Olive Oil Handbook

Volume I

Olive Oil Handbook Volume I

Edited by **Thelma Bosso**

New York

Published by Callisto Reference,
106 Park Avenue, Suite 200,
New York, NY 10016, USA
www.callistoreference.com

Olive Oil Handbook
Volume I
Edited by Thelma Bosso

© 2015 Callisto Reference

International Standard Book Number: 978-1-63239-491-0 (Hardback)

Contents

Preface

This book has been an outcome of determined endeavour from a group of educationists in the field. The primary objective was to involve a broad spectrum of professionals from diverse cultural background involved in the field for developing new researches. The book not only targets students but also scholars pursuing higher research for further enhancement of the theoretical and practical applications of the subject.

The growth of the olive oil industry and health-promoting impacts associated with olive oil has reinforced the quest for novel information, stimulating a broad spectrum of research. This book is a source of currently compiled information. It encompasses a wide spectrum of topics under the section olive oil composition, analysis and quality, which includes quality evaluation and different techniques applied in authenticating the quality of olive-oil. It will be an important reference for food scientists, biotechnologists, nutritionists, researchers, olive oil producers and consumers.

It was an honour to edit such a profound book and also a challenging task to compile and examine all the relevant data for accuracy and originality. I wish to acknowledge the efforts of the contributors for submitting such brilliant and diverse chapters in the field and for endlessly working for the completion of the book. Last, but not the least; I thank my family for being a constant source of support in all my research endeavours.

Editor

Olive Oil Composition, Analysis and Quality

1

Volatile and Non-Volatile Compounds of Single Cultivar Virgin Olive Oils Produced in Italy and Tunisia with Regard to Different Extraction Systems and Storage Conditions

Cinzia Benincasa, Kaouther Ben Hassine,
Naziha Grati Kammoun and Enzo Perri
[1]CRA-OLI Olive Growing and Olive Oil Industry Research Center Rende,
[2]Institut de l'Olivier, Sfax
[1]Italy
[2]Tunisia

1. Introduction

Virgin olive oil has a fundamental role in the markets of alimentary oils because of its unique aroma, its stability and its healthy benefits. In this chapter the attention will be focused on Tunisian and Italian single cultivar olive oils.

The oils under investigation were produced by different extraction systems and characterised for their volatile and non-volatile compounds (Benincasa et al., 2003; Cerretani et al., 2005; Garcia et al., 1996). It is well known that volatile and non-volatile components of products of plant origin are dependent on genetic, agronomic and environmental factors. There are few reports (Angerosa et al., 1996, 1998a, 1998b, 1999; Morales et al., 1995; Solinas et al., 1998) on the evaluation of the relationships between the aroma components of virgin olive oil with the metabolic pathways and varietal factors. Olive ripening process and, to some extent, the fruit growing environment, affect also the composition of the volatile compounds of the oil (Aparicio & Morales, 1998; De Nino et al., 2000; Guth & Grosh, 1993; Montedoro & Garofalo, 1984; Morales et al., 1996). Volatile and non-volatile compounds are retained by virgin olive oils during their mechanical extraction process from olive fruits (Olea europaea L.). Non-volatile compounds such as phenolic compounds stimulate the tasting receptors such as the bitterness perception, the pungency, astringency and metallic attributes. Instead volatile compounds, stimulating the olfactive receptors, are responsible for the whole aroma of the virgin olive oil. The chromatograms of volatile compounds of olive oils were obtained by solid phase micro extraction-gas chromatography/mass spectrometry (SPME-GC/MS) (Hatanaka, 1993; Kataoka et al., 2000; Steffen & Pawliszyn, 1996). The method is based on the assay of the terminal species of the "lipoxygenase pathway" which are present in the volatile fraction of the sampled compounds (Hatanaka, 1993).

2. Materials and methods

2.1 Extraction of olive oil and storage

The olive oils investigated (60 Italian and 60 Tunisian) were single cultivar virgin olive oils (SCVOOs) produced in different regions of Tunisia (Chamlali Cv.) and Italy (Coratina Cv.). Olives were handpicked at the optimal olive ripening degree. Immediately after harvest, olive fruits were transported and cleaned, each fruit sample was divided into three portions of 20 Kg. One portion was extracted using pressure system (see paragraph 2.1.1), the second and the third were extracted by centrifugation systems, three and two phases, respectively (see paragraph 2.1.2 and 2.1.3). The oils obtained were stored in three types of packaging (opaque glass, transparent glass and polyethylene terephtalate PET) and monitored for six months.

2.1.1 Pressure system (PS)

Olives are ground into an olive paste using large millstones. In general, the olive paste stays under the stones for 45–50 minutes. After grinding, the olive paste is spread onto fibre disks, that are easier to clean and maintain, stacked on top of each other and then placed into the press. Afterwards, this pile of disks are put on a hydraulic piston where a pressure of about 400 atm is applied. By the action of this pressure, a olive paste and a liquid phase is produced.

Finally, the liquid phase containing oil and vegetation water is separated by a standard process of decantation.

2.1.2 Two-phase centrifugation (2P)

This system does not need water addition and produces a liquid phase (oil) and a solid waste-water-dampened phase (pomace). The olive paste is kneaded for 60 minutes at 27°C and the oil is extracted with a horizontal centrifugation decanter and separated by means of an automated discharge vertical centrifuge.

2.1.3 Three-phase centrifugation (3P)

This system allows the crushing of olives into a fine paste. This paste is then malaxed for 60 minutes in order to achieve the coalescence of small oil droplets. The aromas are created during these two steps through the action of enzymes. Then, the paste is pumped into an industrial decanter where the phases are separated. Water (500 liters per ton) is added to facilitate the extraction process with the paste. The high centrifugal force created into the decanter separates the phases readily according to their different densities (solid phase pomace, vegetation water, oil). The solid materials is pushed out of the system by the action of a conical drum that rotates with a lower speed. The separated oil and vegetation water are then rerun through a vertical centrifuge, which separates the small quantity of vegetation water still contained in the oil.

2.2 Analytical methods

The physic-chemical and organoleptic analysis of VOO were carried out according to the methods described by the European Union Regulations (UE 61/2011).

Volatile and Non-Volatile Compounds of Single Cultivar Virgin Olive Oils Produced in Italy
and Tunisia with Regard to Different Extraction Systems and Storage Conditions

5

In particular, analysis of fatty acid methyl esters, total phenols, free acidity, peroxide number, conjugated dienes and trienes, sensory analysis and volatile compounds were conducted as described in the following paragraphs.

2.2.1 Fatty acid methyl ester analysis (FAMEs)

FAMEs analysis were carried out after performing alkaline treatment obtained by dissolving the oil (0.05 g) in n-hexane (1 mL) and adding a solution of potassium hydroxide (1 mL; 2 N) in methanol (Christie, 1998). FAMEs were analyzed by gas chromatography by mean of a Shimadzu 17A chromatograph equipped with detector flame ionization and a capillary column. Peaks were identified by comparing their retention times with those of authentic reference compounds.

The fatty acid composition was expressed as relative percentages of each fatty acid calculated considering the internal normalization of the chromatographic peak area.

2.2.2 Total phenols analysis

Total phenols content was determined according to the method developed by Gutfinger (1981). Briefly, an amount of olive oil (2.5 g) was dissolved with hexane (5 mL) and extracted with a solution of methanol and water (5 mL; 60/40). The mixture was then vigorously agitated for 2 minutes. Folin-Ciocalteu reagent (0.5 mL) and bi-distilled water (4.8 mL) were added to the phenolic fraction. The absorbance of the mixture was measured at 725 nm and results were given as mg of caffeic acid per Kg of oil.

2.2.3 Free fatty acids, peroxides, ultra-violet light absorption

Acidity value, peroxide value (PV) and ultra-violet light absorption, conjugated diene (K232) and conjugated trienes (K270), were determined according to the Regulation EEC/2568/91 of the European Union Commission (EEC, 1991).

2.2.4 Sensory analysis

Olive oils were evaluated by a panel according to the official method for the Organoleptic assessment of virgin olive oil referenced COI/T.20/Doc. No 15/Rev. 2.

2.2.5 SPME-GC/MS analysis

Aroma components of products of plant origin are dependent on genetic, agronomic and environmental factors (Benincasa et al., 2003). The complexity of the mass-chromatograms in terms of number of components might represent a drawback when different samples are to be matched. Therefore, in order to consider the minimum set of components that mostly reflect the biogenesis of an oil (Aparicio & Morales, 1998), hexanal (1), 1-hexanol (2), (E)-2-hexenal (3), (E)-2-hexen-1-ol (4) and (Z)-3-hexenyl acetate (5) were chosen as markers of linoleic and linolenic acids specific lipoxygenase oxidation [(path A and B), Fig. 1].

2.2.5.1 Preparation of samples and standard solutions

A solution (200 mg/Kg) was prepared by dissolving 0.04 g of each analytes (see paragraph 2.2.5) in 200 g of commercial seeds oil. In the same manner a solution containing the internal standard (ethyl isobutyrate) was prepared.

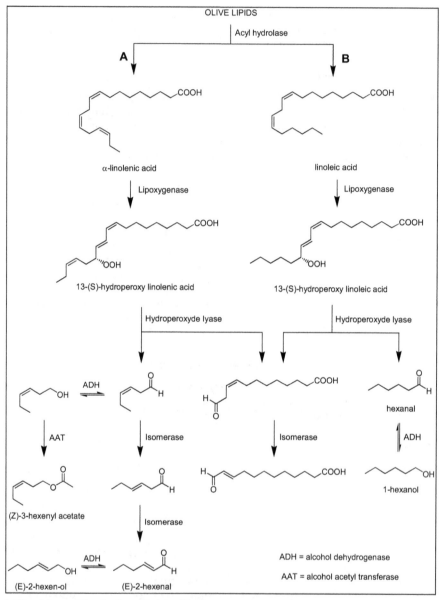

Fig. 1. Linoleic and linolenic acids specific lipoxygenase oxidation.

2.2.5.2 Experimental procedure and instrumentation

The assay of secoiridoid glycosides, such as oleuropein, in virgin olive oil has been proposed as a marker of quality (De Nino et al., 1999, 2005; Perri et al., 1999). With reference to the works previously mentioned, the chromatogram of volatile compounds was considered a useful target. Only the peaks with a certain threshold value (S/N equal to five)

Volatile and Non-Volatile Compounds of Single Cultivar Virgin Olive Oils Produced in Italy
and Tunisia with Regard to Different Extraction Systems and Storage Conditions

7

were taken into account and integrated. Identification of analytes was made by comparison of their mass spectra and retention times with those of authentic reference compounds.

The experimental work was carried out using a Varian 4000 Ion Trap GC/MS system (Varian, Inc. Corporate Headquarters, U.S.A.) equipped with a CP 3800 GC. Volatile components were adsorbed by means of a divilbenzene/carboxen/polydimethylsiloxane (DVB/CAR/PDMS) fiber and separation was obtained by means of a capillary column FactorFOUR (Varian VF-5ms). The ion trap temperature was set at 210 °C with an ionization time of 2 ms, reaction time at 50 ms and scan rate at 1000 ms. The transfer line temperature was set at 230 °C. The column was a 30 m Chrompack CP-Sil 8 CB low bleed/MS (0.25 mm i.d., 0.25 μm film thickness). The GC oven temperature was initially held at 40 °C for 3 min, then ramped at 1 °C/min to 70 °C and finally ramped at 20 °C/min to 250 °C and held for 8 min. The carrier gas was helium at 1 mL/min. Analyses were performed in splitless mode. Mass spectra were collected in EI in positive mode.

2.2.5.3 Quantitative analysis

The calibration curves were obtained by covering two concentration range: 0.4-4 mg/Kg with six steps at 0.4, 0.8, 1.5, 3, 4 mg/Kg for each analyte, with 1.5 mg/Kg of internal standard and 5-150 mg/Kg with six steps at 5, 10, 25, 50, 100, 150 mg/Kg for each analyte, with 40 mg/kg of internal standard. Each experimental value corresponds to the average of three replicates.

The quantitative assay was performed by selecting the area of the ionic species as follows: m/z 41, 56, 67, 72, 82 for hexanal; m/z 55, 56, 69 for 1-hexanol; m/z 55, 69, 83, 97 for (E)-2-hexenal; m/z 57, 67, 82 for (E)-2-hexen-1-ol; m/z 67, 82 for (Z)-3-hexenyl acetate, respectively and m/z 71, 88, 116 for the internal standard.

2.2.5.4 Statistical analysis

The data obtained for each compound were subjected to statistical analysis. Statistical treatment was performed by STATGRAPHICS Plus Version 5.1 (Statistical Graphics Corporation , Professional Edition - Copyrigth 1994-2001). The approach chosen to analyse the set of data obtained was Principal Component Analysis (PCA) and Linear Discriminant Analysis (LDA). Also, in order to check possible differences between the oils, two-way analysis of variance (ANOVA) was performed considering, as main factors, the nationality of the sample and the type of storage container. Moreover, to evaluate significant differences between averages, Tukey test was performed on the oil quality parameters. Differences were considered statistically significant for $P \geq 0.01$ and $P \geq 0.05$. The values obtained for free acidity and FAMEs were analyzed after arcsine transformation in order to meet assumptions for ANOVA.

3. Results and discussions

3.1 FAMEs analysis

VOOs under investigation showed the typical profile of fatty acids of the areas of production. In general, the oils were dominated by palmitic acid (C16: 0), stearic acid (C18: 0), oleic acid (C18: 1) and linoleic acid (C18: 2). The observed values do not show a particular pattern that can indicate the mode of extraction and the type of packing. It is well known, in fact, that fatty acids are dependent on genetic factors, soil and climate (Christie, 1998; Dabbou, et al., 2010; Gharsallaoui, et al., 2011; Manai, et al., 2007).

3.2 Quality parameters

The extraction system has a significant effect on the physical and chemical parameters of the oil: the pressure system can preserve well the colour and the antioxidants of the olive oil, but may affect negatively the sensory profile. From the results obtained, olive oils were characterised by significant differences in free acidity and phenol content (Figures 2 and 3).

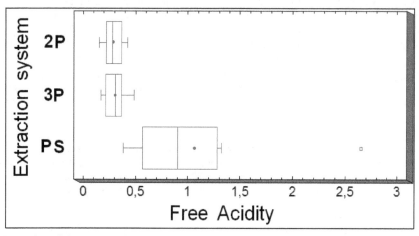

Fig. 2. Box plot of Tunisian VOOs. Free acidity is plotted vs the extraction system.

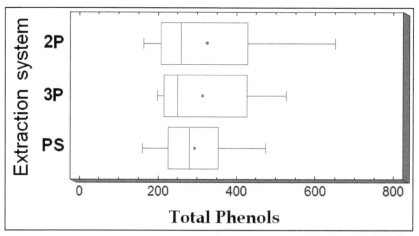

Fig. 3. Box plot of Tunisian VOOs. Total phenols are plotted vs the extraction system.

In general, oils produced with the pressure system have higher free acidity levels than the same oils produced by centrifugation (2P and 3P) and sometime cannot be classified as Extra Virgin Olive Oil (Fig. 2). Moreover, they are often characterised by a lower content of phenols (Fig. 3). In a similar way, the peroxide and K232 and K270 extintion coefficient values were higher than the same oils produced by centrifugation methods.

Volatile and Non-Volatile Compounds of Single Cultivar Virgin Olive Oils Produced in Italy
and Tunisia with Regard to Different Extraction Systems and Storage Conditions

9

3.3 SPME-GC/MS and sensory analysis

Cultivar and extraction systems have a considerable effect on sensory attributes of virgin olive oil. A typical mass chromatogram of the volatile component of one of the analyzed sample is reported in Fig. 4, while the bar chart of Fig. 5 shows the distribution of volatile compounds at five and six carbon atoms that mostly contribute to the olive oil aroma.

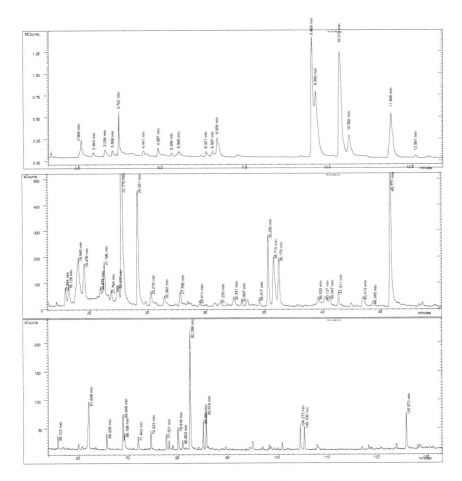

Fig. 4. A typical chromatogram of volatile compounds of one of the analysed samples.

According to the five markers selected as active components of the SPME-GS/MS chromatograms (see paragraph 2.2.5), the distinction of the VOOs under investigation was allowed. Even if both Italian and Tunisian oils were fruity with bitter and pungent characteristics, VOOs of Coratina Cv showed an higher values of fruitiness and bitterness intensity with a clear pungency mainly when they were extracted in centrifugation systems. In fact, these systems can produce olive oils with better organoleptic profiles.

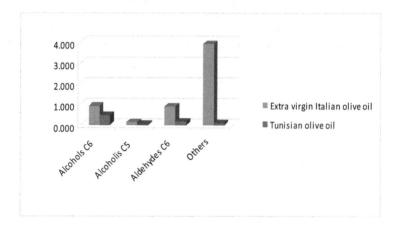

Fig. 5. Bar chart of volatile compounds analysed by SPME-GC/MS. The Cvs under investigation are Coratina and Chamlali from Italy and Tunisia respectively.

Volatile compounds are distributed in a very different concentration in Italian and Tunisian olive oil samples. The flavour of Coratina VOOs was stronger than Chamlali VOOs. In particular, statistical evaluation showed that hexenal (Fig. 6), 1-hexanol (Fig. 7), produced by the lipoxygenase pathway, could discriminate the two VOOs.

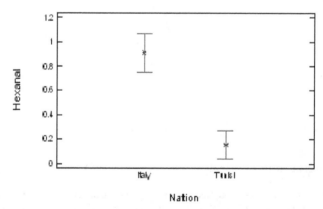

Fig. 6. Biplot of hexenal at 99% confidence level. The genotype was the main factor considered. The Cvs under investigation are Coratina and Chamlali from Italy and Tunisia respectively.

Volatile and Non-Volatile Compounds of Single Cultivar Virgin Olive Oils Produced in Italy
and Tunisia with Regard to Different Extraction Systems and Storage Conditions

11

Means and 95.0 Percent Tukey HSD Intervals

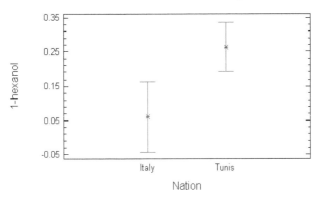

Fig. 7. Biplot of 1.hexanol at 95% confidence level. The genotype was the main factor considered. The Cvs under investigation are Coratina and Chamlali from Italy and Tunisia, respectively.

The VOOs tested by the panelists produced the aromagrams of Figures 8 and 9. According to the panel jury, Coratina olive oils extracted by centrifugation (2P and 3P) were very fruity with a good level of bitterness and astringency. These latter attributes seem to disappear when a pressure system is employed.

Chamlali olive oils extracted by a pressure system were found slightly defected while olive oils extracted by centrifugation systems were fruity with same level of bitterness and astringency. All these results matched those obtained by SPME-GC/MS (see paragraph 3.3).

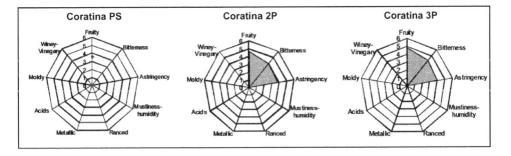

Fig. 8. Sensorial wheels of Italian olive oils of Coratina Cv. extracted by pressure system (PS) and centrifugation two phase and three phase systems (2P and 3P, respectively).

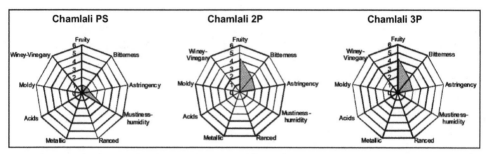

Fig. 9. Sensorial wheels of Tunisian olive oils of Chamlali Cv. extracted by pressure system (PS) and centrifugation two phase and three phase systems (2P and 3P, respectively).

Finally, the organoleptic analysis conducted on custemers demonstrated that consumers prefer olive oils extracted by centrifugation systems rather than olive oils obtained by pressure systems.

3.4 Olive oil storage

Soon after extraction, samples of the sixty Italian and sixty Tunisian VOOs were divided into three groups of twenty and stored in opaque glass, transparent glass and polyethylene terephtalate (PET) bottles. The storage of the oils in opaque glass bottles seemed to be better as it reduced oxidative changes and prolonged shelf life, while polyethylene terephtalate (PET) bottles were the package system that inhibits deterioration to a lesser extent. In fact, free acidity, over the period of six months, became higher when the oils were stored in PET bottles (Fig. 10 and Fig. 11).

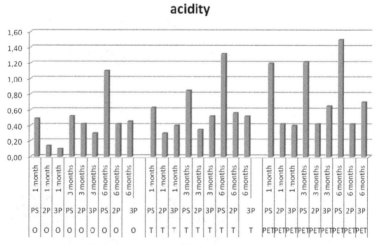

Fig. 10. Bar chart of free acidity of Coratina VOOs during a period of experimentation of six months and depending on the type of packaging utilized. The letters stand for: O opaque glass bottle, T transparent glass bottle, PET polyethylene terephtalate bottle and the extraction system employed: SP pressure system, 2P and 3P centrifugation system at two and three phases respectively.

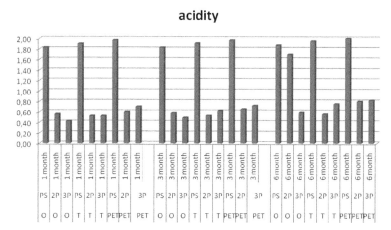

Fig. 11. Bar chart of free acidity of Chamlali VOOs during a period of experimentation of six months and depending on the type of packaging utilized. The letters stand for: O opaque glass bottle, T transparent glass bottle, PET polyethylene terephtalate bottle and the extraction system employed: SP pressure system, 2P and 3P centrifugation system at two and three phases respectively.

Chamlali VOOs were the samples that showed the higher indices of deterioration all over the period. The extraction system plays a key role on the value of the free acidity of an oil. In fact, oils extracted by pressure system have higher free acidity values which increase within the first month (Fig. 12).

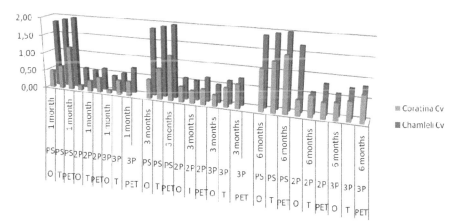

Fig. 12. 3D bar chart of free acidity of Coratina and Chamlali VOOs during a period of experimentation of six months and depending on the type of packaging utilized. The letters stand for: O opaque glass bottle, T transparent glass bottle, PET polyethylene terephtalate bottle and the extraction system employed: SP pressure system, 2P and 3P centrifugation system at two and three phases respectively.

4. Conclusions

The results obtained in this work and discussed in this chapter have shown how important is the method of extraction and the storage of an olive oil (Ben Hassine, et al., 2011; Romano, et al., 2008). A high-quality olive oil can be obtained preferring two phases extraction systems to the classical extraction ones where hydraulic pistons with a pressure of about 400 atm are applied and, storing it in dark glass bottles to better preserve its aroma and phenolic compounds.

5. Acknowledgment

This work was supported by "Riom II – Risorse aggiuntive" project sponsored by the Italian Ministry of Agricultural, Food and Forest Policies (MIPAF); by a grant from the "Institute of Olivier of Sfax" and by the 'Ministère de l'Enseignement Supérieur de la Recherche Scientifique et de la Technologiè (UR03/ES08 Human Nutrition and Metabolic Disorders). We thank M. J. Duff for English revision.

6. References

Angerosa, F.; Di Giacinto, L.; Basti, C. & Serraiocco, A. (1996). Influence of environmental parameters on the virgin olive oil composition. *Riv. Ital. Sostanze Grasse*, Vol.73, pp. 461-467

Angerosa, F.; Camera, L.; d'Alessandro, N. & Mellerio, G. (1998a). Characterization of Seven New Hydrocarbon Compounds Present in the Aroma of Virgin Olive Oils. *J. Agric. Food Chem.*, Vol.46, pp. 648-653

Angerosa, F.; d'Alessandro, N.; Basti, C. & Vito, R. (1998b). Biogeneration of Volatile Compounds in Virgin Olive Oil: Their Evolution in Relation to Malaxation Time. *J. Agric. Food Chem.*, Vol.46, pp. 2940-2944

Angerosa, F.; Basti, C. & Vito, R. (1999). Virgin Olive Oil Volatile Compounds from Lipoxygenase Pathway and Characterization of Some Italian Cultivars. *J. Agric. Food Chem.*, Vol.47, pp. 836-839

Aparicio, R. & Morales, M. T. (1998). Characterization of olive ripeness by green aroma compounds of virgin olive oil. *J. Agric. Food Chem.*, Vol.46, pp. 1116-1122

Benincasa, C.; De Nino, A.; Lombardo, N.; Perri, E.; Sindona, G. & Tagarelli, A. (2003). Assay of aroma active components of virgin olive oils from southern italian regions by SPME-GC/ion trap mass spectrometry, *J. Agric. Food Chem.* Vol.51, pp. 733-741

Ben Hassine, K., Issaaoui, M., Ben Slama, M., Ayadi, M., Lazzez, A., Benincasa, C., Sindona, G., Perri, E., Kammoun, N. & Hammami, M. (2011). Effect of extraction systems on the sensory properties of 'Arbequina' introduced in Tunisia and autochthonous virgin olive oils. Proceeding: "15th Simposium Científico-Técnico Expoliva, EXPOLIVA, May 2007, Jaén (Spain).

Cerretani, L., Bendini, A., Rotondi, A, Lercker, G. & Toschi, T.G. (2005). Analytical comparison of monovarietal virgin olive oils obtained by both a continuous industrial plant and a low-scale mill. *European Journal of Lipid Science and Technology*, Vol.107, pp. 93-100

Christie, WW. (1998). The preparation of derivatives of fatty acids. In: *Gas chromatography and lipids*. Ayr, Scotland: The Oily Press: 1998, pp. 64–84

Dabbou, S.; Chehab, H.; Faten, D.; Dabbou, S.; Esposto, S.; Selvaggini, R.; Taticchi, A.; Servili, M.; Montedoro, GF. & Hammami, H. (2010). Effect of three irrigation regimes on Arbequina olive oil produced under Tunisian growing conditions. *Agric. Water Manag.*, Vol.97, pp. 763-768

De Nino, A.; Mazzotti, F.; Morrone, S. P.; Perri, E.; Raffaelli, A. & Sindona, G. (1999). Characterization of Cassanese olive cultivar through the identification of the new trace components by ionspray tandem mass spectrometry. *J. Mass Spectrom.*, Vol.34, pp. 10-16

De Nino, A.; Mazzotti, F.; Perri, E.; Procopio, A.; Raffaelli, A. & Sindona G. (2000). Virtual freezing of the hemiacetal-aldehyde equilibrium of the aglycones of the oleuropein and ligstroside present in olive oils from Carolea and Coratina cultivars by ionspray ionization tandem mass spectrometry. *J. Mass Spectrom.*, Vol.35, pp. 461-467

De Nino, A.; Di Donna, L.; Mazzotti, F.; Muzzalupo, I.; Perri, E.; Sindona, G. & Tagarelli, A. (2005). Absolute method for the assay of oleuropein in olive oils by atmospheric pressure chemical ionization tandem mass spectrometry, *Anal. Chem.*, Vol. 77, pp. 5961-5964

EEC (1991). Characteristics of olive oil and olive pomace and their analytical methods. Regulation EEC/2568/91 and latter modifications. *Official Journal of European Communities*, L248, pp. 1-82

Garcia, J.M.; Gutierrez, F; Barrera, M.J. & Albi, M.A. (1996). Storage of mill olives on an industrial scale. *J. Agric. Food Chem.*, Vol.44, pp. 590-593

Gharsallaoui, M.; Benincasa, C.; Ayadi, M.; Perri, E.; Khlif, M. & Gabsic, S. (2011). Study on the impact of wastewater irrigation on the quality of oils obtained from olives harvested by hand and from the ground and extracted at different times after the harvesting. *Scientia Horticulturae.*, Vol.128, pp. 23–29

Guth, H. & Grosch, W. (1993). Quantitation of potent odorants of virgin olive oil by stable-isotope dilution assays. *J. Am. Oil Chem. Soc.*, Vol.70, pp. 513-518

Hatanaka, A. (1993). The biogeneration of green odour by green leaves. *Phytochemistry*, Vol.34, pp. 1201-1218

Kataoka, H.; Lord, H.L. & Pawliszyn, J. (2000). Application of solid-phase microextraction in food analysis. *J. Chromatography A*, Vol.880, pp. 35-62

Manaï, M.; Haddada FM.;Trigui, A.; Daoud, D. & Zarrouk, M. (2007). Compositional quality of virgin olive oil from two new Tunisian cultivars obtained through controlled crossings. *J. of the Sc. of Food and Agric.*, 2007, Vol.87, pp. 600–606

Montedoro, G. & Garofolo, L. (1984). The qualitative characteristics of virgin olive oils. The influence of variables such as variety, environment, preservation, extraction, conditioning of the finished product. *Riv. Ital. Sostanze Grasse*, Vol. 61, pp. 157-168

Morales, M. T.; Alonso, M. V.; Rios, J.J. & Aparicio, R. (1995). Virgin olive oil aroma: relationship between volatile compounds and sensory attributes by chemometrics. *J. Agric. Food Chem.*, Vol.43, pp. 2925-2931

Morales, M. T.; Calvente, J.J. & Aparicio, R. (1996). Influence of olive ripeness on the concentration of green aroma compounds in virgin olive oil. *Flavour and Fragrance J.*, Vol.11, pp. 171-178

Perri, E.; Raffaelli, A. & Sindona, G. (1999). Quantitation of oleuropein in virgin olive oil by ionspray mass spectrometry-selected reaction monitoring. *J. Agric. Food Chem.*, Vol.7, pp. 4156-4160

Romano, E., Benincasa, C., Caravita, M.A., Bartoletti, D., De Rose, F., Parise, A., Parise, M.R., Perri, E., Tucci, P. & Rizzuti, B. (2008). Effetti dei processi produttivi su oli provenienti da diverse cultivar dell'Italia meridionale. Proceeding: *"XXI Convegno Nazionale della Divisione di Chimica Analitica della Società Chimica Italiana"*. Arcavacata di Rende (CS), 21-25 September, p. 124

Solinas, M.; Angerosa, F. & Marsilio, V. (1988). Research on some flavour components of virgin olive oil in relation to olives variety. *Rivista Sost. Grasse* 1988, LXV, pp. 361-368

Steffen, A. & Pawliszyn, J. (1996). Analysis of Flavour Volatiles Using Headspace Solid-Phase Microextraction. *J. Agric. Food Chem.*, Vol.44, pp. 2187-2193

Olive Oil Composition: Volatile Compounds

Marco D.R. Gomes da Silva[1], Ana M. Costa Freitas[2],
Maria J. B. Cabrita[2] and Raquel Garcia[2]
*[1]REQUIMTE, Departamento de Química, Faculdade de Ciências e
Tecnologia/ Universidade Nova de Lisboa, Campus da Caparica,
[2]Escola de Ciências e Tecnologia, Departamento de Fitotecnia,
Instituto de Ciências Agrárias e Ambientais Mediterrânicas ICAAM,
Universidade de Évora, Évora,
Portugal*

1. Introduction

In general olive oil is defined on the basis of its sensory characteristics. European Union (EU) regulations establish the organoleptic quality of virgin olive oil by means of a panel test, evaluating positive and negative descriptors (EU regulations). For the organoleptic assessment, several volatile compounds are considered as the main responsible for negative and positive attributes. Volatile compounds, either major or minor, are crucial to olive oil quality; even when present below their olfactory threshold, they can still be important to understand their formation and degradation pathways and provide useful quality marker information.

Volatile composition of olive oils can be influenced by a number of factors, from agronomic and climatic aspects to technological ones. Cultivar, geographic region, ripeness, harvest and processing methods can affect the volatile composition of olive oil. Storage time is also critical for quality. In order to evaluate the volatile profile of olive oil, sensitive analytical techniques as well as extraction procedures were developed. The big issues on aroma analysis are, the loss of compounds during sample preparation steps, and the knowledge that some of the so-called "compounds of interest" (with higher aroma threshold) are, probably, present only in trace amounts. Due to its nature, olive oil is a difficult matrix; for these reasons several methods have been, so far, proposed. The advantages and drawbacks of these methods will be further discussed. One dimension-Gas Chromatography (1D-GC) analysis was, until recently, the most used method to analyze volatiles in different matrices. The increased development of 2D-GC, allowing higher sensitivity and enhanced separation power, is changing the 1D-GC approach. The type of 2D and/or 3D qualitative and quantitative information, provided by 2D-GC systems, promoted the development of powerful chemometrics tools allowing a useful, and potentially easy, way for data interpretation. Fingerprint comparison can be used on a routine basis, providing important and quick information concerning differences among the olive oils produced and, probably most important, also allowing frauds detection.

This work will be divided in four main parts: 1) a brief summary of the composition and biosynthesis of the volatile fraction of olive oil; 2) the role of volatile compounds in olive oil quality: nutritional and sensorial quality; 3) the effect of agronomic and technological practices on olive oil aroma; 4) analytical methodologies for quantification and identification of volatiles compounds: new analytical methods.

2. Composition and biosynthesis of the volatile fraction of olive oil

The wide variety of volatile compounds found in high quality virgin olive oil are produced through biogenic pathways of the olive fruit, namely the lipoxygenase (LOX) pathways (Hatanaka, 1993), and fatty acid or aminoacid metabolism, as depicted in fig.1 (Angerosa et al., 2004; Angerosa et al., 2002). Besides the contribution of several volatile compounds, related with the mentioned pathways, the role of other compounds, especially aldehydes derived from auto-oxidation processes, should also be considered to the final aroma of the olive oils (Angerosa, 2002). Other metabolized products, originated from possible fermentations, conversion of some aminoacids, enzymatic activities of moulds or oxidative processes, are closely related with off-flavour of virgin olive oil. As illustrated in fig. 1, several compounds namely carbonyl compounds, alcohols, esters and hydrocarbons contribute to the aroma profile of olive oil (Angerosa et al., 2004).

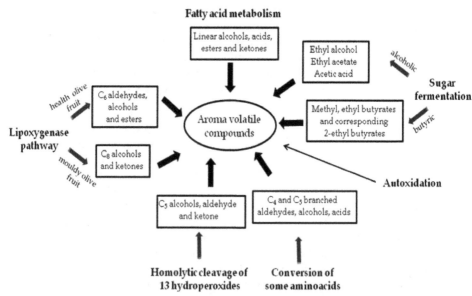

Fig. 1. The main pathways involved in the formation of the volatile profile of high quality virgin olive oils. Adapted from (Angerosa et al. 2004; Angerosa 2002).

The volatile compounds, responsible for virgin olive oil aroma, are usually: low molecular weight (<300 Da); high volatility, sufficient hydrosolubility, fair liposolubility and chemical features to bond with specific proteins (Angerosa et al., 2002).

During crushing and malaxation steps, considerable changes, in olive oil chemical composition occurs accomplished by the activation of olive fruit enzymes due to the inherent disruption of cellular tissues. Consequently, the LOX pathway is initiated by the hydrolysis of triglycerides and phospholipids, mediated by acyl hydrolase (AH), leading to the release of fatty acids. Lypoxygenases, after their release, become immediately active and transform the unsaturated fatty acids, produced by the action of AH, linolenic (LnA) and linoleic (LA) acids, into their corresponding 9- and 13-hydroperoxides, as shown in figure 2. The subsequent cleavage of fatty acids 13-hydroperoxides is catalysed by specific hydroperoxide lyases (HPL) leading to the formation of C_6 aldehydes ((Z)-hex-3-enal and hexanal from linolenic and linoleic acids, respectively) and oxoacids. The unsaturated form of C_6 aldehyde ((Z)-hex-3-enal) undergo rapid isomerisation to the more stable (E)-hex-2-enal. The action of alcohol dehydrogenase (ADH), catalyses the reversible reduction of aliphatic C_6 aldehydes to the corresponding volatile alcohols (Benicasa et al., 2003; Angerosa et al., 1998a). Alcohol species are further transformed into esters due to the catalytic activity of alcohol acetyl transferase (AAT), producing acetates (Kalua et al., 2007) (figure 2). Several factors, such as cultivar and extraction process, including operating temperature, seem to play a relevant role on the improvement of AAT activity (Salas, 2004). When the substrate is LnA, LOX catalyses, besides the hydroperoxide formation, also its cleavage, via an alkoxy radical, increasing the formation of stabilized pent-1,3-diene radicals. These compounds can suffer dimerization leading to the production of C_{10} hydrocarbons (pentene dimmers) or react with a hydroxyl radical present in the medium, leading to C_5 carbonyl compounds (Angerosa et al. 1998b, Pizarro et al., 2011). The most important fraction of volatile compounds, of high quality virgin olive oils, comprises C_6 and C_5 compounds, especially C_6 linear unsaturated and saturated aldehydes. The presence of other volatile compounds, namely C_7-C_{11} monounsaturated aldehydes, C_6-C_{10} dienals, C_5 branched aldehydes and alcohols and some C_8 ketones, in relatively high concentrations, in the aroma of virgin olive oil, is associated with unpleasant notes. The presence, or lack of defects, in the aroma of olive oils is related with the contribution of the various pathways involved on volatiles formation.

When the most active pathway is the LOX cascade the olive oil aroma will not be defective. LOX pathway is predominant in oils of high quality.

3. The role of volatile compounds in olive oil quality: Nutritional and sensorial quality

The International Olive Oil Council (IOOC), European Commission (EC) and Codex Alimentarius have defined the quality of olive oil based on several parameters, such as free fatty acid content, peroxide value, spectrophotometric absorvances in the UV region, halogenated solvents and sensory attributes (Boskou 2006; Kalua et al., 2007; Lopez-Feria et al., 2007). In order to evaluate olive oil quality, the Codex Alimentarius and IOOC include also the insoluble impurities, some metals and unsaponifiable matter determinations (Boskou 2006).

The nutritional value of olive oil arises from high levels of oleic acid and minor components, such as phenolic compounds. It is well recognized that the consumption of some natural antioxidant phenolic compounds produce beneficial health effects. These substances possess strong radical scavenging capacities and can play a relevant role in protecting against oxidative damages and cellular aging. Together with their bioactivity, olive oil phenols have a significant role on the flavour and the bitter taste of olive oil (Boskou 2006; Servili et al. 2002). Sensory quality plays a crucial role in the acceptability of foodstuffs and some characteristics such as colour and flavour are the main sensations which contribute to their acceptability among consumers. Hence, the evaluation of the sensory quality of olive oils involves perception of both favourable and unfavourable sensory attributes.

Olive oil lipids

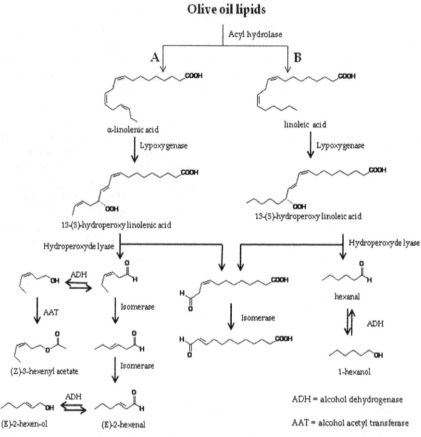

Fig. 2. Lypoxygenase pathway for the formation of major volatile compounds. (Source: Benincasa et al., 2003).

Olive oil possesses a highly distinctive taste and flavor due to specific volatile organic compounds, belonging to several chemical classes, namely aliphatic and aromatic hydrocarbons, aliphatic and triterpenic alcohols, aldehydes, ketones, ethers, esters and furan and thiophene derivatives (Kiritsakis et al. 1998). These compounds, retained by olive oil during the extraction process, stimulate human gustative and olfactive receptors giving rise to olive oil balanced flavour of green and fruity attributes. Such compounds stimulate the free endings of the terminal nerve located in all the palate and in the gustative buds promoting the chemesthetic perceptions of pungency, astringency and metallic attributes. During olive oil tasting, the stimulation of the olfactory epithelium, by a large number of volatile compounds can also occur explaining all other sensations perceived by consumers (Angerosa, 2002). The major volatile compounds of olive oil which contribute for the positive attributes of olive oil aroma (fruity, pungent and bitter) include hexanal, (E)- hex-2-enal, hexan-1-ol and 3-methylbutan-1-ol. Their concentrations, except for (E)-hex-2-enal, varying widely, are generally very low reaching minimum levels of ppb. Thus, volatile compounds, which are responsible for most sensory properties of olive oils, play a significant role on the evaluation of the overall oil quality having a decisive influence on acceptability. The sensory defects are also associated with the volatile composition of the olive oil and are, usually, related with chemical oxidation and exogenous enzymes involved in microbial activity. Chemical oxidation is responsible for the formation of off- flavour compounds, such as pent-2-enal and hept-2-enal The off- flavour compounds associated with unpleasant sensory notes can be assembled in five classes- fusty, moistness- humidity, winey- vinegary, metallic and rancid (Morales et al. 1997; Morales et al., 2005; Escuderos et al., 2007; Faria et al.; Angerosa 2002; Kalua et al. 2007). Moistness-humidity, which possesses the highest sensory significance, is related to the presence of C_8 volatile compounds (e.g. oct-1-en-3-ol and to a lesser extent oct-1-en-3-one) and short chain fatty acids (Morales et al., 2005). Normally they are a characteristic flavour of oils produced from olives infested with fungi and yeasts as a result of an inappropriate storage. Fusty sensory defect is correlated with the presence of ethyl butanoate, propanoic and butanoic acids, a characteristic flavour of oils from olives stored in piles which have undergone an advanced stage of anaerobic fermentation (Morales et al. 2005). Moreover, the presence of acetic acid, ethanol, 3-methylbutan-1-ol and ethyl acetate contributes to winey-vinegary attributes due to the olives fermentation. The rancid negative attribute is due to oils oxidation, characterized by the absence of C_6 aldehydes and alcohols produced from linolenic acid, the absence of esters and the presence of several aldehydes with low odour threshold (Morales et al, 1997). Metallic flavour is associated to oils that have been in prolonged contact with metallic surfaces, during processing, and is characterized by the presence of pent-1-en-3-one; this ketone has been proposed as a useful marker of metallic off-flavour (Venkateshwarlu et al. 2004). The occurrence of pent-1-en-3-one is also positively correlated with bitter and pungency taste while hexanal is negatively correlated with these characteristics, depending on the final amounts. Z-Hex-3-en-1-ol and E-hex-2-enal are negatively correlated with bitter and pungent characteristics, respectively. Other common defects of olive oils, such as muddy sediment and cucumber are related with olive oil preservation.

Poor quality olive oils show remarkable modifications on their sensory basic characteristics, namely the decrease or absence of green, bitter and pungent notes. Generally, the intensity of stimuli elicited by volatile substances is related to their amount. Some other chemical factors, such as volatility and hydrophobic character, size, shape and stereochemistry of volatile molecules, type and position of functional groups as well as external factors, such as matrix effects, seem to affect odour intensity, more than their concentration, due to the influence of these chemical features on the interaction with olfactory and gustative receptors (Angerosa et al., 2004). Odour activity value, evaluated by means of the ratio between its concentration and its odour threshold, constitutes a useful tool to identify the main contributors to the olive oil aroma. The thresholds of several of these compounds are already presented in dedicated literature (Bouskou 2006; Angerosa 2002).

4. The effect of agronomic and technological practices on olive oil aroma

Factors affecting volatile composition of olive oils can be classified into four main groups: environmental (soil and climate); agronomic (irrigation, fertilization); cultivation (harvesting, ripeness) and technological procedures (post-harvest storage and extraction systems) (Aparicio & Luna, 2002). It is generally accepted that volatile profile of virgin olive oils depends on the level and the activity of the enzymes involved in LOX pathway. As previously referred, the major volatile compounds responsible for odour notes of virgin olive oils are the C_6 and the C_5 volatile compounds which emerge from primary or secondary LOX pathway, respectively. The enzymatic levels are determined genetically, so they differ from cultivar to cultivar, but the enzymatic activity is influenced by all factors mentioned above. Apart endogenous plant enzymes, responsible for the positive aroma perception in olive oils, chemical oxidation and microbial activity (associated with sensory defects) should be considered.

4.1 Cultivars

Cultivars and harvest time must be carefully selected in order to correspond to the optimal level of fruit maturity (Esti et al., 1998; Caponio et al., 2001). Olives ripening is quite important for olive oil final composition. The cultivar influence depends on the activity of enzymes and is a genetic characteristic (Tena et al, 2007). The higher LOX activity for linoleic acid than linolenic acid supports the biogenesis of a higher amount of C_6 unsaturated volatile compounds the major constituents of olive oil aroma; usually olive fruits show the highest LOX activity 15 weeks after anthesis; activity decreases during development and ripening periods (Salas et al., 1999). Another enzyme involved is HPL that catalyses the cleavage of fatty acids hydroperoxides producing volatile aldehydes. The highest level of HPL activity is detected in green olive fruits, harvested at the initial development stages. Although there is a slight decrease at maturity, a high activity level is maintained throughout maturation. The decrease in C_6 volatile compounds concentration, in the olive oils of mature olives, is not attributed to HPL activity (Salas & Sanchez, 1999) rather to the availability of substrate. The behaviour of these two enzymes supports the decrease of C_6 volatile compounds content during fruit ripeness. At earlier ripening stages the amount of C_6 aldeydes and alcohols are very similar, and when olive skin colour changes from green to purple most of the C_6 aldehydes reach their maximum concentration (Angerosa & Basti, 2001). With the increase of ripeness a decrease is observed for most of the aldehydes formed from the lipoxygenase pathway, namely E-hex-2-enal (the main volatile compound in most European virgin olive oils), being Z-hex-3-enal an exception (Aparicio & Morales,1998). Kalua et al (2007), however, state that the decrease in C_6 aldehydes, from the lipoxygenase pathway, might not be characteristic of all cultivars.

The olive cultivar influences also fatty acid composition and, particularly, the ratio of oleic to linoleic acid (C18:1/C18:2), triglyceride profile, and phenolic content of olive oil (Aparicio et al., 2002; Tovar et al., 2002; Beltran et al., 2005). Some differences can be found in the fatty acid content of varietal virgin olive oils (Aparicio 2000); they do not vary so much, however, as to be determinant for the volatile profile. In spite, C_6 volatile compounds (aldeydes, alcohols and acetyl esters) formed from 13-hydroperoxides of linoleic and α-linolenic acids, account for 60 to 80% of the total volatile compounds (Aparicio & Luna 2002). The concentration of C_6 volatile compounds, of 36 monovarietal virgin olive oils produced in countries from Mediterranean basin, show that aldehydes (hexanal, Z-hex-3-enal and E-hex-2-enal) and pent-1-en-3-one contribute, distinctly, to the sensory profile of these varietal oils, taking into account the odour thresholds of these volatile compounds (75, 3, 1125 and 50 μg Kg^{-1}, respectively) (Aparicio & Luna 2002). These authors found high concentrations of E-hex-2-enal in Italian cultivars, in accordance with results previously obtained by Solinas et al. (1988), and they all suggest that monovarietal virgin olive oils could be distinguished by this compound. According to Solinas et al. (1988) octanal, nonanal and hex-2-enal contents are a cultivar characteristic; the presence of propanol, amyl alcohols, hex-2-enol, hexan-2-ol and heptanol seems also to be related to the olive cultivar. Nevertheless, olive oils from different cultivars, produced under the same exact conditions (extraction system, ripeness stage, pedoclimatic and agronomic conditions), exhibit different amounts of total volatiles, ranging from 9-83 to 35 mg kg^{-1} (Luna et al., 2006). Baccouri et al. (2008), when studying volatile compounds from Tunisian and Sicilian monovarietal virgin olive oils, found that the overall amounts of C_6 aldehydes were clearly higher than the sum of C_6 alcohols in Chemlali and Sicilian samples, whereas, in Chetoui oils, the sum of C_6 alcohols was generally higher than the C_6 aldehydes. The explanation relays again in the differential activity of the enzymes involved. These authors also reported a decrease, in the amounts of C_5 aldehydes and alcohols, during the maturation. Morales et al. (1996) studied the influence of ripeness on the concentration of green aroma compounds; the total content of volatile compounds decreases with ripeness; there are markers for monovarietal virgin olive oils obtained from unripe (E-hex-2-enal), normal ripe (hexyl acetate) and overripe olives (E-hex-2-enol) regardless of the variety (Aparicio & Morales, 1998)

D'Imperio et al. (2010), when studying the influence of harvest, method and schedule, on olive oil composition, found a remarkable decrease of E-hex-2-enal as was previously reported by Aparicio et al. (1998); an increase of hexanal seems to be related to the use of shakers for harvesting. A decrease in unsaturated fatty acids content was also observed relating these findings to the lipoxygenase pathway.

4.2 Environmental factors

Pedoclimatic factors depends on environmental conditions, soil, type and structure, and/or climatic conditions, temperature and rainfall (Beltran et al., 2005). Cultivars do not always grow at the same altitude, but olive grove zones are spread over a wide range of altitudes, where climatic conditions can be quite different. All these have impact on chemical and sensory profiles of olive oils. Monovarietal olive oils, obtained from olives grown at higher altitudes, are in general sweeter and have a stronger herbaceous fragrance, when compared to the ones produced with olives grown at lower altitudes. Lower temperatures, at higher altitudes, may influence the enzymes from lipoxygenase pathway, since hexanal (green-sweet perception) comes from increased levels of linoleic acid, and E-hex-2-enal (green odour and astringency taste) from lower levels of a-linolenic acids (Aparicio & Luna, 2002). Temime et al. (2006) studied the volatile compounds from Chétoui olive oils, the second variety cultivated in Tunisia, and reported significant differences on volatile compounds when, just, environmental conditions were different. Dabbou et al. (2010) studied the quality and the chemical composition of monovarietal virgin olive oil, from the Sigoise variety, grown in two different locations in Tunisia, a sub-humid zone (Béjaoua, Tunis) and an arid zone (Boughrara, Sfax). Olive oils produced from olives grown at the higher altitude were characterized by higher contents of E-hex-2-enal (11.92 mg kg^{-1}) and hexanal (1.24 mg kg^{-1}), whereas the oils, from the lower altitude, were distinguishable by the higher content of Z-hex-2-en-1-ol (8.78 mg kg^{-1}) and hexan-1-ol (2.17 mg kg^{-1}). The sum of the products of the lipoxygenase oxidation pathways was higher in oils from Béjaoua (15.92 mg kg^{-1}) than in those from Boughrara (15.20 mg kg^{-1}). Among the LOX oxidation products, the amount of hexanal was higher in Béjaoua oils (1.24 mg kg^{-1}), whereas the content of Z-hex-2-en-1-ol was considerably lower.

In a recent study, concerning the behaviour of super-intensive Spanish and Greek olive cultivars grown in northern Tunisia, Allalout et al. (2011) found significant differences between oils; they consider, the majority of the studied analytical parameters, to be deeply influenced by the cultivar-environment interaction.

It seems there is an effect of genotype-environment interaction, responsible for olive oils characteristics.

4.3 Agronomic factors

Irrigation, a practice that has been adequately studied, seems to produce a decrease in the oxidative stability of olive oil volatiles due to a simultaneous reduction in oleic acid and phenolic compounds contents (Tovar et al., 2002).

According to Servili et al. (2007) the olive tree water status has a remarkable effect on the concentration of volatile compounds, such as the C_6-saturated and unsaturated aldehydes, alcohols, and esters. Put simply, deficit irrigation of olive trees appears to be beneficial not only due to its well-known positive effects on water use efficiency, but also by optimizing olive oil volatile quality. Baccouri et al. (2008) reported an enhancement of the whole aroma concentration of Chetouil oils obtained from trees under irrigation conditions when compared to similar ones from non-irrigated trees.

The effect of agronomic practices in oil quality is still controversial: data from Gutierrez et al. (1999) supports the hypothesis that organic olive oils have better intrinsic qualities than conventional ones. These olive oils usually present lower acidity and peroxide index, higher rancimat induction time, higher concentrations of tocopherols, polyphenols, o-diphenols and oleic acid. However, this work was carried out during 1 year, with one olive cultivar only, and results can not be generalized. Ninfali et al. (2008) in a 3-year study, comparing organic *versus* conventional practice did not observe any consistent effect on virgin olive oil quality. Genotype and year-to-year climate changes seem to have a proved influence.

4.4 Technogical factors

Volatile compounds are predominantly generated during virgin olive oil extraction, and are important contributors to olive oil sensory quality. Virgin olive oil quality is intimately related to the characteristics and composition of the olive fruit at crushing. Changes in olive fruit quality during post-harvest is considered determinant to the final sensory quality. Kalua et al. (2008) reported that low-temperature storage of fruits can produce poor sensory quality of the final oil. This decrease in quality might be due to lower levels of E-hex-2-enal and hexanal, associated with a decrease in enzyme activity, and a concurrent increase in E-hex-2-enol, which might indicate a possible enzymatic reduction by alcohol dehydrogenase (Olias et al., 1993,Salas et al. 2000) and reduced chemical oxidation (Morales et al. 1997). Inarejos-Garcia et al. (2010) studied the olive oils from Cornicabra olives stored at different conditions (from monolayer up to 60 cm thicknesses at 10 °C (20 days) and 20 °C (15 days)). E-hex-2-enal showed a Gaussian-type curve trend during storage that can be related to the decrease of hydroperoxide lyase activity. C_6 alcohols showed different trends, during storage, with a strongly decrease of the initial content of Z-hex-3-en-1-ol after 15 and 8 storage days at 20°C and 10°C under the different storage layers, whilst an increase of E-hex-2-en-1-ol was observed (except for mono-layer). Differences might be related to the enhancement of alcohol dehydrogenase activity during storage. Besides the evolution and changes observed in the desirable LOX pathway, C_6 fraction, storage may give rise to undesirable volatile compounds, from metabolic action of yeasts, which was more evident when olive were stored at 20 °C. The effect of the extraction process on olive oil quality is also well documented (Ranalli et al., 1996; Montedoro et al., 1992; Di Giovacchino, 1996; Koutsaftakis et al., 1999; Servili et al., 2004).

Technological operations include several preliminary steps, leaf and soil removal, washing, followed by crushing malaxation and separation of the oil (and water) from the olive paste. This last step can be achieved by pressing (the oldest system), centrifugation (the most widespread continuous system), or percolation (based on the different surface tensions of the liquid phases in the paste).

Ranalli et al. (2008) studied the effect of adding a natural enzyme extract (*Bioliva*) during processing of four Italian olive cultivars (*Leccino, Caroleo, Dritta* and *Coratina*) carried out with a percolation-centrifugation extraction system. The improved rheological characteristics of the treated olive paste resulted in a reduced extraction cycle with good effects concerning olive oil aroma characteristics. Results have shown that enzyme-treated olive pastes always release higher amounts of total pleasant volatiles (*E*-hex-2-enal, *E*-hex-2-en-1-ol, *Z*-hex-3-enyl acetate, *Z*-hex-3-en-1-ol, pent-1-en-3-one, *Z*-pent-2-enal, *E*-pent-2-enal and others). For the individual C_6 metabolites, from the LOX pathway, a similar trend was generally observed, while for the total unpleasant volatiles, *n*-octane, ethyl acetate, isobutyl alcohol, *n*-amyl alcohol, isoamyl alcohol and ethanol, an opposite behaviour was found.

The fundamental step is, however, olive crushing. The release of oil from olives can be achieved by mechanical methods (granite millstones or metal crushers) or centrifugation systems. These different systems affect the characteristics of the pastes and the final oil (Di Giovacchino et al., 2002). Almirante et al. (2006) reported that the oils obtained from de-stoned pastes had a higher amount of C_5 and C_6 volatile compounds, when compared to oils obtained by stone-mills. This increment is due to stones removal, which possess enzymatic activities, metabolizing 13-hydroperoxides other than hydroperoxide lyase, giving rise to a net decrease in the content of C_6 unsaturated aldehydes during the olive oil extraction process. Servili et al. (2007) demonstrate that the enzymes involved in the LPO pathway have different activity in the pulp or in the stone. Stones seem to have a lower hydroperoxide lyase activity and a higher alcohol dehydrogenase activity when compared to the pulp. These authors also found higher amounts of C_6 unsaturated aldehydes olive oils volatiles (VOOs) obtained with the stoning process; the stone presence in traditional extraction procedure increases the concentration of C_6 alcohols (for Coratina and Frantoio cultivars).

The next step is the malaxation. Malaxation is performed to maximize the amount of oil that is extracted from the paste, by breaking up the oil/water emulsion and forming larger oil droplets. The efficiency of this operation depends upon time and temperature. Pressing, percolation, or centrifugation, are finally used to separate the liquid and solid phases. Temperature and time of exposure of olive pastes to air contact (TEOPAC), during malaxation, affect volatile and phenolic composition of virgin olive oil, and consequently its sensory and healthy qualities. Cultivar still plays a fundamental role for the final composition (Servili et al, 2003). These authors showed that TEOPAC can be used to perform a selective control of deleterious enzymes, such as polyphenol oxidase (PPO) and peroxidase (POD), preserving the activity of LPO. High malaxation temperature (> 25 °C) reduces the activity of enzymes, involved in LOP pathway, reducing the formation of C_6 saturated and unsaturated aldehydes. A similar result is described by Tura et al. (2004). These authors found that changes in malaxation time and temperature produces differences in the volatile profile of olive oils. Increasing temperature and decreasing time led to a reduction in the amount of volatiles produced, but they also describe cultivar as the single most important factor in determining volatile profile of olive oils. The decrease of olive oil flavour, produced by high malaxation temperature, is due to the inactivation of hidroperoxide lyase (HPL) rather than lipoxygenase (LOX), as both enzymes have different behaviour regarding temperature (Salas & Sánchez, 1999b). LOX, when assayed with linoleic acid as the substrate, displayed a rather broad optimum temperature around 25 °C and maintained a high activity at temperatures as high as 35 °C, but HPL activity peaked at 15 °C and showed a clear decrease at 35 °C, in assays using 13-hydroperoxylinoleic acid as substrate. Similar results were obtained by Gomez-Rico et al. (2009) who observed a significant increase in C_6 aldehydes, in the final oil, as malaxation time increased; almost no changes in the content of C_6 alcohols were observed. Opposite results were found for the influence of the kneading temperature, where a drop in the C_6 aldehydes content as malaxation temperature increases is observed, especially for E-hex-2-enal and a slight increase in C_6 alcohols, mainly hexan-1-ol and Z-hex-3-en-1-ol.

The final step of olive oil production also affects olive oil quality. Separation of oil from water can be achieved using a two-phase or a three phase centrifugation system. Comparing monovarietal virgin oils obtained by both processes, the oils from two-phase decanters have higher content of E-hex-2-enal and total aroma substances but lower values of aliphatic and triterpenic alcohols (Ranalli & Angerosa, 1996).

Masella et al. (2009), when studying the influence of vertical centrifugation on olive oil quality, observed significant differences both in the total volatile concentration and in the two volatile classes from the LOX pathway involving LnA conversion. The observed decreased of C_6/LnA and C_5/LnA compounds can be explained by the volatiles partition between oil and water phases during vertical centrifugation.

Storage conditions also affect final quality. Light exposure, temperature and oxygen concentration, storage time and container materials are also determinant. A study by Stefanoudaki et al. (2010) evaluating storage under extreme conditions, showed subtle differences, in the pattern of volatile compounds, in bottled olive oils stored indoors or outdoors. When stored with air exposure the levels of some negative sensory components, such as penten-3-ol and hexanal, increased while other positives, like E-hex-2-enal were reduced. Filling the headspace with an inert gas can reduce spoilage.

5. Analytical methodologies for quantitation and identification of volatiles compounds: New analytical methods

5.1 Olive oil volatile compounds

In the volatile fraction of olive oils, approximately three hundred compounds have already been detected and identified by means of gas chromatography/mass spectrometry (GC/MS) methods (Boskou, 2006). Among these compounds, only a small fraction contributes to the aroma of olive oil (Angerosa et al., 2004). The most common olive oil volatiles have 5 to 20 carbon atoms and include short-chain alcohols, aldehydes, esters, ketones, phenols, lactones, terpenoids and some furan derivatives (Reiners & Grosh, 1998; Delarue & Giampaoli, 2000; Kiritsakis, 1992; Boskou, 2006; Vichi et al., 2003a, 2003b, 2003c; Aparicio et al., 1996; Morales et al., 1994; Flath et al, 1973; Morales et al, 1995; Bortolomeazzi et al., 2001; Bentivenga et al., 2002; Bocci et al., 1992; Servili et al., 1995; Fedeli et al., 1973; Fedeli, 1977; Jiménez et al., 1978; Kao et al., 1998; Guth & Grosch, 1991). As all vegetable oils, olive oil comprises a saponifiable and a non-saponifiable fraction and both contribute for the aroma impact. As a result of oxidative degradation of surface lipids (Reddy & Guerrero, 2004) a blend of saturated and mono-unsaturated six-carbon aldehydes, alcohols, and their esters (Reddy & Guerrero, 2004; Matsui, 2006) are produced. As already mentioned they are formed from linolenic and linoleic acids through the LOX pathway, and are commonly emitted due to defence mechanism developed by the plant in order to survive to mechanical damage, extreme temperature conditions, presence of pathogenic agents, among others (Delarue & Giampaoli, 2000; Noordermeer et al., 2001; Pérez et al., 2003; Angerosa et al., 2000; Angerosa et al., 1998b). Volatile phenols are also reported as aroma contributors for olive oil and can play a significant organoleptical role (Vichi et al., 2008; Kalua et al., 2005).

5.2 Analytical methodologies

5.2.1 Sample preparation procedures

When the analysis of a volatile fraction, of complex matrices, is considered sample preparation cannot be underestimated. In biological samples, a wide chemical diversity, in a wide range of concentrations, must be expected (Salas et al., 2005; Wilkes et al., 2000). The chemical nature, and the amount of the detected compounds, strongly depends on the extraction technique used, to remove and isolate them from their matrices. The choice of a suitable extraction methodology depends on sample original composition and target compounds. However, an ideal sampling method does not exist and no single isolation technique produces an extract that replicates the original sample. In order to have enough quantity of each compound to be detected by chromatography, a concentration step must, usually, be considered. Sample preparation can be responsible for the appearance of artefacts, due to the chemical nature of the compounds extracted, and thus detected and quantified, and to a total or partial loss of compounds; this issues can, very strongly, determine the precision, reproducibility, time and cost of a result and/or analysis (Wilkes et al., 2000; Belitz et al., 2004; Buttery 1988; van Willige et al., 2000). These methods are revised in a recent manuscript (Costa Freitas et al.) where sample preparation procedures for volatile compounds are discussed as well as the advantages and drawbacks of each method.

In olive oil analysis, its oily nature strongly influences the choice of the extraction procedure. There are various techniques that can be used for the preparation of the sample analytes in biological material. From those so far applied, liquid extraction with or without the use of ultrasounds (Kok et al., 1987; Fernandes et al., 2003; Cocito et al., 1995) is probably the most used. Besides liquid extraction, simultaneous distillation extraction (SDE) (Flath et al., 1973) has also been widely used. The drawback of these methods is the use of solvents and consequently the need of compounds isolation from the solvent which represents an extra preparation step, as well as the dilutions steps during the extraction procedure. To avoid these steps, supercritical fluid extraction (SFE) (Morales et al., 1998) was also used for the isolation of volatile constituents of olive oil.

The methods based on extraction from the headspace are an elegant choice (Swinnerton et al., 1962). The more often used procedures are the so called "purge and trap" techniques (Morales et al., 1998; Servili et al., 1995; Aparicio & Morales, 1994) in which the compounds of interest are trapped in a suitable adsorbent, from which they can be taken either directly (using a special "thermal desorber" injector) or after retro-extraction into a suitable solvent which, once again, includes an extra extraction step. Another choice is direct injection of the headspace into the injection port of a GC chromatograph. This possibility does not include a concentration step, and consequently, the minor compounds are usually missing or not detected (Del Barrio et al., 1983; Gasparoli et al., 1986). A direct thermal desorption technique can also be applied, avoiding the use of any types of adsorbents, by just heating the target olive oil sample to a suitable temperature in order to promote the simultaneous, extraction, isolation and injection of the volatile fraction into the analytical column (Zunin et al. 2004, de Koning et al., 2008). The main advantage of this technique is its simplicity, although a special injection system is mandatory, which can be expensive. When SPME was introduced (Belardi & Pawliszyn, 1989; Arthur & Pawliszyn, 1990) several authors have focused their attention on adapting the technique for aroma compounds analysis (D'Auria et al., 2004; Vichi et al., 2003; Vichi et al., 2005; Ribeiro et al., 2008). The main advantages of this technique are: a) it does not involve sample manipulations; b) it is an easy and clean extraction method able to include, in just one step, all the steps usually needed for aroma extraction. The extraction step, in SPME, can be made either by headspace sampling or liquid sampling. Headspace sampling (HS) is usually the method of choice for olive oil aroma analysis. The fibre chemical composition is of main interest and   determines the chemical nature of the compounds extracted and further analyzed. There are several coatings commercially available. Polydimethylsiloxane (PDMS) and polyacrylate (PA) coatings extract the compounds by means of an absorption mechanism (Ribeiro et al., 2008) whereas PDMS is a more apolar coating then PA. Polydimethylsiloxane/divinylbenzene (PDMS/DVB), polydimethylsiloxane/carboxene (PDMS/CAR), carbowax/divinylbenzene (CW/DVB), and divinylbenzene/carboxene/polydimethylsiloxane (DVB/CAR/PDMS) extract by an adsorptive mechanism. These second group of fibres have usually a lower mechanic stability but present higher efficiency to extract compounds with low molecular weight (Augusto et al., 2001). In both extraction mechanisms, once the compounds are expelled form the matrix, they will remain in the headspace and a thermodynamic equilibrium is established between these two phases (Zhang & Pawliszyn, 1993). When the fibre is introduced a third phase is present and mass transfer will take place in both interphases (sample matrix/headspace and headspace/fibre). When quantification is a requirement, equilibrium has usually to be achieved. Time and temperature are also very important issues to take in consideration, since they will affect equilibrium (Vas & Vékey, 2004) and thus   extraction efficiency. Methods that consider quantification in non-equilibrium have also been developed (Ai, 1997; Ribeiro et al., 2008). In order to optimize the extraction procedures by HS-SPME, the efficiency, accuracy and precision of the extraction is also directly dependent on operational parameters like extraction time, sample agitation, pH adjustment, salting out, sample and/or headspace volume, temperature of operation, adsorption on container walls and desorption conditions (Pawliszyn, 1997).

5.2.2 Chromatographic methods for the analysis of olive oil volatiles

Capillary gas chromatography (GC) is the most used technique for the separation and analysis of volatile and semivolatile organic compounds (Beesley et al., 2001) in biological samples. GC allows to separate and detect compounds present in a wide range of concentrations in very complex samples, and can be used as a routine basis for qualitative and quantitative analysis (Beesley et al., 2001; Majors, 2003). Enantioselective separations can also be performed when chiral columns are used (Bicchi et al., 1999). The most common detector used is the flame ionization detector (FID), known by its sensitivity and wide linear dynamic range (Scott, 1996; Braithwaite & Smith, 1999). When coupled with Fourier transform infrared spectroscopy (GC/FTIR) or mass spectrometry (GC/MS) (Gomes da Silva & Chaves das Neves, 1997; Gomes da Silva & Chaves das Neves, 1999), compounds tentative identification can be achieved.

The most widely used ionization techniques employed in GC/MS is electron ionization (EI normally at 70 eV) and the more frequently used mass analysers, in olive oil volatile research, are quadrupole filters (qMS), ion traps (ITD) and time of flight instruments (TOFMS). The GC/TOFMS instruments allow the simultaneous acquisition of complete spectra with a constant mass spectral m/z profile for the whole chromatographic peak, while in qMS instruments the skewing effect is unavoidable. This fact enables the application of spectral deconvolution (Smith, 2004), and, potentially, a more accurate use of reference libraries for identification and confirmation of analytes may be possible. Nevertheless, for routine laboratory the development of TOFMS dedicated mass spectral libraries, to complement the libraries now generated by using qMS, should be considered. Spectral matching is usually better when qMS data are compared in some instances (Cardeal et al., 2006; Gomes da Silva et al., 2008).

In an ongoing research in our lab, HS-SPME was performed in order to identify volatile compounds in *Galega Vulgar* variety. Four fibres were used and the HS-SPME-GC/TOFMS system operated with a DB-wax column. In table 1 the complete list of compounds identified (using the four different fibres) is provided as well as fragmentation patterns obtained for those not yet reported in olive oils (table 2). Analysis were performed in two columns: a polar column (DB-WAX), usually recommended for volatiles analysis, and an apolar based column DB-5. The use of these two columns, of different polarity, was also very useful to detect co-elutions, occuring when the polar column was used, and helped the identification task, when associated to mass spectrometric and linear retention indices (LRI) data confrontation. Most identification were performed by comparing retention time and fragmentations patterns, obtained for standards, analysed under the same conditions, or by fragmentation studies, when standards were not available. The differences observed, in the LRI experimentally obtained for the DB-WAX column, compared to the literature were expectable since polar columns are known as being much more unstable, then apolar columns, and cross-over phenomena occur (Mateus et al. 2010). Their retention characteristics varies significantly among different suppliers, which suggest the need of LRI probability regions. This fact explains why few LRI data is available for polar columns. These results aims to fullfill some part of this gap.

Compound name	LRI Experimental [Literature]	SPME Fibres	Compound name	LRI Experimental [Literature]	SPME Fibres
Hexane	n.d. [600]	D-C-P	E-Pent-2-enal	1060 [1127-1131]	D-C-P
Heptane	n.d. [700]	PA D-C-P	p-Xilene	1067 [1133-1147]	PA D-C-P
Octane	800 [800]	PA D-C-P	Butan-1-ol	1074 [1147]	PA D-C-P
Propanone	808 [820]	PA CDVB D-C-P	m-Xilene	1077 [1133-1147]	D-C-P
E-Oct-2-ene	818 [n.f.]	PA	Pent-1-en-3-ol	1093 [1130-1157]	PA D-C-P
Ethyl acetate	832 [892]	D-C-P	2,6-Dimethyl-hepta-1,5-diene (isomer)	1101 [n.f.]	D-C-P
2-Methyl-butanal	850 [915]	D-C-P	Cis-hex-3-enal	1113 [1072-1137]	D-C-P
Dichloromethane	859 [n.f.]	PA CDVB	Heptan-2-one	1123 [1170-1181]	PA CDVB D-C-P
Ethanol	883 [900-929]	PA D-C-P	Heptanal	1126 [1174-1186]	PA CDVB D-C-P
1-Methoxy-hexane	889 [941]	D-C-P	o-Xilene	1128 [1174-1191]	D-C-P
4-Hydroxy-butan-2-one	892 [n.f.]	PA	Limonene	1139 [1178-1206]	PA D-C-P
Pentanal	896 [935-1002]	PA	3-Methyl-butan-1-ol	1141 [1205-1211]	D-C-P
3-Ethyl-octa-1,5-diene (isomer)	907 [n.f.]	D-C-P	2-Methyl-butan-1-ol	1142 [1208-1211]	PA PDMS CDVB D-C-P
3-Methyl-butanal	912 [910-937]	D-C-P	2,2-Dimethyl-oct-3-ene	1144 [n.f.]	D-C-P
Propan-2-ol	918 [n.f.]	PA CDVB D-C-P	E-Hex-2-enal	1160 [1207-1220]	PA CDVB D-C-P
3-Ethyl-octa-1,5-diene (isomer)	930 [1018]	PA D-C-P	Dodecene	1164 [n.f.]	PA D-C-P
Pent-1-en-3-one (isomer)	932 [973-1016]	D-C-P	Ethyl hexanoate	1170 [1223-1224]	PA CDVB D-C-P

Compound name	LRI Experimental [Literature]	SPME Fibres	Compound name	LRI Experimental [Literature]	SPME Fibres
Ethyl butanoate	946 [1023]	PA D-C-P	Pentan-1-ol	1184 [1250-1255]	PA CDVB D-C-P
Toluene	952 [1030-1042]	D-C-P	β-Ocimene	1186 [1242-1250]	CDVB D-C-P
Ethyl 2-methyl-butanoate	963 [n.f.]	D-C-P	Tridec-6-ene (isomer)	1187 [n.f.]	D-C-P
Deca-3,7-diene (isomer)	985 [1077]	D-C-P	Styrene	1199 [1265]	PA CDVB D-C-P
Deca-3,7-diene (isomer)	994 [1079]	D-C-P	Hexyl acetate	1209 [1274-1307]	PA CDVB D-C-P
Hexanal	1000 [1024-1084]	PA CDVB D-C-P	1,2,4-Trimethylbenzene	1223 [1274]	PA PDMS CDVB D-C-P
3-Methylbutyl-acetate	1037 [1110-1120]	D-C-P	Octanal	1231 [1278-1288]	PA PDMS CDVB D-C-P
2-Methyl-propan-1-ol	1054 [1089]	PA	E-4,8-Dimethyl-nona-1,3,7-triene	1247 [1306]	PA PDMS CDVB D-C-P
Ethylbenzene	1056 [1119]	PA CDVB D-C-P	E-Pent-2-en-1-ol	1250 [n.f.]	D-C-P
Z-Hex-3-enyl acetate	1258 [1300-1338]	PA CDVB D-C-P	Hepta-2,4-dienal (isomer)	1453 [1463-1487]	PA CDVB D-C-P
E-Hept-2-enal	1272 [1320]	CDVB D-C-P	Decanal	1456 [1484-1485]	PA CDVB
Z-Pent-2-en-1-ol	1281 [1320]	PA D-C-P	α-Humulene	1472 [n.f.]	PA
6-Methyl-hept-5-en-2-one (isomer)	1285 [1335-1337]	PA CDVB D-C-P	Benzaldehyde	1488 [1513]	PA CDVB D-C-P
Hexan-1-ol	1290 [1316-1360]	PA CDVB D-C-P	α-Terpineol	1493 [1694]	D-C-P
4-Hidroxy-4-methyl-pentan-2-one	1313 [n.f.]	D-C-P	E-Non-2-enal	1494 [1502-1540]	PA D-C-P

Compound name	LRI Experimental [Literature]	SPME Fibres	Compound name	LRI Experimental [Literature]	SPME Fibres
E-Hex-3-en-1-ol	1320 [1356-1366]	PA CDVB D-C-P	Propanoic acid	1495 [1527]	D-C-P
Z-Hex-3-en-1-ol	1322 [1351-1385]	PA D-C-P	Octan-1-ol	1504 [1519-1559]	PA CDVB D-C-P
4-Methyl-pent-1-en-3-ol	1330 [n.f.]	PA D-C-P	2-Diethoxy-ethanol	1565 [n.f.]	PA D-C-P
Methyl Octanoate	1331 [1386]	D-C-P	E,E-Nona-2,4-dienal	1574 [n.f.]	PA
Nonan-2-one	1340 [1382]	PA D-C-P	Methyl benzoate	1587 [n.f.]	D-C-P
Nonanal	1344 [1382-1396]	PA CDVB D-C-P	Butanoic acid	1588 [1634]	PA D-C-P
E-Hex-2-en-1-ol	1348 [1368-1408]	CDVB D-C-P	4-Hydroxybutanoic acid	1593 [n.f.]	D-C-P
Z-2-Hex-2-en-1-ol	1348 [1410-1417]	PA D-C-P	E-Dec-2-enal	1606 [1590]	PA CDVB D-C-P
Oct-3-en-2-one (isomer)	1349 [1455]	D-C-P	Acetophenone	1617 [1624]	D-C-P
Hexa-2,4-dienal (E,E), (E,Z) or (Z,Z)	1349 [1397-1402]	D-C-P	2-Methyl-butanoic acid	1621 [1675]	D-C-P
Ethyl octanoate	1353 [1428]	D-C-P	Nonan-1-ol	1628 [1658]	PA CDVB D-C-P
Hexa-2,4-dienal (isomer)	1360 [1397-1402]	D-C-P	α-Muurolene	1680 [n.f.]	D-C-P
E-Oct-2-enal	1367 [1425]	PA D-C-P	Aromadendrene	1681 [n.f.]	PA PDMS CDVB D-C-P
1-Ethenyl-3-ethyl-benzene	1378 [n.f.]	D-C-P	1,2-Dimethoxy-benzene	1686 [n.f.]	PA PDMS D-C-P
Oct-1-en-3-ol (isomer)	1392 [1394-1450]	PA CDVB D-C-P	4-Methyl-benzaldehyde	1690 [n.f.]	D-C-P
Heptan-1-ol	1400 [n.f.]	PA CDVB D-C-P	Pentanoic acid	1700 [1746]	PA CDVB C-C-P

Compound name	LRI Experimental [Literature]	SPME Fibres	Compound name	LRI Experimental [Literature]	SPME Fibres
Linalool	1403 [1550]	CDVB	Butyl heptanoate	1717 [n.f.]	D-C-P
Acetic acid	1408 [1434-1450]	CDVB D-C-P	E-Undec-2-enal	1726 [n.f.]	PA CDVB D-C-P
Hepta-2,4-dienal (isomer)	1421 [1488-1519]	D-C-P	Methyl salycilate	1758 [1762]	D-C-P
2-Ethyl-hexan-1-ol	1436 [1491]	PA CDVB D-C-P	E, E-Deca-2,4-dienal	1780 [1710]	PA CDVB D-C-P
α-Copaene	1440 [1481-1519]	PA CDVB D-C-P	2-Methoxy-phenol (guaicol)	1836 [1855]	PA CDVB D-C-P
α-Cubebene	1442 [n.f.]	D-C-P	2-Methyl-naphthalene	1839 [n.f.]	D-C-P
Benzyl alcohol	1846 [1822-1883]	PA CDVB D-C-P	Octanoic acid	2047 [2069]	PA D-C-P
Phenylethyl alcohol	1890 [1859-1919]	PA CDVB D-C-P	Nonanoic acid	2198 [n.f.]	PA CDVB D-C-P
Heptanoic acid	1900 [1962]	PA D-C-P	4-Ethyl-phenol	2212 [n.f.]	D-C-P

n.d. denote not determined; n.f. denote not found;
LRI denote linear retention indices for DB-Wax column. LRI between brackets represents the data range found in literature: Angerosa, 2002; Contini & Esti 2006; Flath et al., 1973; Kanavouras et al., 2005; Ledauphin et al,. 2004; Morales et al., 1994; Morales et al., 1995; Morales et al., 2005; Reiners & Grosch, 1998; Tabanca et al., 2006; Vichi et al., 2003a., 2003b; Vichi et al., 2005; Zunin et al., 2004.

Table 1. Compounds identified in olive oil samples of *Galega Vulgar* by means of HS-SPME-GC/TOFMS. The fibres used are polydimethylsiloxane (PDMS), polyacrylate (PA), carbowax/divinylbenzene (CDVB), and divinylbenzene/carboxene/polidimethylsiloxane (D-C-P). The extraction and analysis procedure for all fibres was: 15 g of olive oil sample in 22 mL vial immersed into a water bath at 38 °C. Extraction time was 30 min. Fibre desorption time was 300 seconds into an injection port heated at 260 °C. Splitless time of 1 min. A GC System 6890N Series from Agilent coupled to a Time of Flight (TOF) mass detector GCT from Micromass using the acquisition software MassLynx 3.5, MassLynx 4.0 and ChromaLynx The system was equipped with a 60 m × 0.32 mm i.d. with 0,5 μm d_f DB-Wax column or a 30 m × 0.32 mm i.d. with 1 μm d_f DB-5 column, both purchased from J&W Scientific (Folsom USA). Acquisition was carried out using a mass range of 40-400 u.; transfer line temperature was set at 230 °C; ion source 250 °C. Helium was used as carrier at 100 kPa; Oven temperature was programmed from 50 °C for three minutes and a temperature increase of 2 °C/min up to 210 °C hold for 15 minutes and a rate of 10 °C/min up to 215 °C and hold.

Compound name	LRI Experimental [Literature]	m/z –fragmentation pattern	SPME Fibres
Ethyl pentanoate	1050 [1127]	57(66%); 60(36%); 71(5%); 73(31%); **85(100%)**; 88(87%); 101(30%); 115 (2%) **130 (1%) M+**	D-C-P
2-Methyl-heptan-4-one	1063 [n.f.]	41(41%); 43(45%); 55(10%); **57(100%)**; 69(18%); 71(63%); 85(79%); 95(2%); 100(3%); 113(10%); **128(23%) M+**	PA D-C-P
2,6-Dimethyl-oct-2-ene (isomer)	1181 [n.f.]	41(87%); **55(100%)**; 67(11%); 69(73%); 83(25%); 93(12%); 97(25,74%); 111(16%); 126(9,86%); **140(1%) M+**	D-C-P
3-Methyl-pent-3-en-1-ol (isomer)	1306 [n.f.]	**41(100%)**; 42(16%); 55(52%); 56(12%); 67(93%); 69(49%); 70(19%); 82(72%); 83(4%); **100(3%) M+**	CDVB D-C-P
2,6-Dimethyl-octa-2,4,6-triene (isomer)	1318 [n.f.]	77(15%); 79(38%); 91(3%); 93(22%); 95(10%); 105(55%); **121(100%)**; 122(10%); **136(43%) M+**	D-C-P
1-Methoxy-2-(methoxymethyl)-benzene	1346 [n.f.]	51(15%); 65(18%); 77(33%); 79(20%); **91(100%)**; 21(96%); 137 (17%); 152(6%) **M+**	D-C-P
Hex-4-enyl propanoate (isomer)	1350 [n.f.]	41(42%); 55(29%); 57(25%); **67(100%)**; 82(51%)	PDMS D-C-P
Decan-2-one	1428 [n.f.]	41(11%); 42(10%); 43(82%); 55(4%); 57(6%); **58(100%)**; 59(24%); 60 (6%); 71(24%); 85(2%); 98(4%); 113 (2%); 127(2%); **156(2%) M+**	PA D-C-P
Nonyl acetate	1526 [n.f.]	**43(100%)**; 56(39%); 61(33%); 70(24%); 83(16%); 98(19%); 126(10%)	PA D-C-P
Z-Dec-2-enal	1608 [n.f.]	41(64%); 43(55%); **55(100%)**; 56(98%); 69(71%); 70(94%); 83(57%); 98(34%); 110(5%); 136(2%)	PA D-C-P
Phenyl acetate	1964 [n.f.]	43(39%); 65(22%); 66(28%); 77(8%); 89(16%); **94(100%)**; 95(6%);103(8%); 117(9%); **136(15%) M+**	D-C-P
2-Methyl-phenol	2065 [n.f.]	45(7%); 50(5%); 51(9%); 52(4%); 53(8%); 54(4%); 63(3%); 77(24%); 79(19%); 80(8%); 89(4%); 90(8%); 91(3%); **107(100%)**; **108(98%) M+**; 109(5%)(M+H)+	D-C-P
4-Methyl-byphenyl	2091 [n.f.]	51(6%); 63(5%); 82(10%); 83(12%); 84(11%); 115(10%); 152(21%); 153(17%); 65(32%); 167(71%); **168(100%) M+**; 169(17%)(M+H)+	D-C-P

Table 2. New tentatively identified compound in olive oil samples of *Galela vulgar* by means of HS-SPME-GC/TOFMS. Extraction and analytical conditions according to described in table 1. m/z fragmentation patterns are presented; n.f. denote not found; LRI denotes linear retention indices as in table 1. LRI between brackets represents the data range found in literature, according to table 1.

Co-elutions are often impossible to detect and identify with some GC/MS instruments, in spite of the use of selective single ion monitoring mode (SIM), or complex deconvolution processes. The development of new analytical techniques, that maximize analyte separation, has always been a target. Multidimensional chromatography and comprehensive two-dimensional chromatography (David & Sandra, 1987; Bertsch, 1999) are an example of such achievements. The high complexity of the chromatograms points out new ways of chromatography, such as multidimensional-gas chromatography systems (MD-GC), where the analytes are submitted to two or more independent separation steps, in order to achieve separation. In spite of its efficiency, MD-GC is a time consuming technique, with long analysis times, which does not fit with the demands of routine analysis. Additionally, it is technically difficult to carry out sequential transfers in a narrow window of retention times, since co-elutions are foreseen (Poole, 2003). Nevertheless, MD-GC is a precious tool in peak identification for olive oil analysis when co-elutions occur (Reiners & Grosch, 1998). In 1991, comprehensive two-dimensional gas chromatography (GC × GC) was introduced by Liu & Phillips. The GC × GC system consists of two columns with different selectivities; the first and second dimension columns are serially connected through a suitable interface, usually is a thermal modulator (Phillips & Beens, 1999; Marriott & Shellie, 2002). When performing GC × GC technique the entire sample, separated on the first column, is transferred to the second one, resulting in an enhanced chromatographic resolution into two independent dimensions, where the analytes are separated by two independent mechanisms (orthogonal separation) (Venkatramani et al., 1996; Phillips & Beens, 1999; Marriott & Shellie, 2002; Dallüge et al., 2003). The modulated zones of a peak are thermally focused before the separation on the second column, in a mass conservative process; the resulting segments (peaks), of the modulation, are much narrower with higher S/N ratios, than in conventional GC (Lee et al., 2001; Dallüge et al., 2002), improving the detection of trace analytes and the chromatographic resolution. Fast acquisition TOF spectrometers are the suitable detectors for this technique and have considerably enlarged the application of GC × GC. Few applications are still reported for olive oil analysis, nevertheless, they already showed its potential. GC × GC techniques allowed identification of olive oil key flavour compounds, present in very low concentrations (Adahchour et al. 2005); it has also been used as a flexible technique for the screening of flavours and other classes of (semi-)polar compounds, using the conventional orthogonal approach and the reverse, non-orthogonal approach in order to obtain ordered structures that can simplify the identification task (Adahchour et al. 2004); finally this separation technique can allow easy fingerprint analysis of several olive oil matrices directly, or using image processing statistics (Vaz-Freire et al., 2009).

5.3 Future perspectives for olive oil volatile analysis: Identification tools and fingerprinting

A limitation of electron ionization (EI) in MS analysis is due to the fact that, too often, the molecular ions do not survive fragmentation and, consequently, are not "seen". One way to overcome this problem is to use a complementary technique, that provides "soft" ionization of the molecules, allowing molecular ions detection. Chemical ionization (CI) performs this task (McMaster and McMaster, 1998; Herbert and Johnstone, 2003). The mass spectra obtained by CI are simpler than EI, though most of the interpretable structural information is missing. However the compound´s molecular ions appears as a high intensity fragment and sometimes is the major fragment of the spectra. Thus, molecular weight determination of an analyte becomes possible. Other soft ionization techniques are field ionization (FI) and field desorption (FD). Both produce abundant molecular ions with minimal fragmentation (Herbert and Johnstone, 2003). FI and FD are appliable to volatile and thermally stable samples (Niessen, 2001; Dass, 2007). If high resolution mass analysers are coupled with these ionization techniques, high capability of identification can be achieved. Together with GC × GC a potentially new tool in olive oil compound identification is reachable and desirable.

The application of a multimolecular marker approach to fingerprint allows, in an easy way, the identification of certain sample characteristics. Chromatographic profiles can be processed as continuous and non-specific signals through multivariate analysis techniques. This allow to select and identify the most discriminant volatile marker compounds (Pizarro et al., 2011). The quantity and variety of information, provided by two-dimensional-GC (2D-GC) systems, promoted the increasingly application of chemometrics in order to achieve data interpretation in a usefull and, potentially, easy way. Linear discriminant analysis (LDA) and artificial neural networks (ANN), among other statistical classification methods, can be applied in order to control economic fraud. These applications have been carefully reviewed recently (Cajka et al., 2010). Together with 2D-GC systems the advantage is clear, since, instead of a time consuming trial to determine which variables should be considered for the statistical classification method, the selection may now become as simple as inspecting the 2D contour plots obtained (Cardeal et al 2008, de Koning et al., 2008). Also the use of statistical image treatment, of 2D-GC generated contour plots, can be applied for fingerprint recognitions, precluding the alignment of the contour plots obtained, which already allowed the identification of varieties as well as extraction technologies used to produce high quality Portuguese olive oils (Vaz Freire et al., 2009).

6. Conclusion

A final word should also be addressed to spectral libraries. Commercial spectral libraries are becoming increasingly more complete and specific, making GC/MS one of the most used techniques for routine identifications. However, several compounds are not yet described in library databases and, in spite of better algorithmic calculations, databases are only reliable for target analysis, or when the compounds under study are known, and already characterized with a known mass spectra. Additionally, the full separation of peaks to ensure clean mass spectra, in order to achieve a reliable peak analyte confirmation, is still a necessary goal.

Until now most of the analytical systems used to analyse olive oil volatile compounds are performed in 1D-GC systems with polar or apolar column phases. Since olive oil volatile fraction is very complex, frequent co-elutions occur. Mass spectra obtained are, consequently, not pure, which should preclude the possibility to compare the spectra obtained with the, claimed pure, spectra in the databases. However, tentative identifications are reported in the literature, and it is not rare that some inconsistencies occur, even when linear retention indices LRIs are presented. Because of their nature, the LRIs obtained in apolar columns are more reliable. Nevertheless, a better separation is obtained in 1D-GC systems when polar stationary phases are used, because of the wide chemical variety comprised in the volatile fraction of olive oils. Unfortunately, these columns present a high variability, at least, among different purchasers, which do not facilitate LRIs comparison with literature data. Multidimensional techniques, hyphenated with mass-spectrometry, are now fullfiling this gap also in the separation of optical active compounds, when chiral column phases are used. Clean mass spectra together with compound LRIs in polar, apolar and chiral column phases represents an improved tool in compound identification and thus in olive oil matrices characterization. LRIs considering probability regions in the 2D resulting plot of a GC × GC experiment (with different column set combinations, e.g. polar × apolar, polar × chiral, etc.), can enable comparing standard compounds with the sample compounds retention indices and thus a more reliable peak identification can be achieved, if mass spectrometric data are simultaneously recorded. In the future, for 2D systems, more comprehensive mass spectral libraries should include retention index probability regions for different column sets in order to allow correlation of the results obtained in the used systems with spectral matches and literature LRIs.

7. Acknowledgment

Authors wish to thank Fundação para a Ciência e Tecnologia, Ministério da Ciência, Tecnologia e Ensino Superior and Programa Operacional Ciência e Inovação for financial support (Projects PTDC/AGR-AAM/103377/2008 and PTDC/QUI-QUI/100672/2008).

8. References

Adahchour, M.; Beens, J.; Vreuls, R. J. J.; Batenburg, A. M. & Brinkman, U. A.Th. (2004). Comprehensive Two-Dimensional Gas Chromatography of Complex Samples by Using a 'Reversed-Type' Column Combination: Application to Food Analysis. *Journal of Chromatography A*, Vol. 1054, pp. 47-55

Adahchour, M.; Brandt, M.; Baier, H. U.; Vreuls, R. J. J.; Batenburg, A. M. & Brinkman, U. A. Th. (2005). Comprehensive Two-Dimensional Gas Chromatography Coupled to a Rapid-Scanning Quadrupole Mass Spectrometer: Principles and Applications. *Journal of Chromatography A*, Vol. 1067, pp. 245-254

Ai, J. (1997). Solid Phase Microextraction for Quantitative Analysis in Non-Equilibrium Situations. *Analytical Chemistry*, Vol. 69, pp. 1230-1236

Allalout, A.; Krichène, D.; Methenni, K.; Taamalli, A.; Daoud, D. & Zarrouk, M. (2011) Behavior of super-intensive spanish and greek olivecultivars grown in northern Tunisia *Journal of Food Biochemistry* 35 27–43.

Almirante, P.; Clodoveo, M.L.; Dugo, G.; Leone, A. & Tamborrino, A. (2006). Advance technology in virgin olive oil production from traditional and de-stoned pastes: influence of the introduction of aheat exchanger on oil quality. *Food Chemistry* 98, 797-805.

Angerosa, F. ; d'Alessandro, N. ; Basti, C. & Vito R. (1998a) Biogeneration of volatile compounds in virgin olive oil : their evolution in relation to malaxation time. *Journal of Agricultural and Food Chemistry*, 46, pp. 2940- 2944.

Angerosa F.; Camera, L.; d' Alessandro, N. & Mellerio, G. (1998b). Characterization of Seven New Hydrocarbon Compounds Present in the Aroma of Virgin Olive Oil. *Journal of Agricultural and Food Chemistry* Vol. 46, pp. 648-653

Angerosa F.; Mostallino, R.; Basti, C. & Vito, R (2000). Virgin Olive Oil Odor Notes: Their Relationships With Volatile Compounds From the Lipoxygenase Pathway and Secoiridoid Compounds. *Food Chemistry* Vol. 68, pp. 283-287

Angerosa, F. (2002). Influence of volatile compounds on virgin olive oil quality evaluated by analytical approaches and sensor panels. *European Journal of Lipid Science Technology* Vol. 104, pp. 639-660

Angerosa, F.; Servilli, M.; Selvaggini, R.; Taticchi, A.; Espoto, S. & Montedoro, G. (2004). Volatile Compounds in Virgin Olive Oil: Occurrence and their Relationship with the Quality. *Journal of Chromatography A* Vol. 1054, pp. 17-31

Aparicio, R. & Luna, G. (2002). Characterisation of monovarietal virgin olive oils. *European Journal of Lipid Science and Technology*, 104, 614-627.

Aparicio, R. & Morales, M. (1994). Optimization of a Dynamic Headspace Technique for Quantifying Virgin Olive Oil Volatiles. Relationships Among Sensory Attributes and Volatile Peaks. *Food Quality and Preferences* Vol. 5, 109-114

Aparicio, R. & Morales, M. T. (1998) Characterization of olive ripeness by green aroma compounds of virgin olive oil. *Journal of Agricultural and Food Chemistry* 46 (3), 1116-1122.

Aparicio, R. (2000). Authentication. In J. Harwood & R. Aparicio (Eds.), Handbook of olive oil: analysis and properties (pp. 491-520). Gaithersburg, MD: Aspen Publishers.

Aparicio, R.; Ferreiro, L. & Alonso, V. (1994). Effect of climate on the chemical composition of virgin olive oil. *Analytica Chimica Acta*, 292, 235-241.

Aparicio, R; Morales, M & Alonso, M. (1996). Relationship Between Volatile Compounds and Sensory Attributes of Olive Oils by the Sensory Wheel. *Journal of American Oil Chemists' Society*, Vol. 73, pp. 1253-1264

Arthur, C.& Pawliszyn, J. (1990). Solid Phase Microextraction with Thermal Desorption Using Fused Silica Optical Fibers. *Analytical Chemistry* Vol. 62, pp. 2145-2148

Augusto, F.; Koziel, J. & J. Pawliszyn, J. (2000). Design and Validation of Portable SPME Devices for Rapid Field Air Sampling and Diffusion-Based Calibration. *Analytical Chemistry* Vol. 73, pp. 481-486.

Baccouri, O.; Bendini, A.; Cerretani, L.; Guerfel, M.; Baccouri, B.; Lercker,G.; Zarrouk, M.; Miled, D.D.B. (2008). Comparative study on volatile compounds from Tunisian and Sicilian monovarietal virgin olive oils. *Food Chemistry*, 111, 322-328

Beens, J.; Blomberg, J. & Schoenmakers, P. J. (2000). Proper Tuning of Comprehensive Two-Dimensional Gas Chromatography (GC × GC) to Optimize the Separation of Complex Oil Fractions. *Journal of High Resolution Chromatography*, Vol. 23, pp. 182-188

Beesley, T. E.; Buglio, B. & Scott, R. P. W. (2001). *Quantitative Chromatographic Analysis*. Marcel Dekker, Inc. New York, USA

Belardi, R. P. & Pawliszyn, J. B. (1989). The Application of Chemically Modified Fused Silica Fibers in the extraction of Organics from Water Matrix Samples and their Rapid Transfer to capillary columns. *Water Pollution Research Journal of Canada* Vol. 24, 179-189

Belitz, H. –D.; Grosch, W. & Schieberle, P. (2004). *Food Chemistry*, Springer Verlag, (3rd revised edition) Springer Berlin, Heidelberg, New York, ISBN: 3-540-40817-7

Beltran, G.; Aguilera, M. P.; Del Rio, C.; Sanchez, S.; Martinez, L. (2005) Influence of fruit ripening process on the natural antioxidant content of Hojiblanca virgin olive oils. *Food Chemistry* 89, 207-215.

Benicasa, C. ; De Nino, A. ; Lombardo, N. ; Perri, E. ; Sindona, G. & Tagarelli, A. (2003) Assay of Aroma Active Components of Virgin Olive Oils from Southern Italian

Regions by SPME-GC/Ion Trap Mass Spectrometry. *Journal of Agricultural and Food Chemistry*, 51, pp. 733- 741.

Bentivenga, G.; Dáuria, M.; de Bona, A. & Mauriello, G (2002). On the Flavor of Virgin Oil. *Rivista Italaliana delle Sostanze Grasse* Vol. 79; pp. 101-105

Bertsch, W. J. (1999). Two-Dimensional Gas chromatography. Concepts, Instrumentation, and Applications - Part 1: Fundamentals, Conventional Two-Dimensional Gas Chromatography, Selected Applications. *Journal of High Resolution Chromatography*, Vol. 22, pp. 647-665

Bicchi, C.; D'Amato, A. & Rubiolo, P. J. (1999). Cyclodextrin Derivatives as Chiral Selectors for Direct Gas Chromatographic Separation of Enantiomers in the Essential Oil, Aroma and Flavour Fields. *Journal of Chromatography A*, Vol. 843, pp. 99-121

Bocci, F.; Frega, N. & Lercker, G. (1992). Studio Preliminare sui Componenti di Olio di Oliva Extravirgini. *Rivista Italaliana delle Sostanze Grasse* Vol 69; pp. 611-613

Bortolomeazzi R.; Berno, P.; Pizzale, L. & Conte, L. (2001). Sesquiterpene, Alkene, and Alkane Hydrocarbons in Virgin Olive Oils of Different Varieties and Geographical Origins. *Journal of Agricultural and Food Chemistry* Vol. 49, pp. 3278-3283

Boskou, D. (2006). Olive oil composition. In: *Olive Oil: Chemistry and Technology*. Ed. D. Boskau, 2nd edition AOCS Press, Champaign, IL, USA , pp. 41–7

Braithwaite, A. & Smith, F. J. (1999). *Chromatographic Methods*, 5th edition. Kluwer Academic Publishers, Dordrecht

Buttery, R.; Turnbaugh, J. & Ling, L. (1988). Contribution of Volatiles to Rice Aroma. *Journal of Agricultural and Food Chemistry* Vol. 36, pp. 1006-1009

Cajka, T.; Riddellova, K.; Klimankova, E.; Cerna, M.; Pudil, F. & Hajslova, J. (2010). Traceability of Olive Oil Based on Volatiles Pattern and Multivariate Analysis. *Food Chemistry* Vol. 121, pp. 282–289

Caponio, F.; Gomes, T. & Pasqualone, A. (2001) Phenolic compounds in virgin olive oils: influence of the degree of olive ripeness on organoleptic characteristics and shelf-life. *European Food Research and Technology*, 212, 329–333.

Cardeal, Z. L.; de Souza, P. P.; Gomes da Silva, M. D. R. & Marriott, P. J. (2008). Comprehensive Two-Dimensional Gas Chromatography for Fingerprint Pattern Recognition in *cachaça* Production. *Talanta* Vol. 74, pp. 793–799

Cardeal, Z. L; Gomes da Silva, M. D. R. & Marriott, P. J. (2006). Comprehensive Two-Dimensional Gas Chromatography/Mass Spectrometric Analysis of Pepper Volatiles. *Rapid Communication in Mass Spectrometry*. Vol. 20, pp. 2823–2836

Cocito, C.; Gaetano, G. & Delfini, C. (1995). Rapid Extraction of Aroma Compounds in Must and Wine by Means of Ultrasound. *Food Chemistry* Vol. 52, Vol. 311-320

Contini, M & Esti, M. (2006). Effect of the Matrix Volatile Composition in the Headspace Solid-Phase Microextraction Analysis of Extra Virgin Olive Oil. *Food Chemistry*, Vol. 94, pp. 143-150

Costa Freitas, A. M.; Gomes da Silva, M. D. R. & Cabrita M. J. (accepted). Extraction Techniques and Applications: Food and Beverage. Sampling techniques for the determination of volatile components in grape juice, wine and alcoholic beverages", in *Comprehensive Sampling and Sample Preparation*, J. Pawliszyn (Editor), Elsevier

D'Auria, M. ; Mauriello, G. & Rana, G. (2004). Volatile Organic Compounds from Saffron. *Flavour and Fragrance Journal*, Vol. 19, pp. 17-23

D'Imperio, M.; Gobbino, M.; Picanza, A.; Costanzo, S.; Corte, A.D. & Mannina, L. (2010) Influence of harvest method and period on olive oil composition: an NMR and statistical study. *Journal of Agricultural and Food Chemistry* 58, 11043-11051.

Dabbou, S.; Sifi, S.; Rjiba, I..; Esposto, S.; Taticchi, A.; Servili, M.; Montedoro, G.F & Hammami, M. (2010) Effect of Pedoclimatic Conditions on the Chemical Composition of the Sigoise Olive Cultivar *Chemistry & Biodiversity*. 7 898-908

Dallüge, J.; Beens, J. & Brinkman, U. A. Th.(2003). Comprehensive Two-Dimensional Gas chromatography: A Powerful and Versatile Analytical Tool. *Journal of Chromatography A*, Vol. 1000, pp. 69-108

Dallüge, J.; Vreuls, R. J. J.; Beens, J. & Brinkmann, U. A. Th. (2002). Optimisation and Characterisation of Comprehensive Two-Dimensional Gas Chromatography with Time-of-Flight Mass Spectrometric Detection. *J. Sep. Sci.*, Vol. 25, pp. 201-214

Dass, C. (2007). *Fundamentals of Contemporary Mass Spectrometry*. Wiley-Intersience, John Wiley & Sons, Inc., Hoboken, New Jersey, pp. 28-30

David, F.; Sandra, P. (1987). *Capillary Gas Chromatography in Essential Oil Analysis*, P. Sandra, C. Bicchi (Eds.), Huething Verlag, pp. 387

de Koning, S.; Kaal, E.; Janssen, H.-G.; van Platerink, C. & Brinkman, U. A. Th. (2008). Characterization of Olive Oil Volatiles by Multi-Step Direct Thermal Desorption–Comprehensive Gas Chromatography–Time-of-Flight Mass Spectrometry Using a Programmed Temperature Vaporizing Injector. *Journal of Chromatography A*, Vol. 1186, pp. 228–235

Del Barrio, A.; Gutiérrez, F.; Cabrera, J. & Gutiérrez, R. (1983). Aplicación de la Cromatografía Gas-Líquido, Técnica de Espacio de Cabeza, al Problema del Atrojado de los Aceites de Oliva. II *Grasas y Aceites* 34, pp. 1-6

Delarue, J & Giampaoli, P. (2000). Study of Interaction Phenomena Between Aroma Compounds and Carbohydrates Matrices by Inverse Gas Chromatography. *Journal of Agricultural and Food Chemistry* Vol. 48, pp. 2372-2375

Di Giovacchino L.(1996) Influence of Extraction Systems on Olive Oil Quality, in *Olive Oil, Chemistry and Technology*, Boskou D., ed , AOCS Press, Champaign, IL, 12-51.

Di Giovacchino L.; Sestili, S. & Di Vicenzo, D. (2002) Influence of olive processing on virgin olive oil qualità *European Journal of Lipid Science and Technology* 104, 587–601.

Dlouhy J. (1977) The quality of plant products conventional and bio-dynamic management. *Bio- Dynamics.*, 124, 28–32.

Escuderos, M.E., M. Uceda, S. Sánchez, A. Jiménez. 2007. Instrumental technique evolution for olive oil sensory analysis. *European Journal of Lipid Science and Technology* 109:536–546.

Esti M.; Cinquanta, L. & La Notte, E. (1998) Phenolic compounds in different olive varieties *Journal of Agricultural and Food Chemistry*, 46, 32–35.

Faria, S. L.; Cárdenas, S.; G-Mesa, J. A.; Hernández, A. F-. & Valcárcel, M. (2007). Quantification of the Intensity of Virgin Olive Oil Sensory Attributes by Direct Coupling Headspace-Mass Spectrometry and Multivariate Calibration Techniques. *Journal of Chromatography A*, 1147, pp. 144-152

Fedeli, E. (1977). Caratteristiche Organolettiche dell'Ollio di Oliva. *Rivista Italaliana delle Sostanze Grasse* Vol 54; pp. 202-205

Fedeli, E; Baroni, D. & Jacini, G. (1973). Sui Componenti Odorosi dell'Olio di Oliva – Nota III, *Rivista Italaliana delle Sostanze Grasse* Vol 50; pp. 38-44

Fernandes, L.; Relva, A.; Gomes da Silva, M.D.R. & Costa Freitas, A. M. Different Multidimensional Chromatographic Approaches Applied to the Study of Wine Malolactic Fermentation. *J. Chromatogr, A*, Vol. 995, pp. 161-169

Fischer, A. & Richter, C. (1986) Find more like this influence of organic and mineral fertilizers on yield and quality of potatoes. In Proceedings of the 5th IFOAM international scientific conference, H. Vogtmann Ed, 236–248.

Flath R.; Forrey, R. & Guadagni, D. (1973). Aroma Components of Olive Oil. *Journal of Agricultural and Food Chemistry* Vol. 21, pp. 948-952

Gasparoli, A.; Fedeli, E. & Manganiello, B. (1986). Olio Vergine di Oliva: Valutazione dei Caratteri Organolettici Attraverso Tecniche Strumentali. *Rivista Italaliana delle Sostanze Grasse* Vol. 63, pp. 571-582

Gomes da Silva, M. D. R. & Chaves das Neves, H. J. (1997). Differentiation of Strawberry Varieties Through Purge-and-Trap HRGC-MS, HRGC-FTIR and Principal Component Analysis. *Journal of High Resolution Chromatography* Vol. 20, pp. 275-283

Gomes da Silva, M. D. R. & Chaves das Neves, H. J. (1999). Complementary Use of Hyphenated Purge-and-Trap Gas Chromatography Techniques and Sensory Analysis in the Aroma Profiling of Strawberries (*Fragaria ananassa*). *Journal of Agricultural and Food Chemistry* Vol 47, pp. 4568-4573

Gomes da Silva, M. D. R.; Cardeal, Z. & Marriott, P. J. (2008). Comprehensive Two-Dimensional Gas Chromatography: Application to Aroma and Essential Oil Analysis, in *Food flavour: Chemistry, sensory evaluation and biological activity*, H. Tamura, S. E. Ebeler, K. Kubota, G. R. Takeoka, (Editors), ACS Symposium Series 988, ISBN: 978-0-8412-7411-2

Gomez-Rico, A.; Inarejos-Garcia, A.M.; Salvador, M.D. & Fregapane, G. (2009) Effect of malaxation conditions on phenol and volatile profiles in olive paste and the corresponding virgin olive oils (*Olea europaea* L. Cv. Cornicabra). *Journal of Agricultural and Food Chemistry*, Vol. 57, pp. 3587-3595

Guth H. & Grosch, W. (1991). A Comparative Study of the Potent Odorants of Different Virgin Olive Oils. *Fat Sci. Technol.* Vol. 93, pp. 335-339

Gutierrez, F.; Arnaud, T.; Miguel, A. & Albi, M.A. (1999) Influence of ecological cultivation on virgin olive oil quality. *Journal of the American Oil Chemists' Society*, 76, 617–621.

Gutiérrez, R.; Olías, J.; Gutiérrez, F.; Cabrera, J. & Del Barrio, A. (1975). Los Métodos Organolépticos y Cromatográficos en la Valoración de las Características Aromáticas del Aceite de Oliva Virgen. *Grasas y Aceites* Vol. 26, pp. 21-31

Hatanaka, A. (1993) The biogeneration of green odor by green leaves. *Phytochemistry*, 34, pp. 1201- 1218.

Herbert, C. G. & Johnstone, R. A. W. (2003). *Mass Spectrometry Basics*, CRC Press, Boca Raton, Florida, USA, pp. 23-27

Inarejos-Garcia, A.M.; Gomez-Rico, A.; Salvador, M.D. & Fregapane, G. (2010) Effect of preprocessing olive storage conditions on virgin olive oil quality and composition. *Journal of Agricultural and Food Chemistry* 58, 4858-4865

International Olive Council. 2007. *Sensory analysis of olive oil. Method for the organoleptic assessment of virgin olive oil (COI/T.20/Doc. No 15/Rev. 2)*.

Jiménez, J; Garcia, M; Rosales, F & Quijano, R. G.; (1978). Componentes Volátiles en el Aroma del Aceite de Oliva Virgen. II. Identificación y Análisis Sensorial de los Eluyentes Cromatográficos. *Grassas e Aceites* Vol 29, pp. 211-218

Kalua, C. M.; Allen, M. S.; Bedgood Jr, D. R., Bishop, A. G. & Prenzler, P. D. (2005). Discrimination of Olive Oils and fruits into Cultivars and Maturity Stages Based on Phenolic and Volatile Compounds. *Journal of Agricultural and Food Chemistry*, Vol. 53, pp. 8054-8062

Kalua, C. M.; Allen, S.; Bedgood Jr, D. R.; Bishop, A. G.; Prenzler, P. D. & Robards, K. (2007). Olive Oil Volatile Compounds, Flavour Development and Quality: A Critical Review. *Food Chemistry*, Vol. 100, pp. 273-286

Kalua, C.M.; Bedgood Jr, D.R.; Bishop, A.G. & Prenzler, P.D. (2008) Changes in Virgin Olive Oil Quality during Low-Temperature Fruit Storage. *Journal of Agricultural and Food Chemistry*, 56, 2415-2422

Kanavouras, A; Kiritsakis, A & Hernandez, R. J. (2005). Comparative Study on Volatile Analysis of Extra Virgin Olive Oil by Dynamic Headspace and Solid phase Microextraction. *Food Chemistry* Vol. 90, pp. 69-79

Kao, J.; Wu, X.; Hammond, E & White, E. (1998). The Impact of Furanoid Fatty Acids and 3-Methylnonane-2,4-dione on the Flavor of Oxidized Soybean Oil. *Journal of American Oil Chemists' Society* Vol. 75, pp. 831-835

Kiritsakis A.K., "Flavor components of olive oil- a review" JAOCS, vol. 75, n° 6, (1998), 673-681.

Kiritsakis, A. (1992). *El Aceite de Oliva*. A. Madrid Vicente Ediciones, Madrid 1992.

Kok, M. F.; Yong, F. M. & Lim, G. (1987). Rapid Extraction Method for Reproducible Analysis of Aroma Volatiles. *Journal of Agricultural and Food Chemistry* Vol. 35, 779-781

Koutsaftakis A.; Kotsifaki, F. & Stefanoudaki, E. (1999) Effects of Extraction System, Stage of Ripeness, and Kneading Temperature on the Sterol Composition of Virgin Olive Oils, *Journal of the American Oil Chemists' Society*, 76, 1477-1480.

Koutsaftakis A.; Kotsifaki, F. & Stefanoudaki, E. (2000) A Three-year Study on the Variations of Several Chemical Characteristics and Other Minor Components of Virgin Olive Oils Extracted from Olives Harvested at Different Ripening Stages, *Olivae*, 80, 22-27.

Leclerc, J. Miller, M.L. ; Joliet, E. & Rocquelin, G. (1991) Vitamin and mineral contents of carrot and celeriac grown under mineral or organic fertilization. *Biological Agriculture and Horticulture* 7, 339–348

Ledauphin, J.; Saint-Clair, J.F.; Lablanque, O.; Guichard, H.; Founier, N.; Guichard, E. & Barillier, E. D. (2004). Identification of Trace Volatile Compounds in Freshly Distilled Calvados and Cognac Using Preparative Separations Coupled with Gas Chromatography–Mass Spectrometry.*Journal of Agricultural and Food Chemistry*,Vol. 52, pp. 5124-5134

Lee, A. L.; Bartle, K. L. & Lewis, A. C. (2001). A model of Peak Amplitude Enhancement in Ortoghonal Two Dimensional Gas Chromatography. *Analytical Chemistry*, Vol. 73, pp. 1330-1335

Liu, Z. & Phillips, J. B. (1991). Comprehensive Two-Dimensional Gas Chromatography Using an On-Column Thermal Modulator Interface. *J. Chromatogr. Sci.*, Vol. 29, pp. 227-231

López-Feria, S., S. Cárdenas, J.A. García-Mesa, M. Valcárcel. 2007. Quantification of the intensity of virgin olive oil sensory attributes by direct coupling headspace mass spectrometry and multivariate calibration techniques. *Journal of Chromatography A*. 1147, pp. 144–152.

Luna, G.; Morales, M.T. & Aparício, R. (2006) Characterisation of 39 varietal virgin olive oils *Food Chemistry* 98 243–252

Majors, R. E. (2003). Response to the 2003 Gas Chromatography. User Study - Trends in Column Use and Techniques. *LC•GC North America*, Vol. 21, pp. 960-967

Marriott, P. & Shellie, R. (2002). Principles and Applications of Comprehensive Two-Dimensional Gas Chromatography. *Trends in Analytical Chemistry*, Vol. 21, pp. 573-583

Masella, P.; Parenti, A.; Spugnoli, P. & Calamai, L. (2009). Influence of vertical centrifugation on extra virgin olive oil quality. *Journal of the American Oil Chemists' Society*, 86, 1137-1140

Mateus, E.; Barata, R. C.; , Zrostlíkovác, J.; Gomes da Silva, M.D.R. & Paiva, M. R. (2010). Characterization of the Volatile Fraction Emitted by Pinus spp. by One- and Two-Dimensional Chromatographic Techniques with Mass Spectrometric Detection. Journal of Chromatography A, Vol. 1217, pp. 1845-1855.

Matsui, K. (2006). Green leaf volatiles: hydroperoxide lyase pathway of oxylipin metabolism. *Plant Biology*, Vol. 9, pp. 274–280

McMaster, M. & McMaster, C. (1998). GC/MS: A Practical User's Guide, Wiley-VCH.

Montedoro, G. F.; Garofolo, L. & Bertuccioli, M. (1989) Influence of the Cultivars and Pedoclimatic Conditions on the Virgin Olive Oil Quality. Proceedings of the 6th International Flavor Conference, Rethymnon, Crete, Greece, 5-7 July 1989, Ed. G. Charalambous, Elsevier Science Publishers B.V., Amsterdam (The Netherlands) 881-891.

Morales, M. T. & Aparicio, R. (1999). Effect of extraction conditions on sensory quality of virgin olive oil. *Journal American Oil Chemists Society*, 76, 295–300.

Morales, M. T. & Tsimidou, M. (2000). The role of volatile compounds and polyphenols in olive oil sensory quality. In J. Harwood & R. Aparicio (Eds.), Handbook of olive oil: analysis and properties (pp. 393–458). Gaithersburg, MD: Aspen Publishers.

Morales, M. T.; Aparicio, R. & Calvente, J.J. (1996) influence of olive ripeness on the concentration of green aroma compounds in virgin olive oil. *Flavour and Fragrance Journal* 11, 171-178.

Morales, M. T.; Luna, G. & Aparicio, R. (2005). Comparative Study of Virgin Olive oil sensory Defects. *Food Chemistry* Vol. 91, pp. 293-301

Morales, M. T.; Rios, J. J. & Aparicio, R. (1997) Changes in the volatile composition of virgin olive oil during oxidation: flavors and off flavors. *Journal of Agricultural and Food Chemistry*. 45, 2666–2673.

Morales, M.; Berry, A.; McIntyre, P. & Aparicio, R. (1998). Tentative Analysis of Virgin Olive Oil Aroma by Supercritical Fluid Extraction–High-Resolution Gas Chromatography–Mass Spectrometry. *Journal of Chromatography A*, Vol. 819, 267-275

Morales, M; Alonso, M; Rios, J & Aparicio R. (1995). Virgin Olive Oil Aroma: Relationship Between Volatile Compounds and Sensory Attributes by Chemometrics. *Journal of Agricultural and Food Chemistry* Vol. 43, pp. 2925-2931

Morales, M; Aparicio, R & Rios J. (1994). Dynamic Headspace Gas Chromatograpic Method for Determining Volatiles in Virgin Olive Oil. *J. Chromatogr.* Vol. 668, pp. 455-462

Niessen, W. M. A. (2001). *Current Practice of Gas Chromatography-Mass Spectrometry*. Marcel Dekker, Inc., New York, USA, pp.13.

Ninfali, P.; Bacchiocca, M. ; Biagiotti, E. ; Esposto, S.; Servili, M.; Rosati, A. & Montedoro, G.F. (2008). A 3-year study on quality, nutritional and organoleptic evaluation of organic and conventional extra virgin olives oils. *Journal of the American Oil Chemists' Society* 85; 151-158

Noordermeer, M; Veldink, G. & Vliegenthart, J. (2001). Fatty Acid Hydroperoxide Lyase: A Plant Cytochrome p450 enzyme involved in wound Healing and Pest Resistence. *Chembiochem*, Vol. 2, pp. 494-504

Olias, J. M.; Perez, A. G.; Rios, J. J. & Sanz, L. C. (1993) Aroma of virgin olive oils biogenesis of the green odor notes. *Journal of Agricultural and Food Chemistry*. 41, 2368-2373. on extra virgin olive oil quality. *Journal of the American Oil Chemists' Society*, 86, 1137-1140

Pawliszyn, J. (1997). *Solid Phase Microextraction: Theory and Practice.* Wiley-VCH, Inc., New York, USA, 242 pp..

Pérez A.; Luaces, P.; Ríos, J.; Garcia, J. M. & Sanz, C. (2003). Modification of Volatile Compound Profile of Virgin Olive Oil Due to Hot-water Treatment of Olive Fruit. *Journal of Agricultural and Food Chemistry* Vol. 51, pp. 6544-6549

Petrakis, C. (2006) Olive Oil Extraction In *Olive Oil Chemistry and Technology.* Boskou, D., Ed. Champaign, IL, 191-223.

Phillips, J. B. & Beens, J. (1999). Comprehensive Two-Dimensional Gas Chromatography: A Hyphenated Method with Strong Coupling Between the Two Dimensions. *Journal of Chromatography A*, Vol. 856, pp. 331-347

Pizarro, C.; Tecedor, S. R.; Pérez-del-Notario, N.; Sáiz, J. M. G. (2011). Recognition of Volatile Compounds as Markers in Geographical Discrimination of Spanish Extra Virgin Olive Oils by Chemometric Analysis of Non-Specific Chromatography Volatile Profiles, *Journal of Chromatography A*, Vol. 1218, pp. 518–523

Poole, C. F. (2003). *The Essence of Chromatography.* Elsevier, Boston, USA

Ranalli, A. & Angerosa, F. (1996) Integral centrifuges for olive oil extraction. The qualitative characteristics of products. *Journal of the American Oil Chemists' Society* 73, 417–422.

Ranalli, A.; Lucera, L.; Contento, S.; Simone, N. & Del Re, P. (2004) Bioactive constituents, flavors and aromas of virgin oils obtained by processing olives with a natural enzyme extract. *European Journal of Lipid Science and Technology* 106 187–197

Reddy, G. V. P. & Guerrero, A. (2004). Interactions of Insects Pheromones and Plant Semiochemicals. *Trends in Plant Sci.*, Vol. 9, pp. 253-261

Reiners, J. & Grosch, W. (1998). Odorants of Virgin Olive Oils With Different Flavor Profiles. *Journal of Agricultural and Food Chem*istry Vol. 46, pp. 2754-2763

Ribeiro, L. H.; Costa Freitas , A. M. & Gomes da Silva, M. D. R. (2008). The Use of Headspace Solid Phase Microextraction for the Characterization of Volatile Compounds in Olive oil Matrices, *Talanta* Vol. 77, pp. 110-117

Salas, J. (2004) Characterization of alcohol acyltransferase from olive fruit. *Journal of Agricultural and Food Chemistry*, 52, pp. 3155- 3158.

Salas, J. J.; Sanchez, J.; Ramli, U. S.; Manaf, A. M.; Williams, M.; & Harwood, J. L. (2000) Biochemistry of lipid metabolism in olive and other oil fruits. *Progress in Lipid Research 39*, 151–180.

Salas, J.J. & Sánchez, J. (1999a). The decrease of virgin oil flavour produced by high malaxation temperature is due to inactivation of hydroperoxides lyase. *Journal of Agricultural and Food Chemistry*, Mach 1999, 47 (3) 809-812

Salas, J.J. & Sánchez, J. (1999b). Hydroperoxide lyase from olive (Olea europaea) fruits. *Plant science*, 143(1), 19-26

Salas, J.J.; Williams, M.; Harwood, J.L. & Sánchez, J. (1999). Lipoxygenase activity in olive (Olea europaea) fruits. *Journal of the American Oil Chemists Society*, 76(10), 1163-1168

Salas, J; Sánchez, C.; González, G. D. & Aparicio, R. (2005). Impact of the Suppression of Lipoxygenase and Hydroperoxide Lyase on the Quality of the Green Odor in Green Leaves. *Journal of Agricultural and Food Chemistry* Vol. 53, pp. 1648-1655

Schuphan, W. (1974) Nutritional value of crops as influenced by organic and inorganic fertilizer treatments: results of 12 years' experiments with vegetables (1960-1972). Qualitas plantarum. *Plant Foods for Human Nutrition* 23, 333-350.

Scott, R. P. W. (1996). *Chromatographic Detectors. Design, Function, and Operation.* Marcel Dekker, Inc. New York, USA

Servili M., Montedoro G. (2002). Contribution of phenolic compounds to virgin olive oil quality", *European Journal of Lipid Science Technology*, 104 (2002) 602- 613.

Servili, M.; Esposto, S.; Lodolini, E.; Selvaggini, R.; Taticchi, A.; Urbani, S.; Montedoro, G.; Serravalle, M. & Gucciservili, R.. (2007) Irrigation Effects on Quality, Phenolic Composition, and Selected Volatiles of Virgin Olive Oils Cv. Leccino. *Journal of Agricultural and Food Chemistry* 55, 6609-6618.

Servili, M.; Selvaggini, R. & Esposto, S. (2004) Review, Health and Sensory of Virgin Olive Oil Hydrophilic Phenols: Agronomic and Technological Aspects of Production That Affect Their Occurrence in the Oil, *Journal of Chromatography A*, 1054, 113-127.

Servili, M.; Selvaggini, R.; Taticchi, A.; Esposto, S. & Montedoro, G.F. (2003). Volatile compounds and phenolic composition of virgin olive oil: optimization of temperature and time exposure of olive pastes to ait contact during the mechanical extraction process. *Journal of Agricultural and Food Chemistry* 51, 7980-7988

Servili, M; Conner, J; Piggott, J.; Withers, S. & Paterson, A. (1995). Sensory Characterization of Virgin Olive Oil Relationship with Headspace Composition. *J. Sci. Food Agric.* Vol. 67, pp. 61-70

Servili, M;Taticchi, A.; Esposto, S.; Urbani, S.; Selvaggini, R. & Montedoro, G.F. (2007) Effect of Olive Stoning on the Volatile and Phenolic Composition of Virgin Olive Oil. *Journal of Agricultural and Food Chemistry* 55, 7028-7035

Smith, R. M. (2004). *Understanding Mass Spectra: A Basic Approach.* 2nd edition, Wiley-Interscience, John Wiley & Sons, Inc., USA

Solinas, M.F.; Angerosa, F & Marsilio,V. (1988) Research of some flavour components of virgin olive oil in relation to olive varieties. *Rivista Italiana delle Sostanze Grasse* 65, 361-368.

Stefanoudaki, E.; Williams, M. & Harwood, J. (2010) Changes in virgin olive oil characteristics during different storage conditions. *European Journal of Lipid Science and Technology* 112, 906-914

Swinnerton, J. W.; Linnenbom, V. J. & Cheek, C. H. (1962). Revised Sampling Procedure for Determination of Dissolved Gases in Solution by Gas Chromatography. *Analytical Chemistry* Vol. 34, pp. 1509-1511

Tabanca, N.; Demirci, B.; Ozek, T.; Kirimer, N.; Can Base, K. H.; Bedir, E. ; Khand, I. A. & Wedge, D. E. (2006). Gas Chromatographic–Mass Spectrometric Analysis of Essential Oils from *Pimpinella* Species Gathered from Central and Northern Turkey. *Journal of Chromatography A*, Vol. 1117, pp. 194-205

Temime, S.B.; Campeol, E., Cioni, P.L.; Daoud, D. & Zarrouk, M. (2006). Volatile compounds from Chetoui olive oil and variations induced by growing area. *Food Chemistry* 99, 315-325

Tena, N.; Lazzez, A.; Aparicio-Ruiz, R. & García-González, D. L. (2007) Volatile compounds characterizing Tunisian Chemlali and Chétoui virgin olive oils. *Journal of Agricultural and Food Chemistry*, 55, 7852-7858.

Tovar, M.J.; Romero, M.P.; Girona, J. & Motilva, M.J. (2002) L-phenylalanine ammonia-lyase activity and concentration of phenolics in developing olive (Olea europaea L cv Arbequina) fruit grown under different irrigation regimes. *Journal of the Science of Food and Agriculture* 82, 892–898.

Tura, D; Prenzler, P.D.; Bedgood Jr, D.R.; Antolovich, M. & Robards, K. (2004). Varietal processing effects on the volatile profile of Australian olive oils. *Food Chemistry* 84, 341-349

van Willige, R.; Linssen, J. & Voragen, A. (2000). Influence of Food Matrix on Absorption of Flavour Compounds by Low Density polyethylene: Oil and Real Food Products. *J. Sci. Food Agric.*, Vol. 80, pp. 1790-1797

Vas, G. & Vékey, K. (2004). Solid-Phase Microextraction: A Powerful Sample Preparation Tool Prior to Mass Spectrometric Analysis. *J. Mass Spectrom.*, Vol. 39, pp. 233-254

Vaz-Freire, L. T. ; Gomes da Silva, M. D. R. & Costa Freitas, A. M. (2009). Comprehensive Two-Dimensional Gas Chromatography for Fingerprint Pattern Recognition in Olive Oils Produced by Two different Techniques in Portuguese Olive Varieties *Galega Vulgar, Cobrançoosa* and *Carrasquenha. Anal. Chim. Acta* Vol. 633, pp. 263–270

Venkatramani, C. J.; Xu, J. & Phillips, J. B. (1996). Separation Orthogonality in Temperature-Programmed Comprehensive Two-Dimensional Gas Chromatography. *Analytical Chemistry*, Vol 68, pp. 1486-1492

Vichi, S.; Castellote, A.; Pizzale, L.; Conte, L; Buxaderas, S. & Tamames, E. L. (2003a). Analysis of Virgin Olive Oil Volatile Compounds by Headspace Solid-phase Microextraction Coupled to Gas Chromatography With Mass Spectrometric and Flame Ionization Detection. *J. Chromatogr.* Vol. 983, pp. 19-33

Vichi, S.; Pizzale, L.; Conte, L. S.; Buxaderas, S. & Tamames, E. L. (2005). Simultaneous Determination of Volatile and Semivolatile Aromatic Hydrocarbons in Virgin Olive Oil by Headspace Solid-Phase Microextraction Coupled to Gas Chromatography/mass Spectrometry. *Journal of Chromatography A* Vol. 1090, pp. 146–154

Vichi, S.; Pizzale, L.; Conte, L; Buxaderas, S. & Tamames, E. L. (2003b). Solid-phase Microextraction in the Analysis of Virgin Olive Oil Volatile Fraction: Characterization of Virgin Olive Oils from Two Distinct Geographical Areas of Northern Italy. *Journal of Agricultural and Food Chemistry* Vol. 51, pp. 6572-6577

Vichi, S.; Pizzale, L.; Conte, L; Buxaderas, S. & Tamames, E. L. (2003c). Solid-phase Microextraction in the Analysis of Virgin Olive Oil Volatile Fraction: Modifications Induced by Oxidation and Suitable Markers of Oxidative Status. *Journal of Agricultural and Food Chemistry* Vol. 51, pp. 6564-6571

Vichi, S; Romero, A.; Tous, J. ; Tamames, E. L & Buxaderas, S. (2008). Determination of Volatile Phenols in Virgin Olive oils and their Sensory Significance. *Journal of Chromatography A.*, Vol. 1211, pp. 1-7

Wilkes, J. G.; Conte, E. D.; Kim, Y.; Holcomb, M.; Sutherland, J. B. & Miller, D. W. (2000). Sample Preparation for the Analysis of flavors and Off-Flavors in Foods. *Journal of Chromatography A*, Vol. 880, pp. 3-33

Zhang, Z. & Pawliszyn, J. (1993). Headspace Solid Phase Microextraction. *Analytical Chemistry* Vol. 65, pp. 1843-1852.

Zunin , P.; Boggia, R.; Lanteri, S.; Leardi, R.; Andreis, R. & Evangelisti, F. (2004). Direct Thermal Extraction and Gas Chromatographic–Mass Spectrometric Determination of Volatile Compounds of extra-Virgin Olive Oils *Journal of Chromatography A*, Vol. 1023, pp. 271-276

3

Optical Absorption Spectroscopy for Quality Assessment of Extra Virgin Olive Oil

Anna Grazia Mignani[1], Leonardo Ciaccheri[1*],
Andrea Azelio Mencaglia[1] and Antonio Cimato[2]
[1]CNR – Istituto di Fisica Applicata "Nello Carrara"
[2]CNR – Istituto per la Valorizzazione del Legno e delle Specie Arboree
Italy

1. Introduction

Light travels through space in the form of electromagnetic waves of different wavelengths. The entire wavelength range represents the electromagnetic spectrum. Spectroscopy studies the interaction between light and matter, in order to draw information about the chemical composition inside (Lee et al., 2011). Figure 1 shows the various bands of the electromagnetic spectrum. This chapter refers to measurements performed in the 200-2500 nm band, which is usually subdivided into three portions: the ultraviolet (UV), the visible (VIS) – perceivable by human eyes – and the near-infrared (NIR). They correspond to the 200-400 nm, 400-780 nm, and 780-2500 nm ranges, respectively.

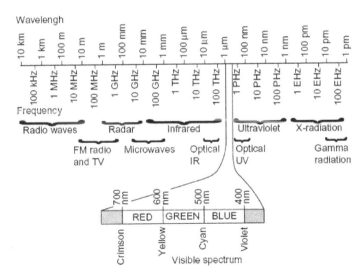

Fig. 1. The electromagnetic spectrum

---

* Corresponding Author

A light beam illuminating an olive oil sample gives rise to reflected, transmitted, and scattered intensities. Optical absorption spectroscopy, as shown in Figure 2, makes use of a broadband UV-VIS-NIR source of intensity I_0 to illuminate the olive oil sample. Then, the transmitted light intensity I, as a function of the illumination wavelength is measured. The change in light intensity, providing the transmittance T, is determined by the molar absorptivity ε, the concentration of absorbing species C, and the optical path L, via the Lambert-Beer relationship, expressed by Equations 1 and 2. T is frequently expressed logarithmically as in Equation 3, to give the so called optical absorbance A, which results linearly dependent on concentration.

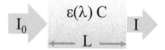

Fig. 2. Optical absorption spectroscopy: the working principle

$$I = I_0 \exp\left(-\varepsilon C L\right) \tag{1}$$

$$T = \frac{I}{I_0} = \exp(-\varepsilon C L) \tag{2}$$

$$A = \log\frac{I_0}{I} = \varepsilon C L \tag{3}$$

This chapter focuses on extra virgin olive oil quality evaluation achieved by means of UV-VIS-NIR absorption spectroscopy. The composition of olive oil is about 98% triglycerides and approximately 2% non glycerid constituents.

- The UV spectrum involves the electronic absorption of fatty acids; in particular, the 230-270 nm band shows high absorption when conjugated dienes and trienes of unsaturated fatty acids are present. For this reason, the absorbances measured at 232 nm and 270 nm, namely K_{232} and K_{270}, provide an official method for olive oil quality control, which is capable of detecting product oxidation and adulteration by means of rectified oils. In addition, the 300-400 nm band provides information about polyphenols (Jiménez Márquez, 2003; Cerretani et al., 2005).
- The VIS spectrum reveals the presence of dyes and pigments (Wrolstad et al., 2005). A and B chlorophylls and their derivatives (pheophytins), carotenoids, and flavonoids such as anthocyanins present distinctive absorption bands in the VIS.
- The wide NIR range is informative for the molecular structure of fats, thanks to the presence of overtones and combinations of vibrational modes of C-H and O-H bonds (Osborne et al., 1993; Ozaki et al., 2007).

In practice, the entire UV-VIS-NIR absorption spectrum can be considered an optical signature, a sort of univocal *fingerprint* of the olive oil. The spectroscopic data can be suitably processed for obtaining a correlation to quality indicators, to the geographic origin of production, to product authenticity as well as to adulteration detection.

2. Instrumentation

Absorption spectroscopy in the UV-VIS-NIR range is one of the most popular measuring methods of conventional analytic chemistry (Mellon, 1950; Bauman, 1962). The most relevant advantages offered are:

- direct measurement: little or no sample preparation is necessary; consequently, the analysis is simple, fast, and does not require manual intervention;
- non-destructive analysis by means of a small quantity of sample;
- compatibility for use in an industrial setting by means of compact instruments.

The conventional instrument for absorption spectroscopy is the double-beam spectrophotometer, the working principle of which is depicted in Figure 3. Since quartz-based optical fibers are transparent in the UV-VIS-NIR range, they are used to equip conventional spectrophotometers by flexible means. Indeed, optical fibers offer the unique possibility of localized probing, a particularly attractive feature for online measurements, which can be carried out in real time without any sample drawing. Moreover, the recent availability of bright LEDs and miniaturized spectrometers further enhances the intrinsic optical and mechanical characteristics of optical fibers and makes it possible to implement compact and moderate-cost instruments.

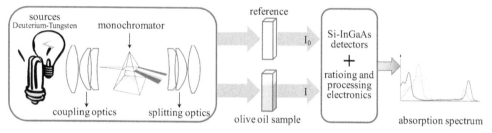

Fig. 3. Working principle of a double-beam spectrophotometer

The conventional spectrophotometer, implemented by means of optical fiber technology, is depicted in Figure 4. In this case, optical fibers are used for both illumination and detection,

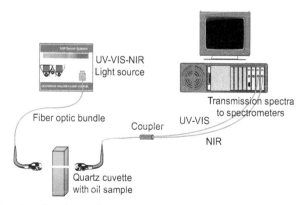

Fig. 4. Fiber optic setup for absorption spectroscopy

and a single cuvette is used. The reference spectrum is measured prior to sample analysis. While the optical fiber strand for illumination can be a single optical fiber or a bundle, the detection is necessarily carried out by means of a bifurcated bundle, or by a coupling device, so as to split the detected light intensity into two spectrometers, for the UV-VIS and the NIR spectroscopic range, respectively.

3. A touch of chemometrics

The chemical information given by an absorption spectrum is contained in the positions and intensities of the absorption bands. Whereas the band positions give information about the appearance and the structure of certain chemical compounds in a mixture, the intensities of the bands are related to the yield of these compounds. Since olive oil contains numerous compounds, the UV-VIS-NIR absorption spectrum shows broad peaks resulting from the convolution of the many overlapping bands, as summarized in Figure 5 (Osborne, 2000).

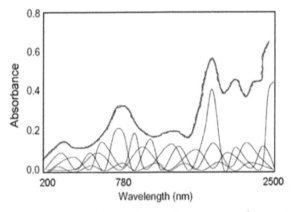

Fig. 5. Absorption spectrum – convolution of absorption spectra of many compounds in a mixture

For qualitative and quantitative analysis, the spectroscopic fingerprinting must be calibrated against reference analytical data from a database of samples representing the best variability in the population. Figure 6 summarizes the steps to follow for achieving a multicomponent analysis from absorption spectroscopy.

- What is needed is a library of representative spectra and relative analytical data to which the spectrum of a test sample may be matched.
- Firstly, a data dimensionality reduction is carried out, which usually leads to a scatter plot where samples are clustered according to the similarity of their spectra. This allows a preliminary inspection of data structure and the detection of what parameters are more likely to be correlated with spectroscopic data.
- Then a more specific analytical tool is chosen according to the type of variable that has to be predicted (quantitative or qualitative). A "Calibration Matrix" is created from which the constituent of interest may be calculated by means of a linear combination of spectroscopic data. The calibration equation has associated statistics which define the closeness of the actual and predicted values. A scatter plot is usually created to detect

any aberrant data. Ideally, the scatter plot should contain data points distributed evenly along a straight line with a narrow confidence limit.

- A validation procedure is then applied for testing the effectiveness of calibration method. While the data dimensionality reduction is usually capable of identifying similarities among products, the correlation to quality indicators always needs the further steps of calibration and validation.

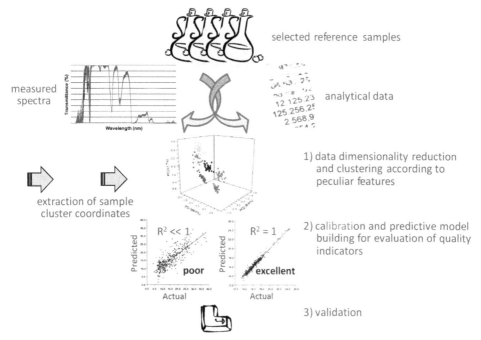

Fig. 6. Steps toward a multicomponent analysis by spectroscopy and chemometrics

Given the nature and complexity of the spectroscopic and analytical data sets involved, many multivariate chemometric techniques have been proposed (Ingle et al., 1988; Mark et al., 2007). The challenge of every multivariate data processing method is to provide excellent classification performance even when few training data are available. Indeed, smart data processing by means of chemometrics makes UV-VIS-NIR optical spectroscopy a rapid and non-destructive method for quality assessment of extra virgin olive oil.

One of the most popular techniques for explorative analysis, data dimensionality reduction, and clustering is the Principal Component Analysis (PCA) that provides the coordinates for identifying the samples in a 2D or 3D map (Jackson, 2003). It linearly combines the spectroscopic data characterizing each oil sample to produce new variables. The coefficients giving the weight of each variable in the linear combination are called *loadings*. The new variables are called Principal Components (PCs), and have the following properties:

- the PCs are mutually uncorrelated (orthogonality);
- the 1st PC (PC1) has the largest variance among all possible linear combination of the starting variables;

- the PCn has the largest variance among all linear combination of the starting variables that are orthogonal to PC1 ... PC(n - 1).

This means that high order PCs have little weight in characterizing the oil samples, and can be disregarded with little loss of information. The loading plots are useful to interpret the score map: they show what variables are important for a given PC. A variable with 0 loading has no importance, a variable with high (positive or negative) loading is important. If a PC has a positive or negative loading at a given wavelength, a sample having high absorbance at that wavelength will tend to have a positive or negative score along that PC.

PCA can also be used for prediction of categorical variables, that is classification, at least in those cases in which they are clearly distinguishable. However, when PCA does not provide satisfactory classification, a more dedicated multivariate data processing tool must be used, such as the Linear Discriminant Analysis (LDA). LDA is a powerful tool that provides both a reduction in dimensionality and automatic object classification. Like PCA, LDA projects a high-dimensional pattern onto a subspace of smaller dimension, but the axes for projection are chosen using a different criterion (Vandeginste et al., 1998).

From the point of view of discriminant analysis, PCA has a drawback: because it weighs the variables in terms of their variance, the features with good discriminating power but limited variance are disregarded. On the contrary, LDA is a tool that is specifically suited for identification, and looks for those variables that show a large spread among different clusters (inter-class variance), but limited variance within each cluster (intra-class variance). Given an N-class problem, the LDA extracts from the data matrix N-1 Discriminating Functions (DFs), which correspond to Principal Components in the PCA, but show a better resolution with regard to poorly-separated clusters. LDA is a *supervised* method. It needs a training set of already classified objects to estimate inter-class and intra-class variances. If all starting variables obey the Gaussian distribution, LDA also provides an easy way to classify unknown patterns. In this case, the points of each cluster in the DF space are distributed following the (N-1)-dimensional Gaussian density function. Thus, classification is achieved simply by evaluating the coordinates of a pattern in the DF space, and then by seeing which of the N density function is higher at that point.

One of the most popular techniques for the prediction of quantitative variables, such as the concentration of an analyte in a multicomponent mixture, is the Partial Least Squares regression (PLS) (Wold et al., 2001). This method is used when the predictor matrix has many collinear variables and the usual Multiple Linear Regression (MLR) cannot be applied. PLS looks for a limited number of PLS factors (PF) which are linear combinations of the original predictors. These new variables are mutually orthogonal (thus uncorrelated) and have the maximum possible covariance with the target variable, among all possible combinations of the original predictors. The idea is that each PF should be linked to a different source of data variance, with the first PF being the most linked to the target variable. The estimation of the optimal number of factors needed to fit the data is a critical issue of PLS. The optimal number of factors is usually assessed by testing each PLS model on the validation set and by minimizing the RMSEP (Root Mean Square Error of Prediction). There are two other fundamental parameters for assessing the goodness of the fit: the RMSEC (Root Mean Square Error of Calibration) and the determination coefficient (R^2), respectively. RMSEC is, like RMSEP, an estimation of the expected prediction error, but is evaluated on the calibration set. The closeness of RMSEC and RMSEP provides an estimation of robustness of the predictive model. R^2 is, instead, the squared correlation coefficients between predicted and reference values, for the calibration set; thus the fit is as better as this value is closer to 1 ($R^2 \leq 1$).

4. Scattered colorimetry – Looking at olive oil as it is

The extra virgin olive oil is a blend of *cultivars* – different varieties of fruit species – which determine the blend's organoleptic properties. In addition to a distinctive taste, each oil blend has a distinctive color and turbidity. Color is mainly determined by the pigment content of the olives, and by the stage of ripening when they are harvested, whereas turbidity is mainly related to the mill type. Moreover, some commercial oils are filtered before bottling, while other oils are bottled unfiltered to provide a natural appearance. Anyway, both color and turbidity provide a means for oil assessment, and standard techniques are used for their independent measurement.

- Color: the intensity of a white-light source crossing the liquid is measured by a spectrometer, giving the transmission spectrum. The chromaticity coordinates L^*, a^*, and b^* of the CIE1976 Chromaticity Diagram, are then computed (Figure 7, top-left) (Billmeyer et al., 1981; Hunt, 1987).
- Turbidity: the intensities of the monochromatic light crossing the liquid along its axis and scattered at 90° are measured. The ratio between the two is the turbidity in nephelometric turbidity units (Figure 7, top-right) (ISO 7027, 1999).

These color and turbidity measurement methods are popular because of their generality, simplicity, and applicability. In the case of color, since the definition of color is independent of the substance, the method can be applied to any liquid. In the case of turbidity, the independence of the test material is attained by reference to an ISO standard turbid material.

However, both methods view the characteristics separately, i.e., the color method never considers the liquid's turbidity, while the turbidity method never considers the color. Usually, in order to avoid mutual interferences, the olive oil is filtered prior to color measurements, while turbidity measurements are performed at a color-independent wavelength – typically at 830 nm. In order to evaluate the olive oil sample as it is, a new technique combining color and turbidity measurements was proposed, which was called scattered colorimetry or spectral nephelometry, since it extended the color and turbidity standards by adding light sources and observation angles, as shown in Figure 7-bottom (Mignani et al., 2003).

Scattered colorimetry makes use of four white light sources, which span the 450-630 nm spectroscopic range, and a miniaturized optical fiber spectrometer as detector. The sources, positioned at different angles with respect to the detector, are sequentially switched on to measure, in addition to the transmitted spectrum, the scattered spectra at the given angles. The transmitted spectrum mainly provides information regarding color, which is also dependent on the turbidity, while the scattered spectra mainly provide information on turbidity, which is also dependent on the color. Measurements on a vial filled by distilled water are carried out prior to sample analysis, so as to obtain reference signals. Then, the spectroscopic data are processed by means of chemometrics, thus providing few coordinates that summarize the combined effects of color and turbidity.

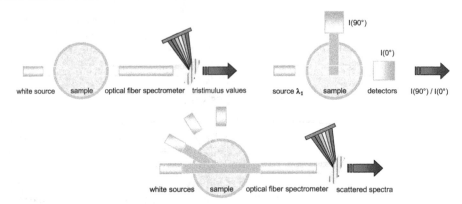

Fig. 7. The working principle of scattered colorimetry (bottom), compared to colorimetry (top-left) and turbidimetry (top-rignt)

Scattered colorimetry showed effectiveness in discriminating the geographic regions of production (Mignani et al., 2005). Figure 8 shows the clustering maps of a couple of experiments carried out by analyzing collections of extra virgin olive oils produced in different geographic regions of the Mediterranean area. Figure 8-left shows the 3D map obtained by LDA processing the spectra of a collection of 236 oils. The collection comprises 115 Tuscan and 53 Calabrian extra virgin olive oils produced by traditional methods, and 68 oils (58 extra virgin and 10 non extra virgin) purchased from retailers. The Tuscan oils are clearly distinguishable from the other extra virgin oils as are the Calabrian oils. As expected, the cluster of non extra virgin olive oils is distinctly separated. Figure 8-right shows another 2D map, again achieved by means of LDA processing of spectroscopic data of a collection of 270 extra virgin olive oils artesanally produced. This collection comprises 213 Italian (90 Tuscan and 123 Sicilian) and 57 Spanish samples. The olive oils of the two countries have intrinsic differences influenced by the diverse cultivars, weather conditions, harvest times, and production methods. An evident clustering according to geographic regions of origin is achieved.

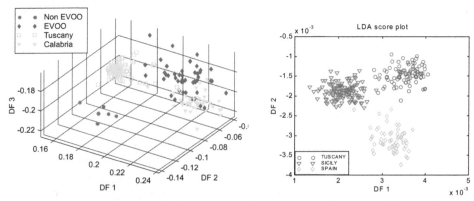

Fig. 8. Effectiveness of scattered colorimetry for discriminating the geographic area of production: Italian extra virgin olive oils from different regions and oils from retailers (left); Spanish and Italian oils (right) (with kind permission of Elsevier and SPIE)

5. UV-VIS-NIR absorption spectroscopy

The little VIS range used in scattered colorimetry only allows to classify oils according to their geographic region of production. In this paragraph, traditional absorption spectroscopy is discussed, which is applied to filtered olive oil samples in order to avoid the influence of turbidity. Indeed, traditional absorption spectroscopy, with or without optical fibers, is certainly the most frequently used spectroscopic technique for olive oil analysis. Since every spectroscopic range is bringing peculiar information, the combination of UV-VIS, or VIS-NIR bands, or the entire UV-VIS-NIR range makes it possible to achieve wider information. Nutraceutic parameters can be predicted, which allow to recognize olive oils of different qualities. In addition, mixtures of extra virgin with lower quality oils can be detected.

Many types of extra virgin olive oils of the Mediterranean region were classified according to their geographic origin by means of absorption spectroscopy combined to chemometrics. The NIR band was used to classify French oils from several regions holding quality labels as registered designation of origin; squalene and fatty acids were also predicted (Galtier et al., 2007). The NIR was also used to classify Spanish oils (Bertran et al., 2000), to predict acidity and peroxide index (Armenta et al., 2007), and to detect and quantify the adulteration with sunflower and corn oil (Özdemir et al., 2007) and other vegetable oils (Christy et al., 2004; Öztürk et al., 2010). Greek oils from Crete, Peloponnese and Central Greece were classified both by the UV-VIS (Kružlicová et al., 2008) and the VIS-NIR bands (Downey et al., 2003), the latter being effective for detecting the adulteration with sunflower oil (Downey et al., 2002). The UV-VIS band was also used to detect the adulteration of extra virgin olive oils mixed with lower quality olive oils (Torrecilla et al., 2010).

The entire UV-VIS-NIR spectrum was exploited to both classify according to geographic region of production, and to predict quality indicators of Italian extra virgin olive oils. The spectra shown in Figure 9-left refer to a collection of 80 extra virgin olive oils produced in four different regions of Italy: Lombardy, Tuscany, Calabria and Sicily. Lombardy is located in the northern part of Italy, Tuscany in the center, while Calabria and Sicily in the south, being Sicily the southest region. A chemometric data processing of these spectra allowed to achieve the regional clustering shown in Figure 9-right, and to predict quality indicators, as shown in Tables 1 and 2 (Mignani et al., 2008).

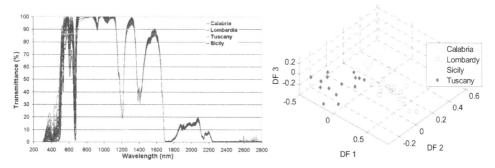

Fig. 9. Transmission spectra in the UV-VIS-VIR ranges of 80 extra virgin olive oil samples from four Italian regions (left), and regional clustering obtained by a chemometric processing of spectroscopic data (right) (with kind permission of SPIE)

Quality parameters	Calibration range	Spectroscopic range (nm)	Number of PLS regressors	R²
Oleic acidity (% oleic acid)	0.12 - 1.555	780 - 2500	3	0.8407
Peroxide value (meq O / kg. oil)	3.76 - 13.98	1000 – 2333	2	0.9628
K_{232}	0.922 - 1.548	1333 – 2222	3	0.9942
K_{270}	0.062 - 0.1178	1333 – 2222	3	0.9825
ΔK	-0.004 - 0.01	1333 - 2222	2	0.4344

Table 1. Prediction of quality parameters of the extra virgin olive oil collection of Figure 9

Fatty acids	Calibration range (%)	Spectroscopic range (nm)	# PLS regressors	R²
Oleic	65.847 - 76.334	1333 - 2222	1	0.9986
Palmitic	9.62 - 17.113	300 - 2300	2	0.9847
Linoleic	4.469 - 10.95	1333 - 2222	1	0.9553
Stearic	2.565 - 4.046	780 - 2500	2	0.9942
Palmiticoleic	0.367 - 1.457	1333 - 2222	2	0.9504
Linolenic	0.646 - 1.066	1000 - 2300	1	0.9822
Arachiric	0.382 - 0.642	1000 - 2222	1	0.9896
Eicosenoic	0.212 - 0.431	1000 - 2300	2	0.9821
Behenic	0.042 - 0.411	300 - 2300	2	0.8892
Heptadecenoic	0.053 - 0.356	300 - 2300	2	0.8081
Heptadecanoic	0.025 - 0.29	1000 - 2300	2	0.8337
Lignoceric	0.026 - 0.205	1333 - 2222	1	0.8532

Table 2. Prediction of fatty acids of the Sicilian extra virgin olive oils of Figure 9

6. Diffuse-light absorption spectroscopy

The scattered colorimetry technique allows for assessing the olive oil by considering both color and turbidity. Indeed, although the intrinsic turbidity of the oil can be regarded as a peculiar characteristic, it has an unstable and non-reproducible influence on absorption measurements because of its time dependent nature. In fact, suspended particles created during production of the olive oil usually settle down in a non-reversible way, because they tend to aggregate at the bottom of the container, creating a sort of sludge. Absorption spectroscopy in the UV-VIS-NIR of filtered samples demonstrated effectiveness to achieve wider quality information. However, sample filtering is not only a time-consuming procedure, but is also an action that alters the composition of the sample. In fact, turbidity is also due to the presence of water, and water removal causes a serious loss of water-soluble compounds–such as polyphenols–that are responsible for the unusual character and authenticity of olive oil.

Diffuse-light absorption spectroscopy, that is, spectroscopy carried out by means of an integrating cavity, is an alternative spectroscopic technique which allows to achieve scattering-free absorption spectra, that is, without caring about the intrinsic turbidity of the olive oil. It has been proposed in the literature as an effective method for overcoming scattering problems in process control (Fecht et al., 1999) and biological applications (Merzlyak et al., 2000), as well as for more general quantitative spectrophotometry (Jàvorfi et al., 2006).

Diffuse-light absorption spectroscopy makes use of an integrating sphere that contains the sample under test. The source and the detector are butt-coupled to the sphere, as shown in Figure 10.

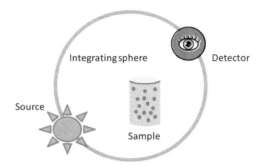

Fig. 10. Setup for diffuse-light absorption spectroscopy by means of an integrating sphere

Almost all the light shining on the sphere surface is diffusely reflected, and the detector can be placed anywhere in the sphere in order to gather the average flux (Elterman, 1970; Fry et al., 1992; Nelson et al., 1993; Kirk, 1995). By inserting an absorbing medium in the cavity, a reduction of the radiance in the sphere occurs. The reduction is related to the absorption of the medium and to its volume, and is independent of non-absorbing objects within it, such as suspended scattering particles. The light intensity detected by means of this measuring setup is described by Equation 4:

$$I = \frac{R\,I_0\,A_d}{S}\,\frac{1}{1 - \dfrac{R}{S}\left(S - A_s - \alpha\,V\right)} \tag{4}$$

where I_0: source power; I: detected power; $\alpha = \varepsilon C$: sample absorption coefficient; V: sample volume; A_d: detector area; A_s: source area; R: cavity power reflectivity; S: cavity surface area.

This technique was used to detect the adulteration of high quality extra virgin olive oils produced in Tuscany caused by lower quality olive oils such as olive pomace, refined olive pomace, refined olive, and deodorized olive oils (Mignani et al., 2011). Mixtures of four original extra virgin olive oils and the four types of adulterants were artificially created at different adulterant concentration. Figure 11-top shows the diffuse-light absorption spectra of high quality oils (left) and the others used as adulterants (right). Then, chemometrics was applied for achieving adulterant discrimination and prediction of relative concentration. Figure 11-bottom shows the discriminating maps: PCA was capable of discriminating samples adulterated by means of deodorized olive oil (left), while a deeper LDA processing was needed for discriminating the other adulterants (right).

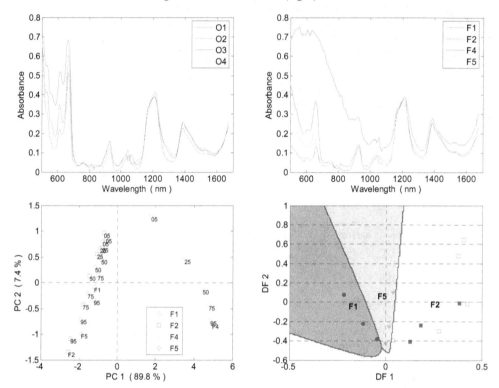

Fig. 11. Top: Diffuse-light absorption spectra of original extra virgin olive oils (left) and other lower quality olive oils commonly used as adulterants (right). Bottom: Discriminating maps obtained by chemometric data processing of absorption spectra (with kind permission of Springer Science+Business Media)

7. Perspectives

Absorption spectroscopy, carried out in the UV-VIS-NIR and combined with chemometrics, can be used for authentication, fraud detection, as well as for quantifying quality indicators. Optical fibers can be used for localized or online analyses.

Absorption spectroscopy for assessing the extra virgin olive oil can be performed also in other spectroscopic bands of the electromagnetic spectrum, especially in the mid-infrared (MIR) and far-infrared (FIR). Recently, NIR and MIR bands were successfully combined and showed effectiveness not only for classifying Italian and French olive oils (Sinelli et al., 2008; Galtier et al., 2011), but also for a classification on the basis of the fruity sensory attribute (Sinelli et al., 2010). The instrumentation for MIR absorption spectroscopy makes use of other types of sources/detectors/optics and measuring schemes. Usually, the attenuated total reflection modality is used, which allows absorption measurements even in high absorption liquids, like olive oil is in the MIR.

Other optical spectroscopic techniques are emerging for the quality assessment of extra virgin olive oils, especially the fluorescence and the Raman spectroscopies. The interested readers can have a look at the most recent bibliography (Ross Kunz et al., 2011; El-Abassy et al., 2010; Paiva-Martins et al., 2010; El-Abassy et al., 2009; Tena et al., 2009; Zou et al., 2009; Sikorska et al., 2008; Poulli et al., 2007).

8. Acknowledgments

Many sponsors and people contributed to the scientific work of the authors of this chapter. The following deserve a special acknowledgement:

- Projects: EC#SMT4-CT97-2157 "Optical Technologies for Intelligent Monitoring Online", EC#003887 "Network of Excellence for Micro-Optics", EC#224014 "Network of Excellence for Biophotonics", Regional Board of Tuscany "Biotecnological characterization and traceability of Tuscan wines and oils", the Spanish Government's Ministry of Science and Technology by means of TEC'2005-08218-C02-02 and TEC'2007-67987-C02-01 projects.
- People: Heidi Ottevaere and Hugo Thienpont, Vrije Universiteit Brussel–Belgium; Cristina Attilio, formerly at CNR IVALSA–Italy; Lanfranco Conte and Milena Marega, Università degli Studi di Udine–Italy; Angelo Cichelli, Università degli Studi di Pescara–Italy; Pilar Beatriz García-Allende and Olga Maria Conde, Universidad de Cantabria–Spain; Debora Tura, Università degli Studi di Milano–Italy.

9. References

Armenta, S., Garrigues, S. & de la Guardia, M. (2007). Determination of edible oil parameters by near infrared spectrometry. *Analytica Chimica Acta*, Vol. 596, pp. 330-337.

Bauman, R.P. (1962). *Absorption Spectroscopy*, John Wiley & Sons Inc., New York.

Bertran, E., Blanco, M., Coello, J., Iturriaga, H., Maspoch, S. & Montoliu, I. (2000). Near infrared spectrometry and pattern recognition as screening methods for the authentication of virgin olive oils of very close geographical origins. *Journal of Near Infrared Spectroscopy*, Vol. 8, pp. 45-52.

Billmeyer, F.W. Jr. & Saltzman, M. (1981). *Principles of Color Technology*, Wiley, New York.

Cerretani, L., Bendini, A., Rotondi, A., Lercker, G. & Gallina Toschi, T. (2005). Analytical comparison of monovarietal virgin olive oils obtained by both a continuous industrial plant and a low-scale mill. *European Journal of Lipid Science and Technology*, Vol. 107, pp. 93-100.

Christy, A.A., Kasemsumra, S., Du, Y. & Ozaki, Y. (2004). The detection and quantification of adulteration in olive oil by near-infrared spectroscopy and chemometrics. *Analytical Sciences*, Vol. 20, pp. 935-940.

Downey, G., McIntire, P. & Davies, A.Y. (2002). Detecting and quantifying sunflower oil adulteration in extra virgin olive oils from the Eastern Mediterranean by visible and near-infrared spectroscopy. *Journal of Agricultural and Food Chemistry*, Vol. 50, pp. 5520-5525.

Downey, G., McIntyre, P. & Davies, A.N. (2003). Geographic classification of extra virgin olive oils from the Eastern Mediterranean by chemometric analysis of visible and near-infrared spectroscopic data. *Applied Spectroscopy*, Vol. 57, pp. 158-163.

El-Abassy, R.M., Donfack, P. & Materny, A. (2009). Visible Raman spectroscopy for the discrimination of olive oils from different vegetable oils and the detection of adulteration. *Journal of the Science of Food and Agriculture*, Vol. 40, pp. 1284-1289.

El-Abassy, R.M., Donfack, P. & Materny, A. (2010). Assessment of conventional and microwave heating induced degradation of carotenoids in olive oil by VIS Raman spectroscopy and classical methods. *Food Research International*, Vol. 43, pp. 694-700.

Elterman, P. (1970). Integrating cavity spectroscopy. *Applied Optics*, Vol. 9, pp. 2140-2142.

Fecht, I. & Johnson, M. (1999). Non-contact scattering independent water absorption measurement using a falling stream and integrating sphere. *Measurement, Science & Technology*, Vol. 10, pp. 612-618.

Fry, E.S., Kattawar, G.W. & Pope, R.M. (1992). Integrating cavity absorption meter. *Applied Optics*, Vol. 31, pp. 2055-2065.

Galtier, O., Dupuy, N., Le Dréau, Y., Ollivier, D., Pinatel, C., Kister, J. & Artaud, J. (2007). Geographic origins and compositions of virgin oils determinated by chemometric analysis of NIR spectra. *Analytica Chimica Acta*, Vol. 595, pp. 136-144.

Galtier, O., Abbas, O., Le Dréau, Y., Rebufa, C., Kister, J., Artaud, J. & Dupuy, N. (2011) Comparison of PLS1-DA, PLS2-DA and SIMCA for classification by origin of crude petroleum oils by MIR and virgin olive oils by NIR for different spectral regions. *Vibrational Spectroscopy*, Vol. 55, pp. 132-140.

Hunt, R.W.G. (1987). *Measuring Color*, Wiley, New York.

Ingle, J.D.J. & Crouch S.R. (1988). *Spectrochemical Analysis*. Prentice Hall, New Jersey.

ISO-International Organization for Standards. (1999). ISO-7027

Jackson, J.E. (2003). *A User's Guide to Principal Components*, J. Wiley & Sons Inc., Hoboken.

Jàvorfi, T., Erostyàk, J., Gàl, J., Buzàdy, A., Menczel, L., Garab, G. & Razi Naqvi, K. (2006). Quantitative spectrophotometry using integrating cavities. *Journal of Photochemistry and Photobiology B: Biology*, Vol. 82, pp. 127-131.

Jiménez Márquez, A. (2003). Monitoring carotenoid and chlorophyll pigments in virgin olive oil by visible-near infrared transmittance spectroscopy. *Journal of Near Infrared Spectroscopy*, Vol. 11, pp. 219-226.

Kirk, J.T.O. (1995). Modeling the performance of an integrating-cavity absorption meter: theory and calculations for a spherical cavity. *Applied Optics*, Vol. 34, pp. 4397-4408.

Kružlicová, D., Mocák, J., Katsoyannos, E. & Lankmayr, E. (2008). Classification and characterization of olive oils by UV-Vis absorption spectrometry and sensorial analysis. *Journal of Food Nutrition Research*, Vol. 47, pp. 181-188.

Lee, C.H. & Hill, T.W. (2011). *Light-Matter Interaction*. Wiley online library, DOI:10.1002/9783527619016.index

Mark, H. & Workman, J. Jr. (2007). *Chemometrics in Spectroscopy*, Academic Press, London.

Mellon, M.G. (1950). *Analytical Absorption Spectroscopy*, John Wiley & Sons Inc., New York.

Merzlyak, M.N. & Razi Naqvi, K. (2000). On recording the true absorption spectrum and the scattering spectrum of a turbid sample: application to cell suspensions of the cyanobacterium Anabaena variabilis. *Journal of Photochemistry and Photobiology B: Biology*, Vol. 58, pp. 123-129.

Mignani, A.G., Smith, P.R., Ciaccheri, L., Cimato, A. & Sani, G. (2003). Spectral nephelometry for making extravirgin olive oil fingerptints. *Sensors and Actuators B*, Vol. 90, pp. 157-162.

Mignani, A.G., Ciaccheri, L., Cimato, A., Attilio, C. & Smith, P.R. (2005) Spectral nephelometry for the geographic classification of Italian extra virgin olive oils. *Sensors and Actuators B*, Vol. 111-112, pp. 363-369.

Mignani, A.G., García-Allende, P.B., Ciaccheri., L., Conde, O.M., Cimato, A., Attilio, C. & Tura, D. (2008). Comparative analysis of quality parameters of Italian extra virgin olive oils according to their region of origin. *Proceedings of SPIE: Optical Sensors 2008*, Vol. 7003, pp. 700326-1/6.

Mignani, A.G., Ciaccheri, L., Ottevaere, H., Thienpont, H., Conte, L., Marega, M., Cichelli, A., Attilio, C. & Cimato, A. (2011). Visible and near-infrared absorption spectroscopy by an integrating sphere and optical fibers for quantifying and discriminating the adulteration of extra virgin olive oil from Tuscany. *Analytical, Bioanalytical Chemistry*, Vol. 399, pp. 1315-1324.

Nelson, N.B. & Prézelin, B.B. (1993). Calibration of an integrating sphere for determining the absorption coefficient of scattering suspensions. *Applied Optics*, Vol. 32 pp. 6710-6717.

Osborne, B.G., Fearn, T. & Hindle, P.T. (1993). *Practical NIR Spectroscopy with Applications in Food and Beverage Analysis*, Longman Scientific & Technical-Wiley, New York.

Osborne, B.G. (2000). Near-infrared spectroscopy in food analysis, In: *Encyclopedia of Analytical Chemistry*, Meyers, R.A. (Ed.), pp. 1-14, J. Wiley & Sons Ltd., Chichester.

Ozaki, Y., McClure, W.F. & Christy, A.A. (2007). *Near Infrared Spectroscopy in Food Science and Technology*, J. Wiley & Sons, Hoboken.

Özdemir, D. & Öztürk, B. (2007). Near infrared spectroscopic determination of olive oil adulteration with sunflower and corn oil. *Journal of Food and Drug Analysis*, Vol. 15, pp. 40-47.

Öztürk, B., Yalçin, A. & Özdemir, D. (2010). Determination of olive oil adulteration with vegetable oils by near infrared spectroscopy coupled with multivariate calibration. *Journal of Near Infrared Spectroscopy*, Vol. 16, pp. 191-201.

Paiva-Martins, F., Rodrigues, V., Calheiros, R. & Marques, M.P.M. (2010). Characterization of antioxidant olive oil biophenols by spectroscopic methods. *Journal of the Science of Food and Agriculture*, Vol. 91, pp. 309-314.

Poulli, K.I., Mousdis, G.A. & Georgiou, A. (2007). Rapid synchronous fluorescence method for virgin olive oil adulteration assessment. *Food Chemistry*, Vol. 105, pp. 369-375.

Ross Kunz, M., Ottaway, J., Kalivas, J.H., Georgiou, C.A. & Mousdis, G.A. (2011). Updating a synchronous fluorescence spectroscopic virgin olive oil adulteration calibration to a new geographical region. *Journal of the Agricultural and Food Chemistry*, Vol. 59, pp. 1051-1057.

Sikorska, E., Khmelinskii, I.V., Sikorski, M., Caponio, F., Bilancia, M.T., Pasqualone, A. & Gomes, T. (2008). Fluorescence spectroscopy in monitoring of extra virgin olive oil during storage. *International Journal of Food Science and Technology*, Vol. 43, pp. 52-61.

Sinelli, N., Casiraghi, E., Tura, D. & Downey, G. (2008). Characterisation and classification of Italian virgin oil by near- and mid-infrared spectroscopy. *Journal of Near Infrared Spectroscopy*, Vol. 16, pp. 335-342.

Sinelli, N., Cerretani, L., Di Egidio, V., Bendini, A. & Casiraghi, E. (2010). Application of near (NIR) infrared and mid (MIR) infrared spectroscopy as a rapid tool to classify extra virgin olive oil on the basis of fruity attribute intensity. *Food Research International*, Vol. 42, pp. 369-375.

Torrecilla, J.S., Rojo, E., Domínguez, J.C. & Rodríguez, F. (2010). A novel method to quantify the adulteration of extra virgin olive oil with low-grade olive oils by UV-Vis. *Journal of Agricultural and Food Chemistry*, Vol. 58, pp. 1679-1684.

Tena, N., García-González, D.L. & Aparicio, R. (2009). Evaluation of virgin oil thermal deterioration by fluorescence spectroscopy. *Journal of the Agricultural and Food Chemistry*, Vol. 57, pp. 10505-10511.

Vandeginste, B.G.M., Massart, D.L., Buydens, L.C.M., De Jong, S., Lewi, D.J. & Smeyers-Verbeke, J. (1998). *Handbook of Chemometrics and Qualimetrics*, Elsevier Science BV, Amsterdam–chapter 33.

Wold, S., Sjöström, M. & Erikkson, L. (2001). PLS-regression: a basic tool for chemometrics. *Chemometrics and Intelligent Laboratory Systems*, Vol. 58, pp. 109-130.

Wrolstad, R.E., Acree, T.E., Decker, E.A., Penner, M.H., Reid, D.S., Schwartz, S.J., Shoemaker, C.F. & Sporns, P. (2005). *Handbook of Food Analytical Chemistry, Pigments, Colorants, Flavours, Texture, and Bioactive Food Components*, J. Wiley & Sons, Hoboken.

Zou, M.Q., Zhang, X.F., Qi, X.H., Ma, H.L., Dong, Y., Liu, C.W., Guo, X. & Wang, H. (2009). Rapid authentication of olive oil adulteration by Raman spectrometry. *Journal of the Agricultural and Food Chemistry*, Vol. 57, pp. 6001-6006.

4

Metal Determinations in Olive Oil

Sema Bağdat Yaşar, Eda Köse Baran and Mahir Alkan
Balıkesir University
Turkey

1. Introduction

It is widely known that trace metals have negative effects on the oxidative stability of olive oil. Natural composition of olive fruit, natural contamination from soil, fertilizers, industrial applications or highways near the plantations are the main sources of metals in olive oils. The olive oil may also be contamined with the metals during the production process and contact with storage materials. The level of trace metals in olive oil is one of the quality parameters and also effective on oil oxidation and human health. Oxidation leading to the development of unfavorable odours and taste is one of the major reasons of deterioration of olive oils. The factors that most affect the rate of oxidation are the degree of unsaturation, the amount of oxygen, temperature, light and the presence of metals (mainly transition metals such as Fe and Cu) (Meira et al., 2011; Sikwese & Duodu, 2007). The trace metals enhance the rate of oxidation of edible oils by increasing the generation of free radicals from fatty acids or hydroperoxides. Benedet & Shibamoto observed that trace amounts of Fe, Cr, Pb and Cd contribute oxidative effects to lipid peroxidation (Benedet & Shibamoto, 2008).

The determination of metals has been a difficult analytical problem because of the hard organic content of the oil matrix. The analytical techniques used for metal determinations in oils are both emission and absorption spectrophotometry. ICP-OES (Allen et al., 1998; Angioni et al., 2006 ; Anthedimis et al., 2005; Costa et al., 2001; De Souza et al., 2005; Murillo et al., 1999; Zeiner et al., 2005), FAAS (Batı & Cesur, 2002; Carbonell et al., 1991; Köse Baran & Bağdat Yaşar, 2010; Nunes et al., 2011), ETA-AAS (Karadjova et al., 1998; Kowalewska et al., 1999 ; Zeiner et al., 2005), GFAAS (Allen et al.,1998 ; Ansari et al., 2009; Calapaj et al., 1988; Chen et al., 1999; Cindric et al., 2007; De Leonardis et al., 2000; Hendrikse et al., 1988, 1991; Lacoste et al., 1999; Martin-Polvillo et al., 1994 ; Matos Reyes & Campos, 2006; Mendil et al., 2009; Nash et al., 1983; Van Dalen, 1996), and ICP-MS (Benincasa et al., 2007 ; Bettinelli et al., 1995; Llorent-Martinez et al., 2011a, 2011b; Pereira et al., 2010; Wondimu et al., 2000) are the most commonly used techniques for the determination of metal contents in oils (Duyck et al., 2007).

In this chapter, recent determination techniques and sample pretreatment methods have been described and compared with each other. Additionally, a novel metal extraction procedure has been introduced in detail. In recent years, scientists have been interested in defining the bioavailable amount of metals more than total metal concentration. Taking into account this, fractionation and speciation analysis of metals in oil samples have also been discussed in the chapter.

2. Sample pre-treatment

The accurate determination of trace metals in olive oil is an analytical challenge due to their low concentration and the difficulties that arise because of the high organic content. Due to the high organic content, sample pretreatment is a critical step and frequently necessary in olive oil analysis. Sample pretreatment step provides the decomposition of organic matrix or the extraction of metals without matrix decomposition. On the other hand, oil sample can be diluted in a suitable solvent or emulsified with an appropriate emulsifier in a rapid pretreatment for direct determinations. The atomic spectrometers are the most commonly used devices but have some problems such as the reduced stability of the analytes in the solution, requirement of organometallic standards, the use of dangerous organic solvents or sample digestion with an acid or acid mixture (Nunes et al., 2011).

2.1 Acid digestion

Digestion procedures are regularly carried out with either open vessels using acid, acid mixture or basic reagents on hot plates or open- and closed-vessel microwave ovens. The decomposition in open system is hard, time consuming and prone to systematic error sources, i.e. contamination or analyte losses. In case of using microwave radiation, the high cost of instrumentation and dilution of the sample can be considered as disadvantages in the microwave assisted digestion system. Although the amount of sample in vessels is limited due to the generation of gaseous reaction products that can increase of pressure, the use of closed high-pressure vessels is appropriate for efficient sample digestion. On the other hand, in the use of open-focused microwave ovens, the advantages are decreasing the risk to the operator, possible introduction of reagents during procedure, opportunity to digest larger amounts of sample and low cooling time (Sant'Ana et al., 2007).

Microwave-assisted digestion has been performed to dissolve the oil sample for elemental analysis in a large number of papers (Angioni et al., 2006; Ansari et al., 2009; Levine et al., 1999; Llorent-Martinez et al., 2011a, 2011b; Mendil et al., 2009; Sant'Ana et al., 2007), while focused microwave assisted digestion for the same purpose has been employed in a few papers (Sant'Ana et al., 2007). As shown in Table 1, some investigation have been done on microwave digestion for olive oil using various procedures.

2.2 Dry ashing

In general, ashing methods may provide lower analyte recovery and exhibit poorer accuracy compared to acid digestion methods. Although dry ashing procedures are effective, they are time consuming and can often result in loss of analyte species that could occur during the preparation of the sample. Oil is decomposed by high-temperature dry ashing, subsequently the ash is dissolved in an acidic aqueous medium and the metal content of the aqueous phase can be measured by various detection techniques such as AAS, adsorptive stripping voltammetry (AdSV) and derivative potentiometric stripping analysis (dPSA) (Abbasi et al., 2009; Carbonell et al., 1991; Lo Coco et al., 2003). There are limited researches for metal determinations in oils after dry ashing of olive oil (Lo Coco et al., 2003).

Microvawe digestion procedure	Reagent	Metals determined	Reference
130 °C 10 min. 140 psi, 150 °C 10 min. 200 psi, 10 min. cooling, 160 °C 20 min. 200 psi	HNO₃	Mg, Ca, Cr, Fe, Mn, Cu, Ni, Zn, Cd, Pb	Bağdat Yaşar & Güçer, 2004
2 min. for 250 W, 2 min. for 0 W, 6 min. for 250 W, 5 min. for 400 W, 8 min. for 550 W, vent.: 8 min.	HNO₃–H₂O₂	Fe, Mn, Zn, Cu, Pb, Co, Cd, Na, K, Ca, Mg	Mendil et al., 2009
250 W 2 min., 0 W 1 min., 250 W 2 min., 600 W 1 min., 400 W 5 min., vent.: 3min.	HNO₃–H₂O₂	Ca, Fe, K, Mg, Na, Zn, Al, Co, Cu, Mn, Ni, Cr, Pb	Cindric et al., 2007
25ᵢ-90f °C 5 min. 700 W, 90ᵢ-90f °C 3 min. 600 W, 90ᵢ-170f °C 10 min. 600 W, 170ᵢ-170f °C 7 min. 600 W	HNO₃	Ag, As, Ba, Be, Cd, Co, Cr, Cu, Fe, Hg, Mn, Mo, Ni, Pb, Sb, Ti, Tl, V	Llorent-Martinez et al., 2011a, 2011b
300 W (83%) for 15 min., 600 W (75%) for 10 min., 1200 W (65%) for 15 min., 300 W (83%) for 5 min.	HNO₃–H₂O₂	Cd, Cu, Pb, Zn	Angioni et al., 2006
750 W 90 °C 6 min., 750 W 90 °C 4 min., 1000 W 180 °C 8 min., 1000 W 180 °C 15 min. (35 bar), vent.: 20 min	HNO₃–H₂O₂	Cu, Fe, Ni, Zn	Nunes et al., 2011
250 W, 4 min., 0 W 4 min., 250 W 5 min., 400 W 7 min., 700 W 6 min., 350 W 5 min.	HNO₃	Be, Mg, Ca, Sc, Cr, Mn, Fe, Co, Ni, As, Se, Sr, Y, Cd, Sb, Sm, Eu, Gd	Benincasa et al., 2007

i: initial ; f: final

Table 1. The summary of microwave digestion procedures for various metals in olive oil.

2.3 Extraction

Sample preparation involves acid extraction (Anwar et al., 2004; De Leonardis et al., 2000; Dugo et al., 2004; Jacob & Klevay, 1975), solid phase extraction (SPE) (Batı & Cesur, 2002) or extraction with special agents (Köse Baran & Bağdat Yaşar, 2010).

After the extraction of metals from oil with nitric acid, hydrochloric acid or acid mixture, the extracts are analyzed. Despite the fact that extraction method has the same advantage both in the separation and preconcentration of metals in oil samples, the recoveries are not satisfactory for many metals in most cases. Batı and Cesur described another method for the preconcentration and separation of copper in edible oils, based on using a solid Pb-piperazine-dithiocarbamate complex for extraction and a potassium cyanide solution for back extraction (Batı & Cesur, 2002).

Anwar et al. reported a simple acid-extraction method for the determination of trace metals in oils and fats. The method has been performed with the use of ultrasonic intensification and successfully applied for accurate determination of iron, copper, nickel and zinc in oils (Anwar et al., 2004). Many extraction procedures are available in literature, the summary of these is given in Table 2.

Extraction method	Metals determined	Detection technique	Notes	Reference
Extraction with 10% HNO₃	Fe, Cu	GF-AAS	[1]Acc.% 94±23 (Cu); 97±12 (Fe)	De Leonardis et al., 2000
Extraction with CCl₄ + 2 N HNO₃ (ultrasonic intensification)	Fe, Cu, Ni, Zn	FAAS	[2]Rec.% 92-98 (Fe); 91-100 (Cu); 92-97 (Ni); 93-101 (Zn)	Anwar et al., 2004
Extraction with 35% H₂O₂ and 36% HCl (30 min., 90 °C)	Cd, Cu, Pb, Zn	dPSA	Rec.% 96.5±2.1 (Cd); 97.0±2.7 (Cu); 95.0±1.8 (Pb); 93.5±1.7 (Zn)	Dugo et al., 2004
Extraction with conc. HCl	Cu	Ad-SSWV	[3]LOD: 0.49 ng mL⁻¹	Galeano Diaz et al., 2006
Extraction with conc. HNO₃ and 6% H₂O₂	Cu, Ni	UV-Vis spec.	Rec.% 90-118 (Cu); 96-100 (Ni)	Hussain Reddy et al., 2003
Extraction with 10% HNO₃ (50 Hz, 60 s)	Cu, Fe, Mn, Co, Cr, Pb, Ni, Cd, Zn	ICP-AES	[4]RSD%: < 10 (Cu); 5 (Fe); 15 (Mn); 8 (Co); 10 (Cr); 20 (Pb); 5 (Cd); 16 (Ni); 11 (Zn)	Pehlivan et al., 2008
Extraction with CCl₄ and 2 N HNO₃ after pretreatment with conc. HNO₃ (ultrasonic bath, 30 °C)	Fe, Cu, Zn, Ni	FAAS	Rec.% 96.5-97.5 (Fe); 96.5-97.1 (Cu); 95.8-97.5 (Ni); 96.0-97.8 (Zn)	Anwar et al., 2003
Pb-piperazinedithiocarbamate SPE and KCN back-extraction	Cu	FAAS	Rec.% 91-97	Batı & Cesur, 2002
Zn-piperazinedithiocarbamate SPE	Cd	FAAS	Rec.% 93.1-100	Yağan Aşçı et al., 2008
Ultrasonic-Assisted extraction with conc. HNO₃ and H₂O₂ (35 kHz)	Cu, Fe, Ni	FAAS and ETAAS	Rec.% 95.9-98.3 (Cu); 95.7-98.2 (Fe); 95.2-97.5 (Ni)	Ansari et al., 2008
Ultrasonic-Assisted extraction with conc. HCl and 30% H₂O₂	Cu, Pb	SCP	Rec.% 82-107 (Cu); 84-105 (Pb)	Cypriano et al., 2008
Extraction with N,N'-bis(salicylidene)-2,2'-dimethyl-1,3-propanediaminato (LDM)	Fe, Cu	FAAS	Rec.% 100.2±5.6 (Fe); 99.4±2.8 (Cu)	Köse Baran & Bağdat Yaşar, 2010

[1]Acc.: Accuracy; [2]Rec.: Recovery; [3]LOD: Limit of detection; [4]RSD: Relative standard deviation

Table 2. The extraction methods used for metal determination in vegetable oils

A Schiff base has been suggested for the extraction of metals from oils as an appropriate chelating agent under the optimum extraction conditions (Köse Baran & Bağdat Yaşar, 2010). In recent analytical applications, Schiff bases have been used in order to form complexes due to their good complexing capacity with metals (Afkhami et al., 2009; Ashkenani et al., 2009; Ghaedi et al., 2009; Khedr et al., 2005; Khorrami et al., 2004; Köse Baran & Bağdat Yaşar, 2010; Kurşunlu et al., 2009; Mashhadizadeh et al., 2008; Shamspur et al., 2003; Tantaru et al., 2002; Ziyadanoğulları et al., 2008). The chemists have attended to the Schiff bases and their metal complexes because of their widespread applications in biological systems and industrial uses (Issa et al., 2005; İspir, 2009; Kurtaran et al., 2005; Li et al., 2007; Mohamed, 2006; Neelakantan et al., 2008; Prashanthi et al., 2008; Sharaby, 2007).

Although most techniques for metal determinations in edible oils require sample digestion, dilution or emulsification, the improved method can be employed for the same purpose without digestion. The procedure is based on efficient extraction of metals from oil to aqueous solution, and the determination of metals in aqueous phase by FAAS. The proposed approach has been applied for Fe, Cu, Ni and Zn successively. This method includes two main steps. Metal complexes with Schiff bases shown in Fig. 1 were investigated spectrophotometrically as a first step. In this step, the investigation of the complexation reaction as a driving force for the extraction is necessary to decide the appropriate pH and the equilibrium time in terms of complexation efficiency.

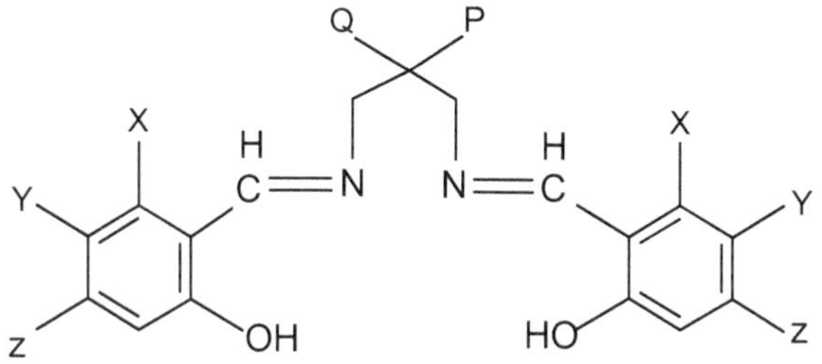

X	-H ; -OCH$_3$
Y	-H ; -OCH$_3$; -Br
Z	-H ; -OCH$_3$; -Br
Q	-H ; -CH$_3$
P	-H ; -CH$_3$; -OH

Fig. 1. Chemical structure of Schiff base used in the extractions

As a second step, the experimental conditions affecting the extraction efficiency of metals should be researched. In the procedure of metal extraction with a Schiff base, the optimization of parameters -the ratio of Schiff base solution volume to oil mass, the stirring time and the temperature- for the metal extractions has been achieved by carrying out central composite design (CCD) as an optimization method.

As shown in Table 3, the CCD consisting of a combination of 2^3 full factorial design and a star design was used, in which three independent factors were converted to dimensionless ones (x_1, x_2, x_3) with the coded values at 5 levels: -1.682, -1, 0, +1, +1.682.

Factors		Levels				
		-1.682	-1	0	+1	+1.682
x_1 (1st factor)	V$_{LDM}$ / m$_{oil}$ ratio (mL g^{-1})	0.159	0.5	1	1.5	1.841
x_2 (2nd factor)	Stirring time (minute)	9.56	30	60	90	110.46
x_3 (3rd factor)	Temperature (°C)	13.18	20	30	40	46.82

Table 3. Variables, levels and the values of levels used in CCD (Köse Baran & Bağdat Yaşar, 2010)

Fifteen experiments should be done in a CCD. Additionally, to estimate the experimental error, replications of factor combinations are necessary at the center point (the level, 0). Experiment at the center point has been repeated five times. The total number of experiments in the CCD with three factors then amounts to 20 (Morgan, 1991; Otto, 1999). Accordingly, 20 experiments given in Table 4 were carried out in the extent of the CCD optimization procedure.

Experiment no.	Coded values of levels		
	V_{LDM} / m_{oil} ratio (mL g⁻¹) x_1	Stirring time (min.) x_2	Temperature (°C) x_3
1	-1	-1	-1
2	+1	-1	-1
3	-1	+1	-1
4	+1	+1	-1
5	-1	-1	+1
6	+1	-1	+1
7	-1	+1	+1
8	+1	+1	+1
9	0	0	0
10	-1,682	0	0
11	+1,682	0	0
12	0	-1,682	0
13	0	+1,682	0
14	0	0	-1,682
15	0	0	+1,682
16	0	0	0
17	0	0	0
18	0	0	0
19	0	0	0
20	0	0	0

Table 4. The coded values of levels for the experiments in the extent of CCD

Organo-metallic standards in oil (Conostan code number; 354770 for iron, 687850 for copper) were used in CCD and metal concentrations of oil standards were fixed to be a certain concentration. The metal concentrations of the extracts gained from each experiment were determined by FAAS. The obtained results were used in order to establish recovery values for the extraction of metals from oil. The response values (y) were calculated from experimentally obtained recovery percentages. The empirical equations were developed by means of response values (Morgan, 1991; Otto, 1999). The following y equations were constructed based on the b values which were calculated by applying to the appropriate matrixes.

$$y = b_1X_1 + b_2X_2 + b_3X_3 + b_{11}X_1^2 + b_{22}X_2^2 + b_{33}X_3^2 + b_{12}X_1X_2 + b_{13}X_1X_3 + b_{23}X_2X_3 + b_{123}X_1X_2X_3 \quad (1)$$

New corresponding equations were obtained by equalization of the derivatives of y equation in terms of x_1, x_2, x_3 to zero and solved using software to provide optimum extraction conditions. Optimum conditions are variable depending on the structure of Schiff base and significant metal. The found optimum conditions are given in Table 5 when LDM (Q and P = CH3; X, Y and Z = H) was used as a Schiff base. The recovery values for the extraction of Cu and Fe from oil under the optimum experimental conditions were found to be 99.4(±2.8) and 100.2(±5.6)%, respectively (n=10). To test the applicability of the improved procedure, it was applied on spiked olive, sunflower, corn and canola oils. The recovery percentages were varied between 97.2-102.1 for Cu and 94.5-98.6 for Fe (Köse Baran & Bağdat Yaşar, 2010).

Metal	Optimum Conditions		
	V_{LDM} / m_{oil} ratio (mL g^{-1})	Stirring time (min.)	Temperature (°C)
Cu	0.76	73	31
Fe	1.19	67	28

Table 5. Optimum extraction conditions for determination of Cu and Fe in edible oils (Köse Baran & Bağdat Yaşar, 2010)

The improved determination strategy after the extraction with Schiff bases has main advantages like independency from hard oil matrix, elimination of explosion risk during decomposition, no requirement for expensive instruments, high accuracy, sensitivity, rapidity and cheapness.

3. Direct determination

The direct determination of metals in oils can be carried out by sample solubilization in an organic solvent, an emulsification procedure in aqueous solutions in the presence of emulsifiers such as Triton X-100 or a solid sampling strategy.

3.1 Dilution with organic solvent

The procedure of the dilution with organic solvents is an easy way to sample pretreatment before detection, but has some requirements: special devices for sample introduction e. g. for FAAS (Bettinelli et al., 1995), the addition of oxygen as an auxiliary gas in ICP-OES or ICP-MS (Costa et al., 2001). The volatile organic solvents have been directly introduced into ICPs for many years, but this can cause plasma instability, less sensitivity, less precision and high cost. Al, Cr, Cd, Cu, Fe, Mn, Ni and Pb contents of olive oil were investigated using diethyl ether, methyl isobutyl ketone (MIBK), xylene, heptane, 1,4-dioxane as solvent and N,N-hexamethylenedithiocarbamic acid, hexamethyleneammonium (HMDC-HMA) salt as a modifier by ETAAS (Karadjova et al., 1998). A transverse heated filter atomizer (THFA) was employed for the direct determination of Cd and Pb in olive oil after sample dilution with n-heptane (Canario & Katskov, 2005). Moreover, Martin-Polvillo et al. (1994) and List et al. (1971) determined trace elements in edible oils based on the direct aspiration of the samples, diluted in MIBK. In another research, the mixture of 2%lecithin-cyclohexane was used to introduce the oil samples to a polarized Zeeman GFAAS (Chen et al., 1999). Van Dalen was also used lecithin and the organopalladium modifier solutions for the injection of the edible oils (Van Dalen, 1996).

3.2 Emulsification

Taking into account parameters such as economy, safety, environment, time, and low risk of contamination, emulsification appears beneficial over microwave assisted acid digestion. On the other hand, optimization of the particle size effect, slurry concentration and homogeneity are necessary in order to obtain good precision and recoveries with slurry techniques. In spite of optimization, complete destruction of the sample matrix in plasma and then liberation of analyte from the sample matrix are not always succeeded, causes unsatisfactory results. An alternative technique for introduction of oil sample directly into ICP is the on-line emulsification (Anthemidis et al., 2005). Direct introduction of oil samples in the form of emulsion into ICP facilitates the spray chamber and plasma torch owing to no need of extra oxygen or sophisticated desolvation device. In such a case, the use of stable emulsions with proper surfactant concentration is very important (Anthemidis et al., 2005).

Emulsification as sample preparation has been performed for the determination of trace metals in vegetable oils by ICP-OES (De Souza et al., 2005; Murillo et al., 1999), ICP-MS (Castillo et al., 1999; Jimenez et al., 2003), FAAS (List et al., 1971) and GF-AAS (Lendinez et al., 2001). Additionally, the use of microemulsion as sample preparation for vegetable oil analysis by High-Resolution Continuum Source FAAS (HR-CS FAAS) has been described by Nunes et al. (2011). The determination of Zn, Cd and Pb in vegetable oils by electrothermal vaporization in combination with ICP-MS (ETV-ICP-MS) was described in literature (Huang et al., 2001).

3.3 Direct solid sampling

Direct introduction of oil samples into the graphite furnace by solid sampling strategy is rarely used, providing an alternative methodology. Due to technical improvements in spectrophotometer and software capabilities of modern instrumentation, this method has not been entirely accepted (Sardans et al., 2010). Direct solid sampling has some advantages such as no sample dilution, satisfactory LOD levels, calibration probability with aqueous analytical solutions, simple analysis and no sample digestion or extraction. Other advantages of this method are reduced time and cost, required little amount of sample and the achievement of high sensitivity. Additionally, it reduces the risk of contamination due to the nonexistence of sample preparation and use of chemical reagents. Some disputes against the method are the difficulty of introducing small sample masses, faulty measurement of the results due to the heterogeneity of some natural samples and the limiting linear working range of AAS (Sardans et al., 2010). Despite these restrictions, direct solid sampling is a reasonable alternative for the determination of the total content of metals in oils, since it needs almost no sample preparation. A method for the direct determination of Ni and Cu in vegetable oils by GFAAS using the solid sampling strategy has been reported without sample dilution by Matos Reyes et al. (2006).

3.4 Flow injection

Various detection techniques like ETAAS, FAAS, ICP-OES, ICP-MS, voltammetry have been utilized for metal determination in oils. However, all of them have the need for sample pretreatment procedures in common like: wet digestion, dry ashing, extraction and dilution with organic solvent in order to eliminate hard organic matrix. In the processing large numbers of samples, flow injection analysis (FIA) systems can be preferred for sample pretreatment. The FIA system for oil analysis is frequently based on the on-line preparation of oil-in-water emulsions by using ultrasonic bath with serious drawbacks in efficient preparation of stable emulsions. By this way, more concentrated emulsions (high oil concentration) can be introduced into the plasma and thereby the LODs were improved. A limited number of researches related to metal determination in oils by FIA systems have been presented. Jimenez et al. succeeded multi-element determination in virgin olive oil by flow injection ICP-MS using with HNO_3 and Triton X-100 as emulsifying agents (Jimenez et al., 2003). A magnetic-stirring micro-chamber has been developed for on-line emulsification and has been successfully employed by Anthemidis et al. to detect Ag, Al, B, Ba, Bi, Ca, Cd, Co, Cr, Cu, Fe, Ga, In, Mg, Mn, Ni, Pb, Tl and Zn in olive oil using flow injection ICP-OES (Anthemidis et al., 2005). The low concentration of analyte in the sample analysed and difficulty of obtaining stable emulsions with rich oil content were reported as the main problems. On-line emulsion preparation procedure was suggested as simpler, more effective, less time consuming, less labor intensive, less matrix interferences and less contamination risk over the other direct sample introducing procedures. The direct determination of Cu and Fe in edible oils based on the flow injection standard addition method by FAAS was performed without sample dilution in a previous study (Carbonell et al., 1991).

As mentioned above, various pretreatment procedure and detection techniques have been employed for the total determination of metals in olive oil. The researchers have dealt with metallic contents of olive oils during last few decades. As can be seen in Table 6, the concentration range of total amount is given for many metals.

Metal	Concentration (μg g^{-1}) (* ng g^{-1})		References
	Minimum	Maximum	
Fe	12.5*	139.0	(Anthemidis et al., 2005; Benincasa et al., 2007; Buldini et al., 1997; Calapaj et al., 1988; Cindric et al., 2007); De Leonardis et al., 2000; Llorent-Martinez et al., 2011a, 2011b; Martin-Polvillo et al., 1994; Mendil et al., 2009; Nunes et al., 2011; Pehlivan et al., 2008; Zeiner et al., 2005)
Cu	1.7*	4.51	(Angioni et al., 2006; Anthemidis et al., 2005; Buldini et al., 1997; Calapaj et al., 1988; Castillo et al., 1999; Cindric et al., 2007; De Leonardis et al., 2000; Galeano Diaz et al., 2006; Jimenez et al., 2003; Karadjova et al., 1998; Llorent-Martinez et al., 2011a, 2001b; Martin-Polvillo et al., 1994; Mendil et al., 2009; Nunes et al., 2011; Pehlivan et al., 2008; Zeiner et al., 2005)
Ni	10.6*	2.26	(Benincasa et al., 2007; Buldini et al., 1997; Calapaj et al., 1988; Castillo et al., 1999; Cindric et al., 2007; Nunes et al., 2011; Zeiner et al., 2005)
Zn	0.6*	4.61	(Angioni et al., 2006; Cindric et al., 2007; Lo Coco et al., 2003; Mendil et al., 2009; Nunes et al., 2011; Zeiner et al., 2005)

Mn	0.7*	0.15	(Anthemidis et al., 2005; Benincasa et al., 2007; Calapaj et al., 1988; Castillo et al., 1999; Cindric et al., 2007; Jimenez et al., 2003; Karadjova et al., 1998; Llorent-Martinez et al., 2011a; Mendil et al., 2009; Pehlivan et al., 2008; Zeiner et al., 2005)
Pb	0.42*	0.032	(Calapaj et al., 1988; Canario & Katskov, 2005; Castillo et al., 1999; Jimenez et al., 2003; Llorent-Martinez et al., 2011a; Mendil et al., 2009; Martin-Polvillo et al., 1994)
Co	0.23*	5.45	(Benincasa et al., 2007; Calapaj et al., 1988; Castillo et al., 1999; Cindric et al., 2007; Mendil et al., 2009; Zeiner et al., 2005)
Cd	0.6*	0.15	(Angioni et al., 2006; Benincasa et al., 2007; Calapaj et al., 1988; Canario & Katskov, 2005; Castillo et al., 1999; Mendil et al., 2009; Yağan Aşçı et al., 2008)
Cr	0.012	2.00	(Anthemidis et al., 2005; Benincasa et al., 2007; Calapaj et al., 1988; Castillo et al., 1999; Llorent-Martinez et al., 2011a)
V	0.005	0.46	(Castillo et al., 1999); (Llorent-Martinez et al., 2011a)
Ge	0.03	0.04	(Castillo et al., 1999)
Zr	0.01	0.04	(Castillo et al., 1999)
Ba	4.9*	0.7	(Castillo et al., 1999; Jimenez et al., 2003; Llorent-Martinez et al., 2011a)
Al	0.030	1.11	(Anthemidis et al., 2005; Cindric et al., 2007; Jimenez et al., 2003; Karadjova et al., 1998; Martin-Polvillo et al., 1994; Zeiner et al., 2005)
Be	0.118*	0.178*	(Benincasa et al., 2007)
Sc	49.94*	747.9*	(Benincasa et al., 2007)
As	1.248*	26.65*	(Benincasa et al., 2007)
Se	1.47*	6.78*	(Benincasa et al., 2007)
Sr	1.52*	48.9*	(Benincasa et al., 2007)
Y	0.082*	0.331*	(Benincasa et al., 2007)
Sb	0.194*	0.411*	(Benincasa et al., 2007)
Sm	0.004*	0.226*	(Benincasa et al., 2007)
Eu	0.004*	0.021*	(Benincasa et al., 2007)
Gd	0.003*	0.094*	(Benincasa et al., 2007)
Sn	0.126	0.159	(Calapaj et al., 1988)
Mg	0.056	4.61	(Bağdat Yaşar & Güçer, 2004; Benincasa et al., 2007; Cindric et al., 2007; Mendil et al., 2009; Zeiner et al., 2005)
Ca	0.63	76.0	(Anthemidis et al., 2005; Benincasa et al., 2007; Cindric et al., 2007; Mendil et al., 2009; Zeiner et al., 2005)
K	0.05	2.14	(Cindric et al., 2007; Mendil et al., 2009; Zeiner et al., 2005)
Na	8.7	38.03	(Cindric et al., 2007; Mendil et al., 2009; Zeiner et al., 2005)

Table 6. The metal levels for olive oils.

4. Speciation and fractionation

Fractionation was defined as "the process of classification of analyte or a group of analytes from a certain sample according to physical (e.g., size, solubility) or chemical (e.g., bonding, reactivity) properties", and speciation of an element was also defined as "distribution of an element amongst defined chemical species in a system" by Templeton et al. (2000). The physicochemical form of an element, i.e. the actual species found in exposure medium and in the different body fractions, is frequently determinant in the evaluation of its bioavailability and toxicity (Flaten, 2002). An element can be found in various species: anionic or cationic inorganic forms, inorganic compounds, complex compounds with protein, peptide etc. Some organometallic compounds are much more toxic than the ions of the corresponding inorganic compounds. Hg, Pb and Sn obey this rule, for example, methyl-Hg and inorganic Hg are both toxic, but methyl-Hg show more toxicity than other (Templeton et al., 2000). In contrast to this, in the case of As and Se, most organo-arsenicals are less toxic than inorganic As species, organic forms of Se are ordinarily less toxic than Se(IV) (Kot & Namiesnik, 2000).

The determination of the total amount of an element in samples cannot give adequate information for understanding its bioavailability or toxicity, that's why the fractionation and speciation of metals in oils are increasingly gaining importance. The fractionation and speciation analysis are more informative than total element determinations for all type of samples.

In general, many works dealing with the total amount of elements in oil samples are reported, but fractionation and/or speciation analysis in vegetable oils are less common in literature. To the best of our knowledge, magnesium fractionation analysis in olive and olive oil was cited firstly in 2004. The improvement of an analytical scheme for fractionation of magnesium in olive products and also the determination of Mg amounts absorbed in stomach and intestine was achieved by Bağdat Yaşar & Güçer (2004). It was reported that 3.37-8.47% of Mg was absorbed in the stomach (ionic and polar groups) and the remaining percentage of Mg was absorbed in the intestine (molecular and complexed structures) in olive oil. As can be seen, the Mg fraction in olive oil is almost absorbable in the intestine. This study can be accepted as a preliminary step for fractionation studies and the fractionation and/or speciation approach for other elements will be described in the future.

5. Detection techniques

Various researchers deal with determination of metals in oils at trace, ultra-trace levels using spectrometric and electrometric techniques. Mentioned detection techniques may be combined with some chromatographic systems. Oils have been analyzed for different metals using atomic absorption spectrometer (FAAS and GFAAS), inductively coupled plasma optical emission spectrometer (ICP-OES), inductively coupled plasma mass spectrometer (ICP-MS). ICP techniques have become more popular since the early 1990s. Although the use of AAS (flame, graphite furnace, hydride generation and cold vapour) has declined during the same period, it is still the most widely used technique (Rose et al., 2001).

Each technique has some special requirements, advantages and disadvantages according to its basic principle. GF-AAS is a sensitive, proper for direct introduction of oil samples in the form of emulsion and does not require a large amount of sample. FAAS and ICP-MS have a requirement of sample pretreatment, but ICP-MS is more sensitive and expensive when compared with FAAS. There are scarce researches dealing with oil samples related to voltammetric and potentiometric techniques such as Ad-SSWV, dPSA (Abbasi et al., 2009; Cypriano et al., 2008; Dugo et al., 2004; Galeano Diaz et al., 2006; Lo Coco et al., 2003).

6. Conclusion

Trace quantities of some metals are naturally present in olive oil. It could be possible to determine the levels of different trace metals with the help of precise and accurate analytical methods. In many cases, a sample pretreatment process is necessary to eliminate the oil matrix prior to the introduction of the sample into the instrument. A direct determination is also possible by sample solubilization in an organic solvent, an emulsification procedure or a solid sampling strategy when ETAAS, GF-AAS or ICP are used for the analysis of edible oils. Microwave-assisted wet digestion sample pretreatment is also employed combined with sensitive detection techniques. An alternative technique can be achieved efficiently and precisely by FAAS after the extraction of metals with a Schiff base ligand.

7. Abbreviations

AAS	Atomic Absorption Spectrometry
FAAS	Flame Atomic Absorption Spectrometry
GF-AAS	Graphite Furnace Atomic Absorption Spectrometry
ETAAS	Electrothermal Atomic Absorption Spectrometry
ICP	Inductively Coupled Plasma
ICP-OES	Inductively Coupled Plasma Optical Emission Spectrometry
ICP-MS	Inductively Coupled Plasma Mass Spectrometry
Ad-SSWV	Adsorptive Stripping Square Wave Voltammetry
dPSA	Derivative Potentiometric Stripping Analysis
SCP	Stripping Chronopotentiometry

8. References

Abbasi, S.; Allahyari, M.; Taherimaslak, Z.; Nematollahi, D. & Abbasi, F. (2009). New Determination of Lead in Edible Oil and Water Samples by High Selective Adsorptive Stripping Voltammetry with SPADNS. *International Journal of Electrochemical Science*, Vol.4, (March 2009), pp. 602-613, ISSN 1452-3981

Afkhami, A.; Abbasi-Tarighat, M. & Khanmohammadi, V. (2009). Simultaneous Determination of Co^{2+}, Ni^{2+}, Cu^{2+} and Zn^{2+} Ions in Foodstuffs and Vegetables with

A New Schiff Base using Artificial Neural Networks. *Talanta*, Vol.77, (July 2008), pp. 995-1001, ISSN 0039-9140

Allen L. B., Siitonen, P. H. & Thompson, H. C. (1998). Determination of Copper, Lead, and Nickel in Edible Oils by Plasma and Furnace Atomic Spectroscopies. *Journal of the American Oil Chemists Society*, Vol.75, No.4, (October 1997), pp. 477-481, ISSN 0003-021X

Angioni, A.; Cabitza, M.; Russo, M. T. & Caboni, P. (2006). Influence of Olive Cultivars and Period of Harvest on the Contents of Cu, Cd, Pb, and Zn in Virgin Olive Oils. *Food Chemistry*, Vol.99, (August 2005), pp. 525-529, ISSN 0308-8146

Ansari, R.; Kazi; T. G.; Jamali, M. K.; Arain M. B.; Sherazi, S. T., Jalbani, N. & Afridi, H. I. (2008). Improved Extraction Method for the Determination of Iron, Copper, and Nichel in New Varieties of Sunflower Oil by Atomic Absorption Spectroscopy. *Journal of AOAC International*, Vol.91, No.2, (November 2008), pp. 400-407, ISSN 1060-3271

Ansari, R.; Kazi; T. G.; Jamali, M. K.; Arain M. B.; Wagan, M. D.; Jalbani, N.; Afridi, H. I. & Shah, A. Q. (2009). Variation in Accumulation of Heavy Metals in Different Verities of Sunflower Seed Oil with The Aid of Multivariate Technique. *Food Chemistry*, Vol.115, (November 2007), pp. 318-323, ISSN 0308-8146

Anthemidis, A. N.; Arvanitidis, V. & Stratis, J. A. (2005). On-line Emulsion Formation and Multi-element Analysis of Edible Oils by Inductively Coupled Plasma Atomic Emission Spectrometry. *Anaylytica Chimica Acta*; Vol.537, (January 2005), pp. 271-278, ISSN 0003-2610

Anwar, F.; Kazi, T.G.; Jakharani, M.A.; Sultana, R. & Sahito, S.R. (2003). Improved Extraction Method fort he Determination of Fe, Cu, Zn and Ni in Fat Samples using Atomic Absorption Spectrophotometer. *Journal of the Chemical Society of Pakistan*, Vol.25, No.3, (April 2003), pp. 210-214, ISSN 0253-5106

Anwar, F.; Kazi, T.G.; Saleem, R. & Bhanger, M.I. (2004). Rapid Determination of Some Trace Metals in Several Oils and Fats. *Grasas y Aceites*, Vol.55, Fasc.2, (September 2003), pp. 160-168, ISSN 0017-3495

Ashkenani, H.; Dadfarnia, S.; Shabani, A. M. H.; Jaffari, A. A. & Behjat, A. (2009). Preconcentration, Speciation and Determination of Ultra Trace Amounts of Mercury by Modified Octadecyl Silica Membrane Disk/Electron Beam Irradiation and Cold Vapor Atomic Absorption Spectrometry. *Journal of Hazardeous Materials*, Vol.161, (March 2008), pp. 276-280, ISSN 0304-3894

Bağdat Yaşar, S. & Güçer, Ş. (2004). Fractionation Analysis of Magnesium in Olive Products by Atomic Absorption Spectrometry. *Analytica Chimica Acta*, Vol.505, (May 2003), pp. 43-49, ISSN 0003-2670

Batı, B. & Cesur H. (2002). Determination of Copper in Edible Oils by Atomic Absorption Spectrometry after Lead Piperazinedithiocarbamate Solid-Phase Extraction and Potassium Cyanide Back-Extraction. *Analytical Sciences*, Vol.18, (August 2002), pp. 1273-1274, ISSN 0910-6340

Benedet, J. A. & Shibamoto T. (2008). Role of Transition Metals, Fe(II), Cr(II), Pb(II), and Cd(II) in Lipid Peroxidation. *Food Chemistry*, Vol.107, (July 2007), pp. 165-168, ISSN 0308-8146

Benincasa, C.; Lewis, J.; Perri, E.; Sindona, G. & Tagarelli, A. (2007). Determination of Trace Element in Italian Virgin Olive Oils and Their Characterization According to Geographical Origin by Statistical Analysis. *Anaytica Chimica Acta*, Vol.585, (December 2006), pp. 366-370, ISSN 0003-2670

Bettinelli, M.; Spezia, S.; Baroni, U. & Bizzari, G. (1995). Determination of Trace Elements in Fuel Oils by Inductively Coupled Plasma Mass Spectrometry after Acid Mineralization of the Sample in a Microwave Oven. *Journal of Analytical Atomic Specrometry*, Vol.10, (April 1995), pp. 555-560, ISSN 0267-9477

Buldini, P. L.; Ferri, D. & Sharma, J. L. (1997). Determination of Some Inorganic Species in Edible Oils and Fats by Ion Chromatography. *Journal of Chromatography A*, Vol.789, pp. 549-555, ISSN 0021-9673

Calapaj, R.; Chiricosta, S.; Saija, G. & Bruno, E. (1988). Method for the Determination of Heavy Metals in Vegatable Oils by Graphite Furnace Atomic Absorption Spectroscopy. *Atomic Spectroscopy*, Vol.9, No.4, pp. 107-109, ISSN 0195-5373

Canario, C. M. & Katskov, D. A. (2005). Direct Determination of Cd and Pb in Edible Oils by Atomic Absorption Spectrometry with Transverse Heated Filter Atomizer. *Journal of Analytical Atomic Spectrometry*, Vol.20, (September 2005), pp. 1386-1388, ISSN 0267-9477

Carbonell, V.; Mauri, A.R.; Salvador, A. & DeLaGuardia, M. (1991). Direct Determination of Copper and Iron in Edible Oils using Flow Injection Flame Atomic Asorption Spectrometry. *Journal of Analytical Atomic Spectrometry*, Vol.6, (June 1991), pp. 581-584, ISSN 0267-9477

Castillo, J.R.; Jimenez, M.S. & Ebdon, L. (1999). Semiquantitative Simultaneous Determination of Metals in Olive Oil using Direct Emulsion Nebulization. *Journal of Analytical Atomic Spectrometry*, Vol.14, (March 1999), pp. 1515-1518, ISSN 0267-9477

Chen, S.S.; Chen, C.M.; Cheng, C.C. & Chou, S.S. (1999). Determination of Copper in Edible Oils by Direct Graphite Furnace Atomic Absorption Spectrometry. *Journal of Food and Drug Analysis*, Vol.7, No.3, (June 1999), pp. 207-214, ISSN 1021-9498

Cindric, I. J.; Zeiner, M. & Steffan, I. (2007). Trace Elemental Characterization of Edible Oils by ICP-AES and GFAAS. *Microchemical Journal*, Vol.85, (April 2006), pp. 136-139, ISSN 0026-265X

Costa, L. M.; Silva, F. V.; Gouveia, S. T.; Nogueira, A. R. A. & Nobrega, J. A. (2001). Focused Microwave-assisted Acid Digestion of Oils: An Evaluation of the Residual Carbon Content. *Spectrochimica Acta Part B*, Vol.56, (May 2001), pp. 1981-1985, ISSN 0584-8547

Cypriano, J.C.; Matos, M.A.C. & Matos, R.C. (2008). Ultrasound-assisted Treatment of Palm Oil Samples for the Determination of Copper and Lead by Stripping Chronopotentiometry. *Microchemical Journal*, Vol.90, (March 2008), pp. 26-30, ISSN 0026-265X

De Leonardis, A.; Macciola, V. & DeFelice, M. (2000). Copper and Iron Determination in Edible Vegetable Oils by Graphite Furnace Atomic Absorption Spectrometry after Extraction with Diluted Nitric Acid. *International Journal of Food Science & Technology*, Vol.35, No.4, (October 1999), pp. 371-375, ISSN 0950-5423

De Souza, R. M.; Mathias, B. M.; da Silveira, C. L. P. & Aucelio, R. Q. (2005). Inductively Coupled Plasma Optical Emission Spectrometry for Trace Multi-element Determination in Vegetable Oils, Margarine and Butter After Stabilization with Ppropan-1-ol and Water. *Spectrochimica Acta Part B*, Vol.60, (January 2005), pp. 711-715, ISSN 0584-8547

Dugo, G.; La Pera, L.; La Torre, G. L. & Giuffrida, D. (2004). Determination of Cd(II), Cu(II), Pb(II), and Zn(II) Content in Commercial Vegetable Oils using Derivative Potentiometric Stripping Analysis. *Food Chemistry*, Vol.87, (December 2003), pp. 639-645, ISSN 0308-8146

Duyck, C.; Miekeley, N.; DaSilveira, C.L.P.; Aucelio, R.Q.; Campos, R.C.; Grinberg, P. & Brandao, G.P. (2007). The Determination of Trace Elements in Crude Oil and Its Heavy Fractions by Atomic Spectrometry. *Spectrochimica Acta Part B*, Vol.62, (April 2007), pp. 939-951, ISSN 0584-8547

Flaten, T. P. (2002). Aluminium in Tea-Concentations, Speciation and Bioavailability. *Coordination Chemistry Reviews*, Vol.228, (February 2002), pp. 385-395, ISSN 0010-8545

Galeano Díaz, T.; Guiberteau, A.; López Soto, M. D. & Ortiz, J. M. (2006). Determination of Copper with 5,5-dimethylcyclohexane-1,2,3-trione 1,2-dioxime 3-thiosemicarbazone in Olive Oils by Adsorptive Stripping Square Wave Voltammetry. *Food Chemistry*, Vol.96, (April 2005), pp. 156-162, ISSN 0308-8146

Ghaedi, M.; Tavallali, H.; Shokrollahi, A.; Zahedi, M.; Montazerozohori, M. & Soylak, M. (2009). Flame Atomic Absorption Spectrometric Determination of Zinc, Nickel, Iron and Lead in Different Matrixes after Solid Phase Extraction on Sodium Dodecyl sulfate (SDS)-Coated Alumina as Their Bis (2-hydroxyacetophenone)-1,3-propanediimine chelates. *Journal of Hazardeous Materials*, Vol.166, (December 2008), pp. 1441-1448, ISSN 0304-3894

Hendrikse, P. W.; Slikkerveer, F. J.; Zaalberg, J. & Hautfenne, A. (1988). Determination of Copper, Iron and Nickel in Oils and Fats by Direct Graphite Furnace Atomic Absorption Spectrometry. *Pure and Applied Chemistry*, Vol.60, No.6, pp. 893-900, ISSN 0033-4545

Hendrikse, P. W.; Slikkerveer, F. J.; Folkersma, A. & Dieffenbacher, A. (1991). Determination of Lead in Oils and Fats by Direct Graphite Furnace Atomic Absorption Spectrometry. *Pure and Applied Chemistry*, Vol.63, No.8, pp. 1183-1190, ISSN 0033-4545

Huang, S.J. & Jiang, S.J. (2001). Determination of Zn, Cd and Pb in Vegetable Oil by Electrothermal Vaporization Inductively Coupled Plasma Mass Spectrometry. *Journal of Analytical Atomic Spectrometry*, Vol.16, (March 2001), pp. 664-668, ISSN 0267-9477

Hussain Reddy, K.; Prasad, N. B. L. & Sreenivasulu Reddy, T. (2003). Analytical Properties of 1-phenyl-1,2-propanedione-2-oxime thiosemicarbazone: Simultaneous Spectrophotometric Determination of Copper(II) and Nickel(II) in Edible Oils and Seeds. *Talanta*, Vol.59, (June 2002), pp. 425-433, ISSN 0039-9140

Issa, R. M.; Khedr, A. M. & Rizk, H. F. (2005). UV-vis, IR and ^{1}H NMR Spectroscopic Studies of Some Schiff Bases Derivatives of 4-aminoantipyrine. *Spectrochimica Acta Part A*, Vol.62, (January 2005), pp. 621-629, ISSN 1386-1425

İspir, E. (2009). The Synthesis, Characterization, Electrochemical Character, Catalytic and Antimicrobial Activity of Novel, Azo-containing Schiff Bases and Their Metal Complexes. *Dyes and Pigments*, Vol.82, (September 2008), pp. 13-19, ISSN 0143-7208

Jacob, R. A. & Klevay, L. M. (1975). Determination of Trace Amounts of Copper and Zinc in Edible Fats and Oils by Acid Extraction and Atomic Absorption Spectrophotometry. *Analytical Chemistry*, Vol.47; No.4; pp. 741-743; ISSN 0003-2700

Jimenez, M. S.; Velarte, R. & Castillo, J. R. (2003). On-line Emulsions of Olive Oil Samples and ICP-MS Multi-Elemental Determination. *Journal of Analytical Atomic Spectrometry*, Vol.18, (May 2003), pp. 1154-1162, ISSN 0267-9477

Karadjova, I.; Zachariadis, G.; Boskou, G. & Stratis, J. (1998). Electrothermal Atomic Absorption Spectrometric Determination of Aluminium, Cadmium, Chromium, Copper, Iron, Manganese, Nickel and Lead in Olive Oil. *Journal of Analytical Atomic Spectrometry*, Vol.13, (November 1997), pp. 201-204, ISSN 0267-9477

Khedr, A. M.; Gaber, M.; Issa, R. M. & Erten; H. (2005). Synthesis and Spectral Studies of 5-[3-(1,2,4-triazolyl-azo]-2,4-dihydroxybenzaldehyde (TA) and Its Schiff Bases with 1,3-diaminopropane (TAAP) and 1,6-diaminohexane (TAAH). Their Analytical Application for Spectrophotometric Microdetermination of Cobalt(II). Application in Some Radiochemical Studies. *Dyes and Pigments*, Vol.67, (November 2004), pp. 117-126, ISSN 0143-7208

Khorrami, A. R.; Naeimi, H. & Fakhari, A. R. (2004). Determination of Nickel in Natural Waters by FAAS after Sorption on Octadecyl Silica Membrane Disks Modified with A Recently Synthesized Schiff's Base. *Talanta*, Vol.64, (October 2003), pp. 13-17, ISSN 0039-9140

Kot, A. & Namiesnik, J. (2000). The role of speciation in analytical chemistry. *Trends in Analytical Chemistry*, Vol.19, Nos.2+3, pp. 69-79, ISSN 0165-9936

Kowalewska, Z.; Bulska, E. & Hulanicki, A. (1999). Oganic Palladium and Palladium-magnesium Chemical Modifiers in Direct Determination of Lead in Fractions from Distillation of Crude Oil by Electrothermal Atomic Absorption Analysis. *Spectrochimica Acta Part B*, Vol.54, (February 1999), pp. 835-843, ISSN 0584-8547

Köse Baran, E. & Bağdat Yaşar, S. (2010). Copper and Iron Determination with [*N,N'*-Bis(salicylidene)-2,2'-dimethyl-1,3-propanediaminato] in Edible Oils Without Digestion. *Journal of the American Oil Chemists Society*, Vol.87, (April 2010), pp. 1389-1395, ISSN 0003-021X

Kurşunlu, A. N.; Guler, E.; Dumrul, H.; Kocyigit, O. & Gubbuk, I. H. (2009). Chemical Modification of Silica-gel with Synthesized New Schiff Base Derivatives and Sorption Studies of Cobalt (II) and Nickel (II). *Applied Surface Science*, Vol.255, (June 2009), pp. 8798-8803, ISSN 0169-4332

Kurtaran, R.; Tatar Yıldırım, L.; Azaz, A. D.; Namlı, H. & Atakol, O. (2005). Synthesis, Characterization, Crystal Structure and Biological Activity of A

Novel Heterotetranuclear Complex: [NiLPb(SCN)$_2$(DMF)(H$_2$O)]$_2$, bis-{[μ-N,N'-bis(salicylidene)-1,3-propanediaminato -aqua-nickel(II)] (thiocyanato) (μ-thiocyanato)(μ-N,N'-dimethylformamide)lead(II)}. *Journal of Inorganic Biochemisrty*, Vol.99, (May 2005), pp. 1937-1944, ISSN 0162-0134

Lacoste, F.; Van Dalen, G. & Dysseler, P. (1999). The Determination of Cadmium in Oils and Fats by Direct Graphite Furnace Atomic Absorption Spectrometry. *Pure and Applied Chemistry*, Vol.71, No.2, pp. 361-368, ISSN 0033-4545

Lendinez, E.; Lorenzo, M.L.; Cabrera, C. & Lopez, M.C. (2001). Chromium in Basic Foods of the Spanish Diet: Seafood, Cereals, Vegetables, Olive Oils and Dairy Products. *The Science of the Total Environment*, Vol.278, (December 2000), pp. 183-189, ISSN 0048-9697

Levine, K.E.; Batchelor, J.D.; Rhoades, C.B. & Jones, B.T. (1999). Evaluation of a High-pressure, High-temperature Microwave Digestion System. *Journal of Analytical Atomic Spectrometry*, Vol.14, (November 1998), pp. 49-59, ISSN 0267-9477

Li, Y.-G.; Shi, D.-H.; Zhu, H.-L.; Yan, H. & Ng, S. W. (2007). Transition Metal Complexes (M = Cu, Ni and Mn) of Schiff-Base Ligands: Syntheses, Crystal Structures, and Inhibitory Bioactivities Against Urease and Xanthine Oxidase. *Inorgica Chimica Acta*, Vol.360, (February 2007), pp. 2881-2889, ISSN 0020-1693

List, G.R.; Evans, C.D. & Kwolek, W.F. (1971). Copper in Edible Oils: Trace Amounts Determined by Atomic Absorption Specroscopy. *Journal of the American Oil Chemists' Society*, Vol.48, No.9, pp. 438-441, ISSN 0003-021X

Llorent-Martinez, E. J.; Ortega-Barrales, P.; Fernandez-de Cordóva, M. L.; Dominguz-Vidal A. & Ruiz-Medina, A. (2011a). Investigation by ICP-MS of Trace Element Levels in Vegetable Edible Oils Produced in Spain. *Food Chemistry*, Vol.127, (January 2011), pp. 1257-1262, ISSN 0308-8146

Llorent-Martinez, E. J.; Ortega-Barrales, P.; Fernandez-de Cordóva, M. L. & Ruiz-Medina, A. (2011b). Analysis of the Legistated Metals in Different Categories of Olive and Olive-pomace Oils. *Food Control*, Vol.22, (July 2010), pp. 221-225, ISSN 0956-7135

Lo Coco, F.; Ceccon, L.; Ciraolo, L. & Novelli, V. (2003). Determination of Cadmium(II) and Zinc(II) in Olive Oils by Derivative Potentiometric Stripping Analysis. *Food Control*, Vol.14, (June 2002), pp. 55-59, ISSN 0956-7135

Martin- Polvillo, M.; Albi, T. & Guinda A. (1994). Determination of Trace Elements in Edible Vegetable Oils by Atomic Absorption Spectrophotometry. *Journal of the American Oil Chemists Society*, Vol.71, No.4, (December 1993), pp. 347-353, ISSN 0003-021X

Mashhadizadeh, M. H.; Pesteh, M.; Talakesh, M.; Sheikhshoaie, I.; Ardakani, M. M. & Karimi, M. A. (2008). Solid Phase Extraction of Copper (II) by Sorption on Octadecyl Silica Membrane Disk Modified with A New Schiff Base and Determination with Atomic Absorption Spectrometry. *Spectrochimica Acta Part B*, Vol.63, (March 2008), pp. 885-888, ISSN 0584-8547

Matos Reyes, M. N. & Campos, R. C. (2006). Determination of Copper and Nickel in Vegetable Oils by Direct Sampling Graphite Furnace Atomic Absorption Spectrometry. *Talanta*, Vol.70, (May 2006), pp. 929-932, ISSN 0039-9140

Meira, M.; Quintella, C. M.; dos Santos Tanajura, A.; da Silva, H. G. R.; Fernando, J. D. S.; da Costa Neto, P. R.; Pepe, J. M.; Santos, M. A. & Nascimento, L. L. (2011).

Determination of the Oxidation Stability of Biodiesel and Oils by Spectrofluorimetry and Multivariate Calibration. *Talanta*, Vol.85, (April 2011), pp. 430-434, ISSN 0039-9140

Mendil, D.; Uluözlü, Ö. D.; Tüzen, M. & Soylak, M. (2009). Investigation of the Levels of Some Element in Edible Oil Samples Produced in Turkey by Atomic Absorption Spectrometry. *Journal of Hazardous Materials*, Vol.165, (October 2008), pp. 724-728, ISSN 0304-3894

Mohamed, G. G. (2006). Synthesis, Characterization and Biological Activity of Bis(phenylimine) Schiff Base Ligands and Their Metal Complexes. *Spectrochimica Acta Part A*, Vol.64, (May 2005), pp. 188-195, ISSN 1386-1425

Morgan, E. (1991). *Chemometrics : Experimental Design*, John Wiley and Sons, ISBN 0 471 92903 4, Chichester, UK

Murillo, M.; Benzo, Z.; Marcano, E.; Gomez, C.; Garaboto, A. & Marin, C. (1999). Determination of Copper, Iron and Nickel in Edible Oils using Emulsified Solutions by ICP-AES. *Journal of Analytical Atomic Spectrometry*, Vol.14, (March 1999), pp. 815-820, ISSN 0267-9477

Nash, A. M.; Mounts, T. L. & Kwolek, W. F. (1983). Determination of Ultratrace Metals in Hydrogenated Vegatable Oils and Fats. *Journal of the American Oil Chemists Society*, Vol.60, No.4, (April 1983), pp. 811-814, ISSN 0003-021X

Neelakantan, M. A.; Rusalraj, F.; Dharmaraja, J.; Johnsonraja, S.; Jeyakumar, T. & Sankaranarayana Pillai, M. (2008). Spectral Characterization, Cyclic Voltammetry, Morphology, Biological Activities and DNA Cleaving Studies of Amino Acid Schiff Base Metal (II) Complexes. *Spectrochimica Acta Part A*, Vol.71, (June 2008), pp. 1599-1609, ISSN 1386-1425

Nunes, L. S.; Barbosa, J. T. P.; Fernandes, A. P.; Lemos, V. A.; dos Santos, W. N. L.; Korn, M. G. A. & Teixeria, L. S. G. (2011). Multi-element Determination of Cu, Fe, Ni and Zn Content in Vegetable Oils Samples by High-resolution Continuum Source Atomic Absorption Spectrometry and Microemulsion Sample Preparation. *Food Chemistry*, Vol.127, (December 2010), pp. 780-783, ISSN 0308-8146

Otto, M. (1999). *Chemometrics : Statistics and Computer Application in Analytical Chemistry*, Wiley-VCH, ISBN 3-527-29628-X, Germany

Pehlivan, E.; Arslan, G.; Gode, F.; Altun, T. & Özcan, M.M. (2008). Determination of Some Inorganic Metals in Edible Vegetable Oils by Inductively Coupled Plasma Atomic Emission Spectroscopy (ICP-AES). *Grasas y Aceites*, Vol.59, No.3, (July 2008), pp. 239-244, ISSN 0017-3495

Pereira, J. S. F.; Moraes, D. P.; Antes, F. G.; Diehl, L. O.; Santos, M. F. P.; Guimarães, R. C. L.; Fonseca, T. C. O.; Dressler, V. L. & Flores; E. M. M. (2010). Determination of Metals and Metalloids in Light and Heavy Crude Oil by ICP-MS After Digestion by Microwave-induced Combustion. *Microchemical Journal*, Vol.96, (December 2009), pp. 4-11, ISSN 0026-265X

Prashanthi, Y.; Kiranmai, K.; Subhashini, N. J. P. & Shivaraj. (2008). Synthesis, Potentiometric and Antimicrobial Studies on Metal Complexes of Isoxazole Schiff Bases. *Spectrochimica Acta Part A*, Vol.70, (July 2007), pp. 30-35, ISSN 1386-1425

Rose, M.; Knaggs, M.; Owen L. & Baxter, M. (2001). A Review of Analytical Methods for Lead, Cadmium, Mercury, Arsenic and Tin Determination Used in Proficiency Testing. *Journal of Analytical Atomic Spectrometry*, Vol.16, (July 2001), pp. 1101-1106, ISSN 0267-9477

Sant'Ana, F. W.; Santelli, R. E.; Cassella A. R. & Cassella R. J. (2007). Optimization of An Open-focused Microwave Oven Digestion Procedure for Determination of Metals in Diesel Oil by Inductively Coupled Plasma Optical Emission Spectrometry. *Journal of Hazardous Materials*, Vol.149, (March 2007), pp. 67-74, ISSN 0304-3894

Sardans, J.; Montes, F. & Penuelas, J. (2010). Determination of As, Cd, Cu, Hg and Pb in Biological Samples by Modern Electrothermal Atomic Absorption Spectrometry. *Spectrochimica Acta Part B*, Vol.65, (November 2009), pp. 97-112, ISSN 0584-8547

Shamspur, T.; Mashhadizadeh, M. H. & Sheikhshoaie, I. (2003). Flame Atomic Absorption Spectrometric Determination of Silver Ion After Preconcentration on Octadecyl Silica Membrane Disk Modified with Bis[5-((4-nitrophenyl)azosalicylaldehyde)] As a New Schiff Base Ligand. *Journal of Analytical Atomic Spectrometry*, Vol.18, (September 2003), pp. 1407-1410, ISSN 0267-9477

Sharaby, C. M. (2007). Synthesis, Spectroscopic, Thermal and Antimicrobial Studies of Some Novel Metal Complexes of Schiff Base Derived from [N¹-(4-methoxy-1,2,5-thiadiazol-3-yl)sulfanilamide] and 2-thiophene carboxaldehyde. *Spectrochimica Acta Part A*, Vol.66, (May 2006), pp. 1271-1278, ISSN 1386-1425

Sikwese, F.E. & Duodu, K.G. (2007). Antioxidant Effect of A Crude Phenolic Extract from Sorghum Bran in Sunflower Oil in the Presence of Ferric Ions. *Food Chemistry*, Vol.104, (November 2006), pp. 324-331, ISSN 0308-8146

Tantaru, G.; Dorneanu, V. & Stan, V. (2002). Schiff Bis Bases :Analylical Reagents. II : Spectrophotometric Determination of Manganese From Pharmaceutical Forms. *Journal of Pharmaceutical and Biomedical Analysis*, Vol.27, (February 2001), pp. 827-832, ISSN 0731-7085

Templeton, D.M.; Ariese, F.; Cornelis, R.; Danielsson, L.G.; Muntau, H.; Van Leeuwen, H.P. & Lobinski, R. (2000). Guidelines for Terms Related to Chemical Speciation and Fractionation of Elements. Definitions, Structural Aspects, and Methodological Approaches. *Pure and Applied Chemistry*, Vol.72, No.8, pp. 1453-1470, ISSN 0033-4545

Van Dalen, G. (1996). Determination of Cadmium in Edible Oils and Fats by Direct Electrothermal Atomic Absorption Spectrometry. *Journal of Analytical Atomic Spectrometry*, Vol.11, (July 1996), pp. 1087-1092, ISSN 0267-9477

Wondimu, T.; Goessler & Irgolic, K. J. (2000). Microwave Digestion of "Residuel Fuel Oil" (NIST SRM 1634b) for the Determination of Trace Elements by Inductively Coupled Plasma-Mass Spectrometry. *Fresenius Journal of Analytical Chemistry*, Vol.367, (December 1999), pp. 35-42, ISSN 0937-0633

Yağan Aşçı, M.; Efendioğlu, A. & Batı, B. (2008). Solid Phase Extraction of Cadmium in Edible Oils using Zinc Piperazinedithiocarbamate and Its Determination by Flame Atomic Absorption Spectrometry. *Turkish Journal of Chemistry*, Vol.32, (December 2007), pp. 431-440, ISSN 1300-0527

Zeiner, M.; Steffan, I. & Cindric, I. J. (2005). Determination of Trace Elements in Olive Oil by ICP-AES and ETA-AAS: A Pilot Study on The Geographical Characterization. *Microchemical Journal,* Vol.81, (December 2004), pp. 171-176, ISSN 0026-265X

Ziyadanoğulları, B.; Cevizici, D.; Temel, H. & Ziyadanoğulları, V. (2008). Synthesis, Characterization and Structure Effects on Preconcentration and Extraction of N, N'-bis-(salicylaldehyde)-1,4-bis-(p-aminophenoxy) butane towards Some Divalent Cations. *Journal of Hazardeous Materials,* Vol.150, (April 2007), pp. 285-289, ISSN 0304-3894

5

Analysis of Olive Oils by Fluorescence Spectroscopy: Methods and Applications

Ewa Sikorska[1], Igor Khmelinskii[2] and Marek Sikorski[3]
[1]Faculty of Commodity Science, Poznań University of Economics,
[2]Universidade do Algarve, FCT, DQF and CIQA, Faro,
[3]Faculty of Chemistry, A. Mickiewicz University, Poznań,
[1,3]Poland
[2]Portugal

1. Introduction

Fluorescence spectroscopy is a well established and extensively used research and analytical tool in many disciplines. In recent years, a remarkable growth in the use of fluorescence in food analysis has been observed (Christensen et al., 2006; Sadecka & Tothova, 2007; Karoui & Blecker, 2011). Vegetable oils including olive oil constitute an important group of food products for which fluorescence was successfully applied. Fluorescence is a type of photoluminescence, a process in which a molecule, promoted to an electronically excited state by absorption of UV, VIS or NIR radiation, decays back to its ground state by emission of a photon. Fluorescence is emission from an excited state, in which the electronic spin is equal to that in the ground state, and typically equal to zero. Such transitions are spin allowed, and occur at relatively high rates, typically 10^8 s^{-1} (Lakowicz, 2006).

A unique feature of fluorescence, distinguishing it from other spectroscopic techniques, is its inherently multidimensional character (Christensen et al., 2006). Excitation of molecules results from absorption of radiation at the energy corresponding to the energy difference between the ground and excited states of a given fluorophore. Subsequently, radiation at a lower energy characteristic for the specific molecule is emitted during its deactivation. Thus, fluorescence properties of every compound are characterized by two types of spectra: excitation and emission. This feature and the fact that not all of the absorbing molecules are fluorescent both contribute to higher selectivity of fluorescence as opposed to absorption spectra.

Another important advantage of fluorescence is its higher sensitivity. In contrast to absorption measurements, the emitted photons are detected against a low background, making fluorescence spectroscopy a very sensitive method. The sensitivity of fluorescence is 100-1000 times higher than that of the absorption techniques, enabling to measure concentrations down to parts per billion levels (Guilbault, 1999).

The fluorescent analysis of olive oils takes advantage of the presence of natural fluorescent components, including phenolic compounds, tocopherols and pheophytins, and their oxidation products. Oils are complex systems and therefore conventional fluorescent techniques, relying on recording of single emission or excitation spectra, are often insufficient if directly applied. In such cases, total luminescence or synchronous scanning fluorescence techniques are used, improving the analytic potential of the fluorescence measurements. With contributions from numerous analytes, the autofluorescence of olive oil exhibits numerous overlapping bands. Such complex spectra should be analyzed using multivariate and multiway methods.

Analytical applications of fluorescence to olive oils include discrimination between the different quality grades, adulteration detection, authentication of virgin oils, quantification of fluorescent components, monitoring thermal and photo-oxidation and quality changes during storage.

In this chapter the application of fluorescence spectroscopy to qualitative and quantitative analysis of olive oils is reviewed. Methodological aspects of fluorescence measurements and analysis of fluorescence spectra are also discussed.

2. Fluorescence of olive oils

2.1 Fluorescence characteristics of olive oil and its components

Conventionally, two basic types of spectra characterize the fluorescent properties: excitation and emission spectra. For a system containing a single fluorophore, the shape and location of the excitation and emission spectra are independent of respective chosen emission and excitation wavelengths. However, for a system containing several fluorescent components, the excitation and emission spectra depend on particular emission and excitation wavelength used for measurements. Therefore, in systems containing several fluorophores, single-wavelength spectra are insufficient for a comprehensive description of fluorescent properties, thus multidimensional measurement methods should be used.

The most comprehensive characterization of a multicomponent fluorescent system is obtained by measurement of an excitation-emission matrix, known also as a total luminescence spectrum or fluorescence landscape. This technique was first introduced by Weber (1961). After the first application to edible oils by Wolfbeis & Leiner (1984), it has been intensively used for exploring oil fluorescence. Total luminescence spectra are usually obtained by measurement of emission spectra at several excitation wavelengths. They may be presented as a three dimensional plot, with the fluorescence intensity plotted in function of the excitation and the emission wavelengths (Ndou and Warner, 1991; Guilbault, 1999). Another representation of the total luminescence is obtained using two-dimensional contour maps, in which one axis represents the emission and another – the excitation wavelength, and the contours are plotted by linking points of equal fluorescence intensity, Fig.1. The total luminescence spectrum gives a comprehensive description of the fluorescent components of the mixture and may serve as a unique fingerprint for identification and characterization of the sample studied. The acquisition of contour maps at sufficient resolution (determined by the number of individual emission spectra recorded) on conventional spectrofluorometers is time-consuming, requiring a large number of scans for each sample (Guilbault, 1999).

Alternatively, multicomponent fluorescent systems may be investigated by the synchronous fluorescence techniques, proposed by Lloyd, (1971). This technique involves simultaneous scanning of both excitation and emission wavelengths, keeping a constant difference between them. Synchronous scanning fluorescence spectroscopy is very useful for the analysis of mixtures of fluorescent compounds, because both excitation and emission characteristics are included into a single spectrum. Although it provides less information than the excitation-emission matrix, it may still present a viable alternative to the total luminescence measurements due to its inherent simplicity and rapidity. A set of synchronous spectra recorded at different wavelength intervals may be concatenated into a total synchronous fluorescence spectrum. In such spectra fluorescence intensity is plotted as a function of the excitation wavelength and the wavelength interval. Both single wavelength interval and total synchronous fluorescence spectra were used for studies of olive oils (Sikorska et al. 2005a; Poulli et al. 2005). The relation between various kinds of fluorescence spectra of a virgin olive oil is presented in Fig. 1.

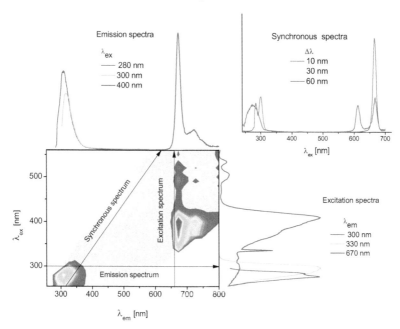

Fig. 1. Different types of fluorescence spectra; fluorescence spectra of a diluted sample of extra virgin olive oil (1%, v/v, in n-hexane) are shown as an example.

Numerous factors affect measured fluorescence intensity and spectral distribution. These factors are related to the nature and the concentration of fluorophores, their molecular environment, and scattering and absorption effects. They may be immeasurably important in complex natural systems, such as oils, and have to be taken into account when measuring and interpreting the fluorescence spectra. Fluorescence intensities are proportional to the concentration over only a limited range of optical densities (Lakowicz, 2006). To obtain proportionality between the fluorescence intensity and the fluorophore concentration, the absorbance at the excitation wavelength should be below 0.05 and close to zero in the emission spectral region. At higher concentrations, the inner filter effects have to be taken into account. These effects may decrease the observed fluorescence intensity by either reducing the intensity of the excitation or by absorbing the emitted radiation. To avoid the inner filter effects due to the high optical densities or sample turbidity, appropriate geometry of sample illumination should be used. The most common geometry using right-angle observation of the center of a centrally illuminated sample is only appropriate for diluted solutions with low optical densities. For opaque samples, front-face illumination is achieved using either triangular or square cuvettes oriented at 30 or 60° relative to the incident beam.

The effects of concentration and sample geometry on oil spectra were addressed by several authors. Zandomeneghi et al. (2005) in the very detailed studies compared the emission fluorescence spectra of undiluted extra virgin olive oil obtained with the traditional setup (right-angle fluorescence) and using front face fluorescence. The absorption of undiluted olive oil samples was from 2 up to 12 absorbance units, on passing from 325 to 260 nm, and therefore the inner filter phenomena affected the right angle spectra considerably. Excluding the long-wavelength chlorophyll region, significant differences in the number, shape, intensity, and position of the bands in spectra of the same undiluted oil obtained with right angle and front face geometry were observed, Fig. 2. The right angle fluorescence spectra showed considerable distortions, even after the mathematical corrections for the inner filter effects due to the absorption of both the excitation and emission radiation were applied. The front-face fluorescence spectra were less affected by self-absorption and thus provided reliable information about type of fluoropores and their concentration. It was also demonstrated that analysis of spectra affected by inner filter effects may lead to spectral misinterpretation and invalid assignments of origin of some fluorescent bands (M. Zandomeneghi & G. Zandomeneghi, 2005; Zandomeneghi et al., 2006).

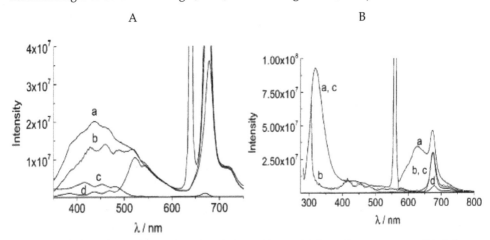

Fig. 2. A. Emission spectra of olive oil with λ_{ex} = 320 nm. (a) front face spectrum; (b) right angle spectrum with the absorption correction; (c) absorption spectrum of the oil multiplied by 10^6; (d) right angle spectrum without the absorption correction. B. Emission spectra of olive oil with λ_{ex} = 280 nm. (a) front face spectrum; (b) right angle spectrum with the absorption correction and multiplied by 3.7; (c) front face spectrum, second-order corrected; (d) right angle spectrum. Reprinted with permission from Zandomeneghi et al. 2005. Copyright 2005 American Chemical Society.

Typical fluorescence spectra of extra virgin and refined olive oils are shown in Fig. 3, (Sikorska et al., 2011). The fluorescence depends on sample concentration; therefore spectra for diluted and intact samples are shown. Both total fluorescence and total synchronous spectra are presented for the same oils, to enable comparison.

Based on the published data, one may conclude that the fluorescence properties depend on the quality grade of olive oils (Kyriakidis&Skarkalis, 2000; Poulli et al., 2006; Guimet et al., 2004a). For a selected quality category, the spectra may show minor differences between samples, however, the general features remain similar, permitting identification and authentication of oil samples.

The total fluorescence spectrum of diluted extra virgin olive oils, measured with the use of right angle geometry, exhibits two intense bands, one with excitation at about 270–330 nm and emission at about 295–360 nm and the second with excitation at about 330–440 nm and emission at about 660–700 nm, Fig. 3. An additional band appears in spectra of refined olive oil, located in the intermediate range, with excitation at 280-330 nm and emission at 372-480 nm. The long-wavelength band has a lower intensity in refined as compared to virgin olive oil (Sikorska et al., 2011).

The spectra of the same undiluted oils measured with the front face geometry method show a clearly different fluorescence pattern. The spectra are not affected by the inner filter effect, because front face geometry was used for measurement of undiluted samples. Additional bands are observed in the spectra of extra virgin olive oil at about 310–390 nm in excitation and 440–580 nm in emission. The ratio of fluorescence intensity of short- and long-wavelength bands is lower as compared to the spectrum of the diluted sample. The spectrum of undiluted refined olive oil exhibits a broad band with emission at 350-600 nm, two maxima at 320/420 nm and 365/450 nm in excitation/emission, and a long wavelength emission at 650-700 nm. Only a trace of the short-wavelength emission is observed with the maximum at 300/331 nm (Sikorska et al., 2011).

The differences in the spectra between diluted and undiluted samples may result from the high fluorophore concentrations in the intact oil samples and a variety of molecular interactions, such as quenching and energy transfer, which alter fluorescence characteristics. The effect of concentration on the total and synchronous fluorescence spectra of vegetable oils including olive oils was reported by Sikorska et al. (2004; 2005b).

The total synchronous fluorescence spectra of undiluted oils showed dependence of spectral shape and intensity on the wavelength interval ($\Delta\lambda$) used in the measurements, with the presence of particular bands dependent on $\Delta\lambda$. At lower values of $\Delta\lambda$ the bandwidths are reduced and the spectrum is simplified as compared to the total fluorescence spectra. Appearance of new bands or splitting of existing bands is typically observed with increasing $\Delta\lambda$. Emission bands are present in the excitation region below 310 nm, 310-350 nm, 350-380 nm, and above 550 nm in spectra of virgin olive oils (Sikorska et al., 2011). Similar spectral characteristics for virgin olive oil were reported by Poulli et al., (2006). The bands in total synchronous fluorescence spectra were observed in the 270-325, 347-365 and 602-685 nm excitation wavelength ranges with the respective wavelength intervals of 20–120, 30–50 and 20–76 nm.

Refined oils are characterized by a relatively weak band between 290-320 nm, a very broad band spreading to about 500 nm, and a band above 550 nm. All of these bands equally appear in the total fluorescence spectra (Sikorska et al., 2011).

The identification of origin of the particular emission bands relies mainly on comparison to the spectra of chemically pure fluorescent components. The fluorescence properties of compounds occurring in oils or suggested to contribute to their emission are listed in Table 1. The short wavelength band in total fluorescence spectra, which covers the region of 270–330 nm in excitation and 295–360 nm in emission, corresponds to the band at 280–310 nm in the total synchronous fluorescence spectra and is assigned to tocopherols and phenols. This assignment has been confirmed by several observations. Firstly, it was shown that a similar band appears in various vegetable oils, either cold-pressed or refined, and not only in olive oils (Sikorska et al., 2004). Olive oils contain considerable amounts of phenolic compounds, with their concentrations significantly reduced in refined oils. This observation seems to confirm that tocopherols also contribute to the emission observed in this wavelength range. In fact, tocopherols are present in most vegetable oils in widely variable amounts, from 70 to 1900 mg/kg (Cert et al., 2000). The vitamin E group includes four natural tocopherols (α-, β-, γ-, δ-) and four tocotrienols (αT3, βT3, γT3, δT3), all − in the R-configuration at the three double bonds in the side-chain of tocotrienols. Due to their structural similarity, all of these compounds exhibit very similar UV-absorption spectra and have similar fluorescence properties, Table 1. Of all tocopherols, α-tocopherol is predominant in olive oils. Indeed, the band in olive oil spectra being discussed is similar to the one in the total luminescence spectrum of α-tocopherol dissolved in *n*-hexane. Moreover, conventional excitation and emission spectra of the olive oils in the wavelength range mentioned are also similar to those of α-tocopherol, and the excitation spectra are in good agreement with the absorption spectrum of α-tocopherol in *n*-hexane (Sikorska et al. 2004). Still, the detailed analysis of excitation and emission spectra suggests contributions from several other fluorophores.

There still remain some inconsistencies concerning the assignment of vitamin E (tocopherol) bands in olive oil spectra. In one of the pioneering papers, where the emission spectra of various oils were reported, it has been suggested that the bands in the emission spectrum (λ_{ex}=365 nm) with the maximum at 525 nm may partly originate from compounds of the vitamin E group, or their derivatives formed upon oxidation (Kyriakidis & Skarkalis, 2000). However, this interpretation is based on spectra of undiluted olive oils measured using right angle geometry, and therefore strongly affected by inner filter effects, and in some cases referring to the spectral region where no emission of tocopherols is present (Zandomeneghi et al. 2005). It should be underlined that the emission of vitamin E in *n*-hexane has its maximum at about 320 nm, with a similar maximum appearing in the spectra of oils. Moreover, it has been stated (Zandomeneghi et al. 2006) that the known products of oxidation of R-α-, β-, γ-, δ-tocopherols, the R-α-, β-, γ-, δ-tocopherolquinones, are all nonfluorescent substances (Pollok & Melchert, 2004).

Note that a considerable number of minor components belonging to different classes of phenolic compounds such as phenolic acids, phenolic alcohols, hydroxyisochromans, secoiridoids, lignans, and flavonoids are present in virgin olive oils (Servili et al., 2004). Most of polyphenols are fluorescent substances, absorbing in the 260-310 nm range and emitting in the near-UV range, with their bands centered at 310-370 nm (M. Zandomeneghi & G. Zandomeneghi, 2005). These phenolic compounds can be detected by fluorescence after separation by HPLC, using excitation/emission wavelengths of 264/354, 310/430 or 280/320 nm (Dupuy et al., 2005). Fluorescence typical for phenolic components of olive oils was reported recently by Tena et al., (2009), using excitation at 270 nm with the fluorescence maxima appearing in the 362-420 nm range, Table 1.

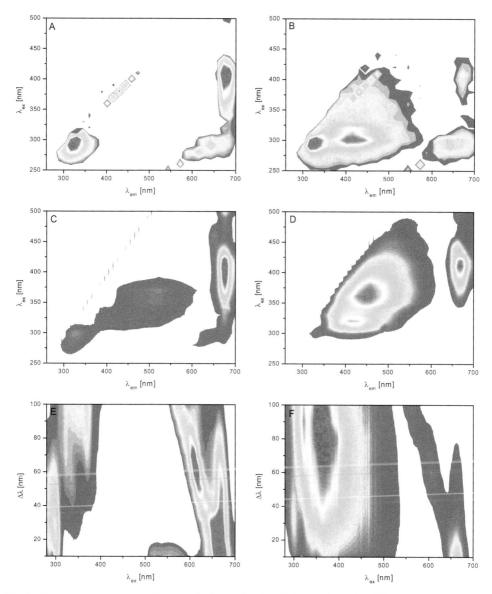

Fig. 3. Fluorescence spectra of extra virgin and refined olive oil: A. Total fluorescence spectrum of diluted olive oil, 1% in n-hexane, B. Total fluorescence spectrum of diluted refined olive oil, 1% in n-hexane, C. Total fluorescence spectrum of undiluted extra virgin olive oil, front face geometry, D. Total fluorescence spectrum of undiluted refined olive oil, front face geometry, E. Total synchronous fluorescence spectrum of undiluted extra virgin olive oil, front face geometry, F. Total synchronous fluorescence spectrum of undiluted refined olive oil, front face geometry (Sikorska et al., 2011).

	Solvent	λ_{abs} [nm]	ε [dm³ mol⁻¹ cm⁻¹]	Solvent	λ_{ex} [nm]	λ_{em} [nm]
Vitamin E						
α-Tocopherol [a]	ethanol	292	3265	n-hexane	295	320
β- Tocopherol [a]	ethanol	296	3725	n-hexane	297	322
δ- Tocopherol [a]	ethanol	298	3515	n-hexane	297	322
γ- Tocopherol [a]	ethanol	298	3809	n-hexane	297	322
α-Tocotrienol [a]	ethanol	292	3652	n-hexane	290	323
β- Tocotrienol [a]	ethanol	292	3540	n-hexane	290	323
δ- Tocotrienol [a]	ethanol	297	3403	n-hexane	292	324
γ- Tocotrienol [a]	ethanol	297	3737	n-hexane	290	324
Chlorophylls						
Chlorophyll a [b,c,d]	acetone	430	94700	ether	436	668
		663	75000	acetone	405	669
				9:1 acetone/water	430[e]	669
Chlorophyll b [b,c,d]	acetone	455	131000	ether	436	648
		645	47100	acetone	405	652
				9:1 acetone/water	458 [e]	653
Pheophytin a [b,c,d]	acetone	409	101800	ether	436	673
		666	44500	9:1 acetone/water	406 [e]	671
Pheophytin b [b,c,d]	acetone	434	145000	ether	436	661
		654	27800	9:1 acetone/water	435[e]	658
				-	-	
Pheophorbide a [b,c,d]	acetone	409	119200	-	-	-
		667	55200	-	-	-
Phenolic compounds						
Oleuropein	ethanol/n-hexane	282	-	ethanol/n-hexane	270	310
Vanillic acid [e]				methanol	270	349
Syringic acid [e]					270	361
Gallic acid [e]					270	382
p-Coumaric acid [e]						416
o-Coumaric [e]					270	426
Cinnamic acid [e]					270	420
Tyrosol [e]					270	420
Caffeic acid [e]					270	457

λ_{abs}, λ_{exc}, λ_{em} – absorption, excitation and emission maxima, ε - molar absorption coefficient, [a] (Eitenmiller et al., 2008), [b] (Ward et al., 1994), [c] (Undenfriend, 1962), [d](Diaz et al., 2003), [e](Tena et al., 2009).

Table 1. Fluorescence properties of olive oil components.

The fluorescence spectra of a vitamin E standard in hexane and an oil polyphenol extract in methanol–water at the same concentration as found in the extra virgin olive oil were investigated separately (Cheikhousman et al., 2005). The maximums in the fluorescence excitation spectrum measured at the 330 nm emission wavelength of oil polyphenol extract and vitamin E were observed at 284 and 290 nm, respectively. The spectral contribution of both the tocopherols and phenolic compounds to the fluorescence of extra virgin olive oil was confirmed by the similarity between the reconstructed spectrum of the mixture and the spectrum of extra virgin oil (Cheikhousman et al., 2005). Recently the fluorescence intensity at 280/320 nm in excitation/emission was successfully used to determine phenol contents in methanol/water extracts of olive oils (Papoti & Tsimidou, 2009).

Thus, both tocopherols and phenolic compounds contribute to the short-wavelength emission of the olive oils, with the tocopherol contribution dominating in refined oils. The exact positions of the maxima of the short-wavelength emission vary slightly between various oils, which may result from differences in the respective tocopherol composition. Note that in oils obtained by physical methods (cold pressing), including olive, linseed and rapeseed oils, this fluorescence maximum was blue-shifted as compared to refined oils, pointing out the difference between fluorescence of refined and cold-pressed oils in this spectral region (Sikorska et. al., 2004).

Due to the similar fluorescence properties of tocopherols and some phenolic compounds (Table 1), their emission appears as a single broad band, therefore a detailed study of excitation and emission spectra in this region is required to reveal presence of various species (Sikorska et al., 2008a). The excitation and emission spectra of virgin olive oil measured respectively at λ_{em}=330 nm and λ_{ex}=295 nm agree very well with the respective spectra of α-tocopherol, Fig. 4. On the other hand, the excitation and emission spectra measured respectively at λ_{em}=300 nm and λ_{ex}=280 nm are blue shifted as compared to α-tocopherol, being attributed to the phenolic compounds, according to Cheikhousman et al. (2005).

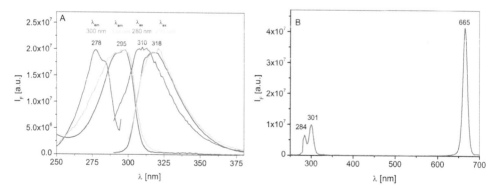

Fig. 4. (A) Excitation spectra (λ_{em}=300 and 330 nm) and emission spectra (λ_{ex}=280 and 295 nm) of extra virgin olive oil (green and blue) and tocopherol (red); the respective excitation and emission wavelengths are shown in the picture; (B) synchronous fluorescence spectrum ($\Delta\lambda$ = 10 nm) of extra virgin olive oil, (Sikorska et al., 2011).

Synchronous fluorescence spectroscopy offers a superior solution to the problem of the overlapping spectral bands, by reducing their spectral widths. The synchronous fluorescence spectrum of virgin olive oil measured at $\Delta\lambda$=10 nm shows two separate bands, with the band at 301 nm attributed to tocopherol. Linear correlation was found between the fluorescence intensity at 301 nm and the total tocopherol concentration as determined chromatographically, for a set of different diluted vegetable oils (1% v/v, in n-hexane) (Sikorska et. al., 2005b).

The band at 284 nm could originate from phenolic compounds such as phenolic aglycons, based on the molecules of tyrosol and hydroxytyrosol, derived from phenolic glycosides present in the olives. Oleuropein aglycons are present in high amounts in virgin olive oil. The synchronous spectrum ($\Delta\lambda$=10 nm) of pure oleuropein exhibits a maximum at about 289 nm and is shifted by about 5 nm as compared to the fluorescence band observed in the virgin olive oil. This shift may result from the solvent, as the phenolic compounds are poorly soluble in nonpolar solvents, the spectrum was recorded in the n-hexane – ethanol mixture. On the other hand, the emission observed in the oil may originate from oleuropein derivatives, with slightly different emission properties (Sikorska et al., 2008).

The band splitting, observed in synchronous spectra only at small $\Delta\lambda$ values, seems to be characteristic for virgin olive oils and has been not observed for refined oils (Sikorska et. al., 2005a). Synchronous fluorescence spectra acquired for virgin olive and sunflower oils at $\Delta\lambda$=20 nm were reported to have one double band at around 275 and 297 nm, and one more band at 660 nm, in contrast to sunflower oil that shows an intense band at around 300 nm and a weak one at 325 nm (Poulli et al., 2006).

The synchronous fluorescence spectra of phenolic compounds (tyrosol, p-coumaric and caffeic acids) present in virgin olive oils were measured at $\Delta\lambda$=30 nm (Dupuy et al., 2005). The spectra present a fluorescence band in the 275-350 nm spectral range, with the maxima for tyrosol and tocopherol observed respectively at 276 and 295 nm. Interestingly, it was shown that the mathematically calculated emission spectrum from a mixture of tyrosol, p-coumaric and caffeic acids, and α-tocopherol was very similar to the experimentally obtained spectrum of olive oil. Therefore, we once more conclude that the fluorescence emission between 275 and 400 nm has contributions from both tocopherols and phenolic compounds (Dupuy et al., 2005).

A long-wavelength band is observed in the olive oil spectra, with excitation at about 350–420 nm and emission at about 660–700 nm, corresponding to the band above 550 nm in total synchronous fluorescence spectra. This band was attributed to pigments of chlorophyll group, based on its excitation and emission characteristics (Zandomeneghi et al., 2005; Diaz et al., 2003). This group includes chlorophylls *a* and *b*, and pheophytins *a* and *b*, derived from chlorophylls by loss of magnesium. The emission spectra of these four chlorophyll derivatives are very similar, with their maxima in the range of 653 to 671 nm in 9:1 acetone/water (Diaz et al., 2003). Their presence is characteristic for virgin olive oils, being reduced to trace amounts in refined oils.

The origin of the emission in the intermediate region for both virgin and refined olive oils is unclear. It seems that various components may be responsible for bands appearing in this region. Wolfbeis & Leiner (1984), suggested that in addition to tocopherols and chlorophylls, parinaric acid, a conjugated 18:4 (n-3) acid, may also contribute to vegetable oil emission.

Recently, fluorescence from conjugated fatty acids including α-eleostearic acid, *cis*-parinaric acid and 8,10,12,14,16-octadecapentaenoic acid, was reported for Borage oil (Smyk et al., 2009). The fluorescence spectrum of methyl ester of the *cis*-parinaric acid has the maximum at 416 nm. It was also suggested that the low intensity emission with the maximum at 524 nm may originate from riboflavin (vitamin B_2) in virgin olive oils. Riboflavin is a polar, water soluble compound and no quantitative data regarding the presence of riboflavin in olive oils are available. This vitamin was found in olive pulp (Zandomeneghi et al., 2005). Additional emission in the intermediate region has been detected as a result of oxidation (Cheikhousman et al., 2005; Poulli et al., 2009a, 2009b; Tena et al., 2009; Sikorska et al., 2008).

2.2 Methods of analysis of fluorescence data

In past decades improvements in both spectroscopic instruments and computers contributed to the extensive application of fluorescence spectroscopy in food analysis, including olive oils. Although even the simple conventional analysis of fluorescence spectra may produce valuable data, most of the successful applications rely on multivariate methods for extracting useful analytical information from the measured fluorescence signals.

According to Christensen et al. (2006), several conditions should be met in an ideal system for fluorescence measurements: 1) the concentration of the fluorophores must be sufficiently low, to be approximately linearly related to the fluorescence intensity, 2) signals from each of the fluorescent components must beindependent of each other, 3) the signal contribution from interfering species must be insignificant compared to the target fluorophore signal. Most of these conditions, however, are routinely violated in intact food samples. Therefore, due to the complex character of the spectra, they are rarely used for direct analysis, being rather used as spectral patterns or fingerprints of particular samples. The vast amount of spectral information contained in such fingerprints could be used in qualitative and quantitative analysis.

Multivariate and multiway methods are specifically suited for treatment of such complex spectral data. The multivariate analysis has several advantages over the univariate approach. It enables analysis of nonselective signals in the presence of spectral interferences, providing diagnostic tools for detection of the outliers. Their application to spectral data has provided important tools for food analysis, where they can be used for exploration, classification and calibration purposes (Christensen et al., 2006).

Traditional multivariate analysis of fluorescence data is performed on a series of emission, excitation or synchronous spectra arranged into a matrix. It starts usually with data exploration that is aimed at discovering structures in the data set, clustering of objects and outlier detection. This analysis does not require any prior knowledge of the explored data, employing unsupervised pattern recognition methods, including principal component analysis (PCA). Other methods used to explore the food fluorescence data include hierarchical cluster analysis, non-negative matrix factorization, common components and specific weights analysis, and canonical correlation analysis (Sadecka & Tothova, 2007).

Three-way models are used for analysis of sets of fluorescence excitation-emission matrices, including parallel factor analysis (PARAFAC) and the Tucker model. The PARAFAC model decomposes the fluorescence data into a number of components. These components correspond to the distinct fluorophores present in the samples. The analysis provides relative concentrations of each of the fluorophores in the mixture, accompanied by the respective excitation and emission loadings, which correspond to the respective excitation and emission spectra, facilitating identification of the fluorescent constituents. This approach is called mathematical chromatography, enabling qualitative and quantitative analysis of the individual mixture components (Bro, 2003; Christensen et al., 2006).

The analytical problem of the food quality assessment often involves assignment of a particular product to a specific category. To perform such classification, supervised pattern recognition methods are used. In these methods the information about the class membership of the samples in a certain category is used to derive classification rules, which are next applied to classify new samples into correct categories on the basis of patterns present in their measurements (Berrueta et al., 2007). A number of classification techniques were used for analysis of food fluorescence data in the supervised mode: linear discriminate analysis, factorial discriminate analysis, k - nearest neighbors, discriminate partial least squares regression (DPLS), soft independent modeling of class analogy, and artificial neural networks.

Multivariate calibration is the most important and successful combination of chemometrics with spectral data used in analytical chemistry. The calibration consists of building a relationship between a desired chemical, biological or physical property of a sample, and its spectrum. The advantage of such approach is the replacement of the wet chemical measurements of a concentration, which are usually slow and expensive, by the spectral measurements, which are nondestructive and fast, requiring little or no sample preparation and producing no waste chemicals. The multivariate regression methods most frequently used in fluorescence analysis are partial least-squares regression (PLS) and principal component regression (PCR). N-way partial least-squares regression (N-PLS) is used for calibration analysis of fluorescence excitation-emission matrices (Geladi, 2003).

3. Application of fluorescence in olive oil analysis

Application of fluorescence to quality assessment of olive oils was proposed already in the beginning of the 20th century. From 1925, when mercury lamp with the Wood's filter became commercially available, visual observation of oil fluorescence induced by UV light was utilized to detect adulteration of extra virgin olive oils. It was shown that extra virgin olive oils exhibit characteristic yellow fluorescence, due to chlorophylls, while fluorescence of refined oils was blue due to the changes in chlorophyll content during the refining process. This method allowed detecting adulteration of extra virgin olive oils at the level of 5% with refined oils (Sidney & Willoughby, 1929; Glantz, 1930). The use of Wood's lamp was accepted as the U.S. official method for detection of olive oil adulteration. (Kyriakidis & Skarkalis, 2000). The authors of the first papers that reported spectral properties of fluorescence of vegetable oils also point out practical applications of fluorescence spectra as fingerprints in oil analysis (Wolfbeis & Leiner, 1984; Kyriakidis & Skarkalis, 2000).

3.1 Discrimination between quality grades of olive oil

Olive oil is an economically important product and its quality control and detection of possible fraud are of great interest. Olive oils are classified and priced according to acidity. The most expensive is the high-quality extra-virgin olive oil. This oil may be subject of both mislabeling and adulteration. Refined olive oil is obtained from virgin olive oil using refining methods that do not alter the initial glyceridic structure; pure olive oil (or simply olive oil) consists of a blend of virgin and refined olive oil. The potential of fluorescence to discriminate olive oils of different quality was the subject of several studies. Both total luminescence and synchronous fluorescence spectra combined with various chemometric approaches were successfully used for this purpose.

Scott et al. (2003) used total luminescence spectra of four different types of edible oils: extra virgin olive, non-virgin olive, sunflower and rapeseed oils. The spectra of undiluted oil samples were measured in the excitation range from 350 to 450 nm with 10 nm intervals and in the emission range from 400 to 720 nm with 5 nm interval. Three supervised neural network algorithms were used for sample classification: simplified fuzzy adaptive resonance theory mapping, traditional back propagation and radial basis function. The 100% correct classification was obtained using all methods.

Guimet et al. (2004a) in a series of studies investigated possibility of application of total fluorescence spectra for discrimination between various quality grades of olive oils. The excitation-emission matrices of undiluted oils were measured using right angle geometry. The hierarchical agglomerative clustering method with the Euclidean distance as a similarity measure and the average linkage method were applied to discriminate between three classes of commercial Spanish olive oils (virgin olive oils, pure olive oils, and olive-pomace oils). To optimize the sample grouping into clusters, different preprocessing methods and two spectral ranges were tested, which either included or not the fluorescence peak of chlorophylls. The oils were distinguished using the unfolded excitation–emission fluorescence matrices in the 300-400 nm excitation range and 400-600 nm emission range, thus excluding the chlorophyll band, Fig. 5. The large variations in the chlorophyll band intensity, even between samples of the same type, tend to deteriorate oil discrimination. The optimal preprocessing included normalization of the unfolded spectral excitation–emission fluorescence matrices, followed by column autoscaling. The comparison of the results obtained from the excitation–emission fluorescence matrices to those from a single emission (λ_{ex}=345, 360, 390 nm) and excitation (λ_{em}=410 nm) fluorescence spectrum analysis showed the advantage of the total fluorescence data, which result in a significantly better discrimination.

Other studies used unfold PCA and PARAFAC to explore the excitation–emission fluorescence matrices of virgin and pure olive oils (Guimet, 2004b). The spectral ranges studied were λ_{ex}=300-400 nm, λ_{em}=400-695 nm and λ_{ex}=300-400 nm, λ_{em}=400-600 nm. The first range included chlorophylls, whose peak was much more intense than those of the other components. The second range did not include the chlorophyll peak, being limited to the fluorescence spectra of the oxidation products and vitamin E. The three-component PARAFAC model on the second range (chlorophylls excluded) was found to produce the most useful results. With this model, it was possible to distinguish well between the two groups of oils and to calculate the underlying fluorescent spectra of the three families of compounds. Both unfold PCA and PARAFAC applied to the excitation–emission matrices showed clear differences between fluorescence of the two main groups of olive oils (virgin and pure). Chlorophylls had a strong influence on the models because of their high fluorescence intensity and high variability. Differentiation between the two types of oils was better when the chlorophyll fluorescence region was excluded from the models. The oxidation products are the species that most contribute to the separation between the two groups. PCA was calculated from the emission spectra of oils between λ_{em}=400 and 695 nm measured at λ_{ex}=365 nm (Guimet, 2004b).

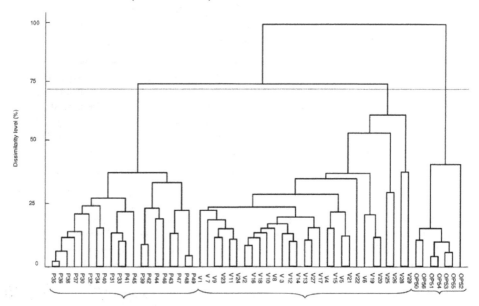

Fig. 5. Dendrogram of the 56 x 861 unfolded matrix (λ_{ex} =300-400 nm, λ_{em} = 400-600 nm) using the Euclidean distance as similarity measure and the average linkage method. The distance is expressed as a percentage of dissimilarity (normalized and autoscaled spectra): virgin V, pure, P; and olive-pomace, OP oils. Reprinted with permission from Guimet et al., 2004a. Copyright 2004, American Chemical Society.

Non-negative matrix factorization with Fisher's linear discriminant analysis were applied for discriminating between different types of olive oils: 1) discrimination between commercial Spanish olive oils of different quality (virgin, pure, and olive-pomace oil); 2) discrimination between virgin oils from two "Siurana" (Protected Denomination of Origin) regions; 3) discrimination between the original "Siurana" virgin olive oils and oils adulterated with olive-pomace oil at 5% (w/w) level (Guimet et al., 2006). In all cases, classifications at above 90% confidence were achieved. The proposed method was also compared to PARAFAC and discriminant N-PLS regression. The classification results were better with non-negative matrix factorization than PARAFAC for two data sets out of three. Non-negative matrix factorization combined with Fisher's linear discriminant analysis was also comparable with discriminant N-PLS regression, giving better classifications for the second data set, but slightly worse results for the other two. The main advantage of non-negative matrix factorization with respect to discriminant N-PLS regression is that its basis functions are more interpretable than the regression loadings, because they are positive and correspond to parts of the spectra than can be directly related to the fluorescent components of oils.

The fluorescence excitation-emission matrices (λ_{ex}=300-390 nm and λ_{em}=415-600 nm) were used in studies of the Spanish extra virgin, virgin, pure, and olive pomace oils, to investigate the relationship between oil fluorescence and the conventional quality parameters, including peroxide value, K_{232}, and K_{270} (Guimet et al., 2005c). Multiway methods were applied to the data analysis: PARAFAC with multi-linear regression and N-PLS regression. Better regression fits and lower prediction errors were obtained using N-PLS. The best results were obtained for prediction of K_{270}. The detection of extra virgin olive oils was highly degraded at early stages (with high peroxide value) and little oxidized pure olive oils (with low K_{270}).

Synchronous fluorescence spectroscopy combined with multi-dimensional chemometric techniques was applied to the classification of virgin olive oils according to their quality by Poulli et al. (2005). They studied the fluorescence of virgin olive oils and lampante oils, using total fluorescence, synchronous and total synchronous fluorescence spectra. Total luminescence spectra recorded in the 350–720 nm range while exciting in the 320 to 535 nm range showed different shapes and intensities for the two classes of oils. Lampante olive oil had a broad emission with its maximum at 450-500 nm in addition to the 685-690 nm peak. Total synchronous fluorescence spectra measured at $\Delta\lambda$=20-180 nm had emission peaks between 500-700 nm, depending on $\Delta\lambda$, for both classes of oils. However, lampante oil had additional fluorescence in the of 360-500 nm range, which is not observed for edible virgin olive oil. Classification of virgin olive oils based on their synchronous fluorescence spectra ($\Delta\lambda$=80 nm) was performed by hierarchical cluster analysis and PCA using the 429-545 nm spectral range. The authors conclude that the fluorescence in the 429-545 nm range, which they used for data analysis, originates from oleic acid. PCA provided 100% correct discrimination between the two classes, while hierarchical cluster analysis allowed 97.3% correct classification.

3.2 Adulteration detection of olive oils

A few papers were published in recent years on the use of fluorescence to assess adulteration detection of virgin olive oils. Adulteration of virgin olive oils has been a common fraud practice that involves addition of cheaper oils, including olive oils of lower quality or other plant oils. The most common adulterants found in virgin olive oil are refined olive oil, pomace oil, residue oil, synthetic olive oil–glycerol products, seed oils, and nut oils. The current analytical standards for olive oil enable detection of the presence of almost all of the possible adulterants; however, they require the measurement of several parameters established by the EU Regulations: (EEC) No 2568/91 and (EC) No 796/2002.. Thus, rapid methods to detect olive oil adulteration are important for quality control purposes (Karoui & Blecker, 2011).

Hazelnut oil is chemically similar to virgin olive oil; its presence is difficult to detect at low concentration levels using standard methods. A different approach was tested to detect this type of adulteration using fluorescence (Sayago et al., 2004; Sayago et al., 2007). The emission spectra of undiluted olive oil mixtures with virgin and refined hazelnut oils with excitation at 350 nm were measured (Sayago et al., 2004). The spectra were subjected to mathematical treatment by calculation of the first derivative. One-way analysis of variance was used for the selection of suitable wavelengths to differentiate oil samples. The response to the addition of adulterant, as evaluated by multiple linear regression, was linear for virgin olive and refined hazelnut oil mixtures (R^2=0.99), and for virgin olive and virgin hazelnut oil mixtures (R^2=0.98). Stepwise linear discriminant analysis used to discriminate genuine from adulterated olive oil samples allowed 100% correct classifications for each mixture separately, and also for the entire set of samples. Another work explored application of the fluorescence spectroscopy to differentiate between refined hazelnut and refined olive oils (Sayago et al., 2007). Classification of these oils based on their excitation (in 300-500 nm spectral range, using λ_{em}=655 nm) and emission spectra (in the 650-900 nm range, using λ_{ex}=350 nm) was performed, using PCA and artificial neural networks. Both methods provided good discrimination between the refined hazelnut and olive oils. Using the artificial neural networks model, the presence of refined hazelnut oils in refined olive oils was robustly detected at levels exceeding 9%.

Several studies devoted to the detection of of adulteration of virgin olive oil with sunflower oil. Poulli et al. (2006) applied total synchronous fluorescence to differentiate virgin olive from sunflower oil and synchronous fluorescence combined with PLS regression for quantitative determination of olive oil adulteration. Total synchronous fluorescence spectra were acquired in the 270-720 nm range, using the wavelength interval variable from $\Delta\lambda$=20 to 120 nm. The emission band at around 660 nm was only observed in virgin olive oil, attributed to pigments of the chlorophyll group. For sunflower, in contrast to virgin olive oil, a fluorescence band in the 325-385 nm excitation range is observed. This band was attributed to linoleic acid, however, there are no published data on fluorescence of this compound. In contrast, virgin olive oil has only small signals in this range if scanned at 30 to 50 nm wavelength interval. Synchronous fluorescence spectra of virgin olive oil recorded at $\Delta\lambda$=20 nm show a double band at 275 and 297 nm and a single band at 660 nm, in stark contrast to sunflower oil that has an intense band at around 300 nm and a weak one at 325 nm. For quantification of the adulteration, the PLS regression model was used for analysis of synchronous fluorescence spectra of mixtures of virgin olive oil and sunflower oil at $\Delta\lambda$=20 and 80 nm, Fig. 6. The detection limits were 3.6% and 3.4% (w/v) when using the 20 and 80 nm wavelength intervals, respectively.

The potential of fluorescence spectroscopy for detecting adulteration of extra virgin olive oil with olive oil has been investigated recently (Dankowska & Małecka, 2009). Synchronous fluorescence spectra were collected in the 240-700 nm range, using $\Delta\lambda$=10, 30, 60 and 80 nm. A narrow band at around 300 nm appeared in the synchronous fluorescence spectrum at $\Delta\lambda$=10 nm, attributed to tocopherols, and an intense band with a peak at around 665 nm, attributed to compounds of the chlorophyll group. The raw spectra were subject to calculation of the first and second derivatives to find the maximum or the intersection point. Five wavelengths at each of the wavelength intervals were chosen for further analysis. Multiple regression analysis was applied separately to the data acquired at each of the wavelength intervals. The ability to detect olive oil in extra virgin olive oil was better at the wavelength interval of 60 or 80 nm, rather than 10 or 30 nm. Using the spectra acquired at 60 and 80 nm wavelength intervals, the lowest detection limits of adulteration were 8.9% and 8.4% at 350 and 302 nm, respectively.

Fluorescence was used to detect adulteration of virgin olive with others oils (Poulli et al., 2007). Synchronous fluorescence spectra of virgin olive, olive-pomace, corn, sunflower, rapeseed, soybean and walnut oils at 20 nm wavelength interval were used for analysis. Virgin olive oil shows a double band in the 275-297 nm range and a single band at 660 nm, in contrast to other oils that show a strong band around 300 nm and a weak to moderate band near 325 nm. Total synchronous fluorescence spectra were acquired for the excitation wavelength in the 250-720 nm range and the wavelength interval $\Delta\lambda$ in the 20 to 120 nm range. Total synchronous fluorescence spectra for olive oils show a spectral band at around 660 nm, attributable to pigments of chlorophyll group. Moreover, all studied oils save the virgin olive oil show a band at above 315 nm when using $\Delta\lambda$=20 nm. This band could be attributed to linoleic acid. It has been suggested that differentiation of virgin olive oil from low quality oils can be achieved using this wavelength region. The PLS regression model was used to quantify adulteration using 20 nm synchronous fluorescence spectra. This technique enabled detection of olive-pomace, corn, sunflower, soybean, rapeseed and walnut oil in virgin olive oil at levels of 2.6, 3.8, 4.3, 4.2, 3.6, and 13.8% (w/w), respectively (Poulli et al., 2007).

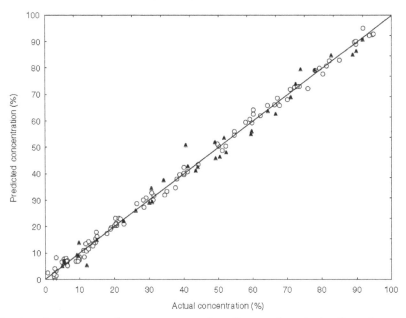

Fig. 6. Predicted versus actual concentrations of sunflower oil in virgin olive oil at a wavelength interval of 80 nm: open circles - calibration samples, filled triangles - validation samples. Reprinted with permission from Poulli et al., 2006. Copyright 2006, Springer.

For adulteration detection of extra virgin olive oil with rapeseed oil a neural network method – a simplified fuzzy adaptive resonance theory mapping - was found to be very efficient, resulting in a total of 99.375% correctly classified oil samples at the 5% v/v adulteration level (Scott et al., 2003). It was shown for extra virgin olive oil adulterated by rapeseed oil that the percentage of adulteration may be described by either a radial basis functional network (2.435% RMSE) or a simple Euclidean distance relationship of the PCA scores (2.977% RMSE).

3.3 Authentication of virgin olive oils geographical origin

Classification of virgin olive oils according to variety and/or geographical origin is of great importance for the producers, importers, and consumers. Dupuy et al. (2005) developed a method to discriminate olive oils according to their geographic origin. Samples of virgin olive oil from five French registered designations of origin (RDOs), namely, Nyons, Vallé des Baux, Aix-en-Provence, Haute-Provence, and Nice, were discriminated by applying multivariate regression methods to synchronous fluorescence spectra of oils. The synchronous fluorescence spectra were collected in the 250-700 nm range at a constant wavelength difference of 30 nm between the excitation and emission wavelengths. The PLS regression analysis of synchronous fluorescence spectra allowed to determine the origin of the oils with satisfactory results, despite the similarity between two denominations of origin (Baux and Aix) that are composed by some common cultivars (Aglandau and Salonenque). The interpretation of the regression coefficients shows that RDOs are correlated to chlorophylls, pheophytins, tocopherols, and phenolic compounds, present in different amount for each origin (Dupuy et al., 2005).

Guimet et al. (2005a) developed a method based on excitation-emission matrices and three-way methods for detecting adulterations of pure olive oils in extra virgin olive oil from the protected denomination of origin (PDO) "Siurana", which is a prestigious distinction given to the extra virgin olive oils produced in a specific area in the south of Catalonia, Spain. Excitation and emission ranges were λ_{ex}=300–390 nm and λ_{em}=415–600 nm, respectively. Unfold PCA and PARAFAC were used for exploratory analysis. Discrimination between non-adulterated and adulterated samples was performed using Fisher's linear discriminant analysis and discriminant N-PLS regression. Using discriminant N-PLS regression, 100% correct classification was obtained. Adulteration at around 5% level was quantified, with a prediction error of 1.2% (Guimet et al., 2005a). In another study, Guimet et al. (2005b) achieved discrimination between oils from the two PDO "Siurana" regions by means of discriminant unfold PLS regression, giving correct classification for 94% of samples for "Siurana-Camp" and 100% for "Siurana-Montsant" oils.

3.4 Quantification of fluorescent components in olive oils

The olive oil autofluorescence is attributed to minor components, species such as tocopherols, phenols and chlorophylls, thus fluorescence spectroscopy has been used to analyze these compounds in olive oils. Diaz et al. (2003) used fluorescence for determination of chlorophylls a and b and pheophytins a and b in olive oil samples. The analysis was accomplished by PLS multivariate calibration using the three types of spectra (excitation, emission, and synchronous spectra of these solutions). The best results were obtained for the excitation spectra. The optimum wavelength range to record the excitation spectra (λ_{em}=662 nm) was selected to minimize the contribution of pheophytin a and to maximize the contribution of the other pigments, which are the minor constituents in olive oil. To perform the PLS calibration, a set of samples with final concentration ranges varying from 140 to 560 ng mL^{-1} for pheophytin a, from 10 to 40 ng mL^{-1} for chlorophyll a and pheophytin b, and from 20 to 80 ng mL^{-1} for chlorophyll b was used as the calibration matrix. The oil samples were diluted in acetone. Recovery values from olive oil, spiked with chlorophylls a and b and pheophytins a and b, were in the ranges of 70-112, 71-111, 76-105, and 82-109%, respectively.

Fluorescence was proposed as an alternative to the Folin-Ciocalteu assay for estimation of the total phenol content in virgin olive oil, olive fruit and leaf polar extracts (Papoti & Tsimidou, 2009). Phenol content in olive oils was determined by measuring the fluorescence intensity of methanol/water extract, with the excitation/emission wavelengths set at 280/320 nm. The method was shown to be more sensitive (limit of detection and limit of quantification values 10-fold lower) and three times faster then the Folin-Ciocalteu assay. Good correlation was found with the results of colorimetric assay (r = 0.69, n = 65) for virgin olive oil extracts.

Fluorescence combined with PLS regression was used to determine tocopherol homologues (α-, β-, γ-, and δ- tocopherol) in the quaternary mixture and the oils (Diaz et al., 2006). The calibration set that included mixtures of tocopherols dissolved in hexane: diethyl ether (70:30 v/v) was constructed based on the central composite plus a full factorial plus a fractionated factorial design. PLS regression was applied to analyze matrices of fluorescence excitation and emission spectra and with fluorescence excitation, emission, and synchronous spectra. For analysis of synthetic samples, recoveries around 100% were obtained. For the analysis of the oils, the samples were diluted in hexane, cleaned in silica cartridges and then tocopherols were eluted with hexane:diethyl ether (90:10 v/v). The method was applied to different edible oils giving satisfactory results for α-, β-, and γ-, but not for δ- tocopherol.

PLS regression was utilized to develop calibration models between front face and right angle synchronous fluorescence spectroscopy for the characterization of edible oils and total tocopherol content as determined by HPLC (Sikorska et al., 2005b). The studies were performed on commercially available edible oils: olive, grapeseed, rapeseed, soybean, sunflower, peanut, and corn oils were analyzed. The regression models showed a good ability to predict tocopherol content. The best fitting results were obtained for 1% v/v diluted oils and for bulk samples using the entire spectrum, yielding the regression coefficient of 0.991, and the root mean square error of cross-validation of 8%.

3.5 Monitoring thermal and photo-oxidation of olive oils

The studies of thermal deterioration of oils are important because changes during oxidation involve degradation of oil constituents and formation of new products that alter quality attributes and nutritional profile, as the oxidation products are potentially toxic. Fluorophores in olive oils are compounds that can participate in oxidation, thus fluorescence spectroscopy can serve as a tool for better understanding of oil oxidation. The fluorescence was compared to other spectroscopic techniques (NIR/VIS, FT-IR and FT-Raman) and chemical and physical methods in determining the deterioration of frying oils, collected from a commercial Chinese spring roll plant, (Engelsen, 1997). Fluorescence has been measured by using five selected excitation wavelengths varying from 395 to 530 nm. Data analysis was performed using PCA and PLS regression. Overal, fluorescence provided the best models for the anisidine value, oligomers, iodine value, and vitamin E concentration, among the spectroscopic techniques used.

Fluorescence spectrometry and PLS regression were used as a rapid technique for evaluating the quality of heat-treated extra virgin olive (Cheikhousman et al., 2005). Two commercial extra virgin olive oils were heated at 170°C for 3 h. Changes in excitation spectra were correlated with changes in concentrations determined by other methods. The fluorescence excitation band emitting at 330 nm was attributed to vitamin E and some fluorescent polyphenols. This fluorescence decreased during the heating process, with the exponential decay constant similar to that obtained chromatographically. Fluorescence excitation spectra with the emission wavelength at 450 nm were inversely correlated with the hydroperoxide content of oil. Indeed, the degradation products generated during heating, particularly the compounds formed by reaction between amino-phospholipids and aldehydes, fluoresce in this wavelength range.

Thermal deterioration of extra virgin olive oils was studied by Tena et al. (2009). The sample of virgin olive oil was heated at 190°C for 94 h in cycles of 8 h per day. The fluorescence intensity in the spectral region between 290 and 400 nm decreased during the oxidation and a bathochromic shift of the maximum from 350-360 to around 420-440 nm was observed. The fluorescence observed in the 300-390 nm range was assigned to tocopherols together with polyphenols; the information collected from the spectra was compared to the results of the HPLC analysis of these compounds. The observed changes in the spectral profile were explained by the decrease of the tocopherols and phenols and the increase of the oxidation products of vitamin E homologues correlated to K_{232} and K_{270}, and hydrolysis products. The intensity of the band between 630 and 750 nm, associated with chlorophylls and pheophytins, decreased exponentially with the thermal oxidation time.

The fluorescence intensity recorded at 350 nm and at the wavelength of the spectral maximum occurring in the range of 390-630 nm allowed to explain the increase of the percentage of polar compounds during the experiment. It was stated that the spectra of the undiluted heated oils with maxima at 490 nm or higher correspond to polar compounds exceeding 25%, which is the maximum percentage acceptable for edible oils used in frying.

Poulli et al. (2009b) studied the effect of heating to 100, 150 and 190° C on extra virgin olive, olive pomace, sesame, corn, sunflower, soybean, and a commercial blend of oils. The changes in fluorescence were assessed by measuring total synchronous fluorescence spectra, in the 250-720 nm excitation range, with the wavelength interval, $\Delta\lambda$, from 20 to 120 nm at 20 nm step. The synchronous fluorescence intensities below 315 nm recorded at $\Delta\lambda$ =80 nm decreased during heating, presumably due to the consumption of phenolic antioxidants by the lipid radicals generated. The decrease of the fluorescence bands in the 250-350 and 350-400 nm ranges for extra virgin olive and olive pomace oil, respectively, was in accordance with the percentage of trolox equivalent antioxidant capacity reduction. The bands in the total synchronous fluorescence spectra at below 350 nm disappeared during heating, with those at 600-700 nm also decreasing, probably due to the decay of antioxidant compounds and chlorophyll, respectively. The bands in the 400-450 nm range increased, probably due to the formation of secondary oxidation products. PCA of synchronous fluorescence spectra obtained at $\Delta\lambda$=80 nm allowed oil discrimination according to the degree of oxidation. For extra virgin olive, olive pomace, and sesame oil the spectral range of 300-500 nm was used for classification, while the 320-520 nm range was more appropriate for corn, soybean, and sunflower oil, and a commercial blend of oils. Spectroscopic changes are indicative of oxidative deterioration as measured through wet chemistry methods: peroxide value, p-anisidine value, totox value, and radical-scavenging capacity (Poulli, 2009a).

Extra virgin olive oil is very stable in the dark; it is susceptible to oxidation under UV light. An accelerated thermal and photooxidation under UV light was studied by Poulli et al. (2009a, 2009b) on samples of extra-virgin, regular-quality and pomace olive oils. Synchronous fluorescence spectra were collected using the 250–720 nm excitation range at $\Delta\lambda$=80 nm. Extra virgin olive oil bands in the 300–330 nm range decreased during oxidation, while the fluorescence in the 350–550 nm range increased during the initial 8 h and then remained almost constant for up to 12 h. Regular quality olive oil exhibited fluorescence in the 300–550 nm range. The bands in the 300–370 nm range decreased during oxidation, whereas fluorescence bands in the 370–550 nm range increased during the initial period and remained almost constant afterwards. Also, the fluorescence bands of pomace oil in the 350–550 nm range decreased during the initial period of the experiment and then a small additional increase was observed. All olive oils show fluorescence bands in the 550–700 nm range, attributed to chlorophyll pigments, intensive in extra virgin olive oil and with very low intensity in olive-pomace oil. These bands decreased significantly due to deterioration of the chlorophyll pigments involved in photo-oxidation. Total synchronous fluorescence spectra were obtained by scanning the excitation wavelength in the same spectral range and changing the wavelength interval from 20 to 120 nm at 20 nm steps. These spectra showed considerable changes during oxidation for all of the oils studied, Fig. 7. Fluorescence intensity in the 600–720 nm range, attributed to chlorophylls, decreased significantly. In contrast, the fluorescence bands in the low-wavelength range expanded up to 590 nm. PCA applied to the synchronous fluorescence spectra recorded at $\Delta\lambda$=80 nm in the 300–500 nm range reveals five different classes of oils depending on their oxidation degree.

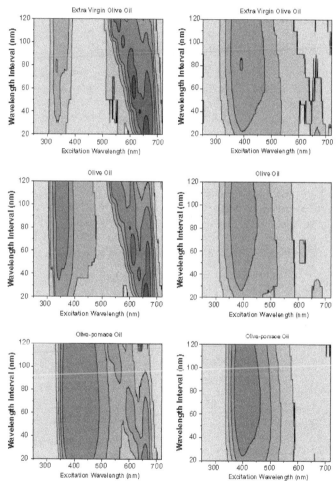

Fig. 7. Total synchronous fluorescence spectra: contour plots of olive oils before (left) and after 12 h (right) exposure to UV light at 80°C. Grayscale indicates fluorescence intensities. Reprinted from *Food Chemistry*, (2009) Vol.117, No.3, Poulli K. I.; Mousdis G. A. & Georgiou C. A., "Monitoring Olive Oil Oxidation Under Thermal and UV Stress Through Synchronous Fluorescence Spectroscopy and Classical Assays", pp. 499-503, Copyright (2009), with permission from Elsevier.

An interesting study on thermal oxidation of extra virgin olive oils has been published recently (Navarra et al., 2011). Among different experimental techniques (including FTIR and rheology) time-resolved luminescence was used to investigate early steps of the thermally induced oxidative process. The oxidation process was followed at three different heating temperatures (30, 60 and 90°C) as a function of time for up to 35 days. The chlorophyll fluorescence lifetime increased from 6.0 ± 0.1 ns, measured before, to 6.3 ± 0.1 ns, measured after 35 days of experiment. These changes were in agreement with the observed viscosity rise, resulting from formation of polar molecules with propensity to form hydrogen bonds. The viscosity increase reduced the frequency of collisions between the chromophore and its environment, consequently lowering the non-radiative contribution to the luminescence decay.

3.6 Assessing quality changes of olive oil during storage

Fluorescence spectroscopy was applied to monitoring changes in virgin olive oil during storage (Sikorska et al. 2008b). The extra virgin olive oil samples were stored for the period of 12 month in different conditions: in clear and green glass bottles exposed to light, and in darkness. Changes occurring in olive oil during storage were assessed by total fluorescence and synchronous scanning fluorescence spectroscopy techniques. In the total fluorescence spectra the intensity of emissions ascribed to tocopherols and chlorophyll pigments decreased during storage, depending on the storage conditions. Additional bands appeared in oils exposed to light in the intermediate range of excitation and emission wavelengths. Bands attributed to tocopherols, chlorophylls and those tentatively ascribed to phenolic compounds were observed in the synchronous scanning fluorescence spectra, allowing monitoring of the storage effects on these constituents. PCA of the synchronous fluorescence spectra revealed systematic changes in the overall emission characteristics dependent on the storage conditions, such as exposure to light, and packaging, Fig. 8.

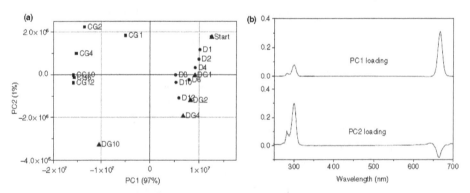

Fig. 8. (a) Scores plot for the two most significant principal components, PC1 vs. PC2, of a PCA of the synchronous scanning fluorescence (SSF) spectra ($\Delta\lambda$ = 10 nm) of virgin olive oil samples. A nonstored sample (Start), and samples stored in different conditions: in darkness (D), in green glass bottles (DG) and in clear glass bottles (CG). The samples are numbered according to the months of storage. The values in brackets describe the fraction of the total variation explained by each of the PC. Each point represents an average of the spectra obtained from three replicates . (b) One-vector loading plots for the PC1 and PC2 principal components. Reprinted with permission from Sikorska et al., 2008b. Copyright 2008, John Wiley and Sons.

4. Conclusions

Fluorescence spectra of olive oils contain information about fluorophores (tocopherols, phenolic compounds, and chlorophylls) that are important for oil quality. The spectra may be used to monitor either selected constituents or to determine overall sample characteristics, which may serve as the spectral fingerprint. The analytic potential of fluorescence is enhanced by application of multivariate data analysis methods for the analysis of spectra.

The review of literature data demonstrates that fluorescence measurements conducted directly on olive oil samples with subsequent multivariate data analysis can be efficiently used for qualitative and quantitative analysis as a valid alternative to conventional, chemical methods of quality assessment. These methods can be used for oil discrimination and for quantitative determination of fluorescent components after an appropriate calibration. Further studies are needed to resolve various issues that are important for practical application of the fluorescence techniques, among which are the method verification for specific kinds of oil and identification and quantification of other fluorescent oil constituents.

5. Acknowledgment

Grant NN312428239, 2010-2013, from the Polish Ministry of Science and Higher Education is gratefully acknowledged.

6. References

Berrueta L. A.; Alonso-Salces R. M. & Heberger K. (2007). Supervised Pattern Recognition in Food Analysis. *Journal of Chromatography* A, Vol.1158, No.1-2, pp. 196-214, ISSN 0021-9673

Bro R. (2003). Multivariate Calibration. What is in Chemometrics for the Analytical Chemist?. *Analytica Chimica Acta*, Vol.500, No.1-2, pp. 185-194, ISSN 0003-2670

Cert A.; Moreda W. & Perez-Camino M. C. (2000). Chromatographic Analysis of Minor Constituents in Vegetable Oils. *Journal of Chromatography* A, Vol.881, No.1-2, pp. 131-148, ISSN 0021-9673

Cheikhousman R.; Zude M.; Bouveresse D. J. R.; Leger C. L.; Rutledge D. N. & Birlouez-Aragon I. (2005). Fluorescence Spectroscopy for Monitoring Deterioration of Extra Virgin Olive Oil During Heating. *Analytical and Bioanalytical Chemistry*, Vol.382, No.6, pp. 1438-1443, ISSN 1618-2650

Christensen J.; Norgaard L.; Bro R. & Engelsen S. B. (2006). Multivariate Autofluorescence of Intact Food Systems. *Chemical Reviews*, Vol.106, No.6, pp. 1979-1994, ISSN 1520-6890

Dankowska A. & Małecka M. (2009). Application of Synchronous Fluorescence Spectroscopy for Determination of Extra Virgin Olive Oil Adulteration. *European Journal of Lipid Science and Technology*, Vol.111, No.12, pp. 1233-1239, ISSN 1438-9312

Díaz T. G.; Merás I. D.; Correa C. A.; Roldan B. & Cáceres M. I. R. (2003). Simultaneous Fluorometric Determination of Chlorophylls A and B and Pheophytins A and B in Olive Oil by Partial Least-Squares Calibration. *Journal of Agricultural and Food Chemistry*, Vol.51, No.24, pp. 6934-6940, ISSN 1618-2650

Díaz, T. G.; Durán-Merás, I.; Cáceres, M. I. R.; Murillo, B. R. (2006). Comparison of Different Fluorimetric Signals for the Simultaneous Multivariate Determination of Tocopherols in Vegetable Oils. *Applied Spectroscopy*, Vol.60, No.2, pp. 194-202, ISSN 1943-3530

Dupuy N.; Le Dreau Y.; Ollivier D.; Artaud J.; Pinatel C. & Kister J. (2005). Origin of French
    Virgin Olive Oil Registered Designation of Origins Predicted by Chemometric
    Analysis of Synchronous Excitation-Emission Fluorescence Spectra. *Journal of
    Agricultural and Food Chemistry*, Vol.53, No.24, pp. 9361-9368, ISSN 1618-2650
Eitenmiller R. R.; Ye L. & Landen W. O. (2008). Vitamin E: Tocopherols and Tocotrienols. In
    Vitamin Analysis for the Health and Food Sciences, 2 edn CRC Press, New York, ISBN
    0849326680
Engelsen S. B. (1997). Explorative Spectrometric Evaluations of Frying Oil Deterioration.
    Journal of the American Oil Chemists Society, Vol.74, No.12, pp. 1495-1508, ISSN 1558-
    9331
Geladi P. (2003). Chemometrics in Spectroscopy. Part 1. Classical Chemometrics.
    Spectrochimica Acta Part B, Vol.58, No.5, pp. 767-782, ISSN 0584-8547
Glantz A. L. (1930). Fluorescence of Olive Oil Under Ultra-Violet Light. *Industrial &
    Engineering Chemistry Analytical Edition*, Vol.2, No.3, pp. 256-258, ISSN 0096-4484
Guilbault G. G. (1999). *Practical Fluorescence*, Marcel Dekker, ISBN 0824712633, New York
Guimet F.; Boque R. & Ferre J. (2004a). Cluster Analysis Applied to the Exploratory Analysis
    of Commercial Spanish Olive Oils by Means of Excitation-Emission Fluorescence
    Spectroscopy. *Journal of Agricultural and Food Chemistry*, Vol.52, No.22, pp. 6673-
    6679, ISSN 1618-2650
Guimet F.; Ferre J.; Boque R. & Rius F. X. (2004b). Application of Unfold Principal
    Component Analysis and Parallel Factor Analysis to the Exploratory Analysis of
    Olive Oils by Means of Excitation-Emission Matrix Fluorescence Spectroscopy.
    Analytica Chimica Acta, Vol.515, No.1, pp. 75-85, ISSN 0003-2670
Guimet F.; Ferre J. & Boque R. (2005a). Rapid Detection of Olive-Pomace Oil Adulteration in
    Extra Virgin Olive Oils From the Protected Denomination of Origin "Siurana"
    Using Excitation-Emission Fluorescence Spectroscopy and Three-Way Methods of
    Analysis. *Analytica Chimica Acta*, Vol.544, No.1-2, pp. 143-152, ISSN 0003-2670
Guimet F.; Boque R. & Ferre J. (2005b). Study of Oils From the Protected Denomination of
    Origin "Siurana" Using Excitation-Emission Fluorescence Spectroscopy and Three-
    Way Methods Of Analysis. *Grasas y Aceites*, Vol.56, No.4, pp. 292-297, ISSN 1988-4214
Guimet F.; Boque R. & Ferre J. (2006). Application of Non-Negative Matrix Factorization
    Combined With Fisher's Linear Discriminant Analysis for Classification of Olive
    Oil Excitation-Emission Fluorescence Spectra. *Chemometrics and Intelligent
    Laboratory Systems*, Vol.81, No.4, pp. 94-106, ISSN 0169-7439
Guimet F.; Ferre J.; Boque R.; Vidal M. & Garcia J. (2005c). Excitation-Emission Fluorescence
    Spectroscopy Combined with Three-Way Methods of Analysis as a
    Complementary Technique for Olive Oil Characterization. *Journal of Agricultural
    and Food Chemistry*, Vol.53, No.24, pp. 9319-9328, ISSN 1618-2650
Karoui R. & Blecker Ch. (2011). Fluorescence Spectroscopy Measurement for Quality
    Assessment of Food Systems—A Review. *Food and Bioprocess Technology*. Vol.4,
    No.3, pp. 364-386, ISSN 1935-5149
Kyriakidis N. B. & Skarkalis P. (2000). Fluorescence Spectra Measurement of Olive Oil and
    Other Vegetable Oils. *Journal of AOAC International*, Vol.83, No.24, pp. 1435-1439,
    ISSN 1944-7922
Lakowicz J. R. (2006). *Principles of Fluorescence Spectroscopy*, Third edn. Kluwer
    Academic/Plenum Publishers, ISBN 0387312781, New York

Lloyd J. B. F. (1971). Synchronized Excitation of Fluorescence Emission Spectra. *Nature (London) Physical Science*, Vol.231, pp. 64-65, ISSN 1476-4687

Navarra G.; Cannas M.; D'Amico M.; Giacomazza D.; Militello V.; Vaccaro L. & Leon M. (2011). Thermal Oxidative Process in Extra-Virgin Olive Oils Studied by FTIR, Rheology and Time-Resolved Luminescence, *Food Chemistry*, Vol.126, No.3, pp. 1226-1231, ISSN 0308-8146

Ndou T. & Warner I. M. (1991). Applications of Multidimensional Absorption and Luminescence Spectroscopies in Analytical Chemistry. *Chemical Reviews*, Vol.91, No.4, pp. 493-507, ISSN 1520-6890

Papoti V. T. & Tsimidou M. Z. (2009). Looking Through the Qualities of a Fluorimetric Assay for the Total Phenol Content Estimation in Virgin Olive Oil, Olive Fruit or Leaf Polar Extract, *Food Chemistry*, Vol.112, No.1, pp. 246-252, ISSN 0308-8146

Pollok, D. & Melchert, H. U. (2004). Determination of alpha-Tocopherolquinone in Human Serum Samples by Liquid Chromatography with Fluorescence Detection and On-Line Post-Column Derivatization. *Journal of Chromatography* A, Vol.1056, No. 1-2, pp. 257-262, ISSN 0021-9673

Poulli K. I.; Mousdis G. A. & Georgiou C. A. (2005). Classification of Edible and Lampante Virgin Olive Oil Based on Synchronous Fluorescence and Total Luminescence Spectroscopy. *Analytica Chimica Acta*, Vol.542, No.2, pp. 151-156, ISSN 0003-2670

Poulli K.I.; Mousdis G. A. & Georgiou C. A. (2006). Synchronous Fluorescence Spectroscopy for Quantitative Determination of Virgin Olive Oil Adulteration with Sunflower Oil. *Analytical and Bioanalytical Chemistry*. Vol.386, No.5, pp. 1571-1575, ISSN 1618-2650

Poulli K. I.; Mousdis G. A. & Georgiou C. A. (2007). Rapid Synchronous Fluorescence Method for Virgin Olive Oil Adulteration Assessment. *Food Chemistry*, Vol.105, No.1, pp. 369-375, ISSN 0308-8146

Poulli K. I.; Chantzos N. V.; Mousdis G. A. & Georgiou C. A. (2009a). Synchronous Fluorescence Spectroscopy: Tool for Monitoring Thermally Stressed Edible Oils. *Journal of Agricultural and Food Chemistry*, Vol.57, No.18, pp. 8194-8201, ISSN 1520-5118

Poulli K. I.; Mousdis G. A. & Georgiou C. A. (2009b). Monitoring Olive Oil Oxidation Under Thermal and UV Stress Through Synchronous Fluorescence Spectroscopy and Classical Assays, *Food Chemistry*, Vol.117, No.3, pp. 499-503, ISSN 0308-8146

Sadecka J. & Tothova J. (2007). Fluorescence Spectroscopy and Chemometrics in the Food Classification - A Review. *Czech Journal of Food Sciences*, Vol.25, No.4, pp. 159-173, ISSN 1212-1800

Sayago A.; Morales M.T. & Aparicio R. (2004). Detection of Hazelnut Oil in Virgin Olive Oil by a Spectrofluorimetric Method. *European Food Research and Technology*, Vol.218, No.5, pp. 480-483, ISSN 1438-2385

Sayago A.; Garcia-Gonzalez D. L.; Morales M.T. & Aparicio R. (2007). Detection of the Presence of Refined Hazelnut Oil in Refined Olive Oil by Fluorescence Spectroscopy. *Journal of Agricultural and Food Chemistry*, Vol.55, No.6, pp. 2068-2071, ISSN 1520-5118

Scott S. M.; James D.; Ali Z.; O'Hare W. T. & Rowell F. J. (2003). Total Luminescence Spectroscopy With Pattern Recognition for Classification of Edible Oils. *Analyst*, Vol.128, No.7, pp. 966-973, ISSN 0003-2654

Servili M.; Selvaggini R.; Esposto S.; Taticchi A.; Montedoro G. & Morozzi G. (2004). Health and Sensory Properties of Virgin Olive Oil Hydrophilic Phenols: Agronomic and Technological Aspects of Production That Affect Their Occurrence in the Oil. *Journal of Chromatography* A, Vol.1054, No.1-2, pp. 113-127, ISSN 0021-9673

Sidney M. S. & Willoughby C. E. (1929). Olive Oil Analytical Method. Part II The Use of the Ultraviolet Ray in the Detection of Refined in Virgin Olive Oil, *Journal of the American Oil Chemists' Society*, Vol.6, No.8, pp. 15-16, ISSN 0003-021X

Sikorska E.; Romaniuk A.; Khmelinskii I. V.; Herance R.; Bourdelande J. L.; Sikorski M. & Kozioł J. (2004). Characterization of Edible Oils Using Total Luminescence Spectroscopy. *Journal of Fluorescence*, Vol.14, No.1, pp. 25-35, ISSN 1573-4994

Sikorska E.; Górecki T.; Khmelinskii I. V.; Sikorski M. & Kozioł J. (2005a). Classification of Edible Oils Using Synchronous Scanning Fluorescence Spectroscopy. *Food Chemistry*, Vol.89, No.2, pp. 217-225, ISSN 0308-8146

Sikorska E.; Gliszczyńska-Świgło A.; Khmelinskii I. & Sikorski M. (2005b). Synchronous Fluorescence Spectroscopy of Edible Vegetable Oils. Quantification of Tocopherols. *Journal of Agricultural and Food Chemistry*, Vol.53, No.18, pp. 6988-6994, ISSN 1520-5118

Sikorska E. (2008a). Metody fluorescencyjne w badniach żywności, AE Poznań. ISBN 978-83-7417-360-5, Poznań, Poland

Sikorska E.; Khmelinskii I. V.; Sikorski M.; Caponio F.; Bilancia M. T.; Pasqualone A. & Gomes T. (2008b). Fluorescence Spectroscopy in Monitoring of Extra Virgin Olive Oil During Storage. *International Journal of Food Science and Technology*, Vol.43, No.1-2, pp. 52-61, ISSN 1365-2621

Sikorska E.; Khmelinskii I. V. & Sikorski M. (2011). Unpublished results

Smyk B.; Amarowicz R.; Szabelski M.; Gryczynski I. & Gryczynski Z. (2009). Steady-State and Time-Resolved Fluorescence Studies of Stripped Borage Oil, *Analytica Chimica Acta*, Vol.646, No.1-2, pp. 85-89, ISSN 1520-5118

Tena N.; Garcia-Gonzalez D. L. & Aparicio R. (2009). Evaluation of Virgin Olive Oil Thermal Deterioration by Fluorescence Spectroscopy. *Journal of Agricultural and Food Chemistry*, Vol.57, No.22, pp. 10505-10511, ISSN 1520-5118

Undenfriend S. (1962). *Fluorescence Assay in Biology and Medicine*. Academic Press, New York

Ward K.; Scarth R.; Daun J. K. & Thorsteinson C. T. (1994). Comparison of High-Performance Liquid Chromatography and Spectrophotometry to Measure Chlorophyll in Canola Seed and Oil. *Journal of the American Oil Chemists' Society*, Vol.71, No.9, pp. 931-934, ISSN 0003-021X

Weber G. (1961). Enumeration of Components in Complex Systems by Fluorescence Spectrophotometry. *Nature*, Vol.190, No.4770, pp. 27-29, ISSN 1476-4687

Wolfbeis O. S. & Leiner M. (1984). Characterization of Edible Oils via 2D-Fluorescence, *Mikrochimica Acta*, Vol.1, pp. 221-233, ISSN 0026-3672

Zandomeneghi M.; Carbonaro L. & Caffarata C. (2005). Fluorescence of Vegetable Oils: Olive Oils. *Journal of Agricultural and Food Chemistry*, Vol.53, No.3, pp. 759-766, ISSN 1520-5118

Zandomeneghi, M. & Zandomeneghi, G. (2005). Comment on Cluster Analysis Applied to the Exploratory Analysis of Commercial Spanish Olive Oils by Means of Excitation-Emission Fluorescence Spectroscopy. *Journal of Agricultural and Food Chemistry*, Vol.53, No.14, pp. 5829-5830, ISSN 1520-5118

Zandomeneghi M.; Carbonaro L. & Zandomeneghi G. (2006). Comment on Excitation-Emission Fluorescence Spectroscopy Combined with Three-Way Methods of Analysis as a Complementary Technique for Olive Oil Characterization. *Journal of Agricultural and Food Chemistry*, Vol.54, No.14, pp. 5214-5215, ISSN 1520-5118

6

Sensory Analysis of Virgin Olive Oil

Alessandra Bendini, Enrico Valli,
Sara Barbieri and Tullia Gallina Toschi*
Department of Food Science, University of Bologna
Italy

1. Introduction

Virgin olive oil (VOO) is the supernatant of the fresh juice obtained from olives by crushing, pressure and centrifugation, without additional refining. Its flavour is characteristic and is markedly different from those of other edible fats and oils. The combined effect of odour (directly via the nose or indirectly through a retronasal path, via the mouth), taste and chemical responses (as pungency) gives rise to the sensation generally perceived as "flavour".

Sensory analysis is an essential technique to characterize food and investigate consumer preferences. International cooperative studies, supported by the International Olive Oil Council (IOOC) have provided a sensory codified methodology for VOOs, known as the "COI Panel test". Such an approach is based on the judgments of a panel of assessors, conducted by a panel leader, who has sufficient knowledge and skills to prepare sessions of sensory analysis, motivate judgement, process data, interpret results and draft the report. The panel generally consists of a group of 8 to 12 persons, selected and trained to identify and measure the intensity of the different positive and negative sensations perceived. Sensory assessment is carried out according to codified rules, in a specific tasting room, using controlled conditions to minimize external influences, using a proper tasting glass and adopting both a specific vocabulary and a profile sheet that includes positive and negative sensory attributes (Dec-23/98-V/2010). Collection of the results and statistical elaboration must be standardized (EEC Reg. 2568/91, EC Reg. 640/08). The colour of VOO, which is not significantly related to its quality, may produce expectations and interferences in the flavour perception phase. In order to eliminate any prejudices that may affect the smelling and tasting phases, panelists use a dark-coloured (blue or amber-coloured) tasting glass.

Many chemical parameters and sensory analyses (EEC Reg. 2568/91 and EC Reg. 640/08), with the latter carried out by both olfactory and gustatory assessments, can classify oils in different quality categories (extra virgin, virgin, lampant). Extra virgin olive oil (EVOO) extracted from fresh and healthy olive fruits (Olea europaea L.), properly processed and adequately stored, is characterized by an unique and measurable combination of aroma and taste. Moreover, the category of EVOO should not show any defects (e.g. fusty, musty, winey, metallic, rancid) that can originate from incorrect production or storage procedures.

---

* Corresponding Author

Positive or negative sensory descriptors of VOO have been related to volatile and phenol profiles, which are responsible for aroma and taste, respectively.

The characteristic taste of VOO, and in particular some positive attributes such as bitterness and pungency that are related to important health benefits, is not completely understood or appreciated by consumers. In this respect, it is interesting to consider the degree of acceptability of VOO in several countries based on literature data. In this way, it is possible to lay the foundations for correct instruction of the sensory characteristics of EVOO. The main chemical, biochemical and technological processes responsible for the positive and negative (defects) descriptors of VOO are summarized in this chapter. An overview on the sensory methodologies proposed, applied and modified during the last 20 years is also presented.

2. Flavours and off-flavours of virgin olive oil: The molecules responsible for sensory perceptions

VOOs are defined by the European Community as those "...oils obtained from the fruit of the olive tree solely by mechanical or other physical means under conditions that do not lead to alteration in the oil..." (EEC Reg. 2568/91). This production method renders VOO different from other vegetable oils that undergo refining, which leads to loss of most of the minor components such as volatile molecules and "polar" phenolic compounds.

Many authors (Angerosa et al., 2004; Kalua et al., 2007) have clarified that several variables affect the sensory characteristics and chemical composition of an EVOO. These include environmental factors, cultivation and agronomic techniques, genetic factors (cultivar), ripening degree of drupes, harvesting, transport and storage systems of olives, processing techniques, storage and packaging conditions of the oil.

The sensory attributes of EVOO mainly depend on the content of minor components, such as phenolic and volatile compounds. The independent odours and tastes of different volatile and phenolic compounds that contribute to various and typical EVOO flavours have been extensively studied; the sensory and chemical parameters of EVOO have been correlated in a large number of investigations (Bendini et al., 2007; Cerretani et al., 2008).

Each single component can contribute to different sensory perceptions. It is well established that specific phenolic compounds are responsible for bitterness and pungency (Andrewes et al., 2003; Gutiérrez-Rosales et al., 2003; Mateos et al., 2004). Few individuals, except for trained tasters of EVOO, know that the bitterness and pungency perceived are considered positive attributes. These two sensory characteristics, more intense in oils produced from olives at the start of crop year, are strictly related to the quali-quantitative phenolic profile of EVOO.

Even in small quantities, phenols are fundamental for protecting triacylglycerols from oxidation. Several authors (Gallina Toschi et al., 2005, Carrasco-Pancorbo et al., 2005; Bendini et al., 2006; Bendini et al., 2007) have reported their importance as antioxidants as well as nutracetical components. The major phenolic compounds identified and quantified in olive oil belong to five different classes: phenolic acids (especially derivatives of benzoic and cinnamic acids), flavones (luteolin and apigenin), lignans ((+)-pinoresinol and (+)-

acetoxypinoresinol), phenyl-ethyl alcohols (hydroxytyrosol, tyrosol) and secoiridoids (aglycon derivatives of oleuropein and ligstroside). The latter are characteristic of EVOOs.

Several investigations (Gutiérrez-Rosales et al., 2003; Mateos et al., 2004) have demonstrated that some phenols, and in particular secoiridoid derivatives of hydroxytyrosol, are the main contributors to the bitterness of olive oil; other phenolic molecules such as decarboxy-methyl-ligstroside aglycone, which seems to be a key source of the burning sensation, can stimulate the free endings of the trigeminal nerve located in the palate and gustative buds giving rise to the chemesthetic perceptions of pungency and astringency (Andrewes et al., 2003). Using a trained olive oil sensory panel, some investigators (Sinesio et al., 2005) have studied the temporal perception of bitterness and pungency with a time-intensity (TI) evaluation technique. It has been shown that the bitterness curves had a faster rate of increase and decline than the pungency curves. It was also demonstrated that differences in kinetic perception are linked to the slower signal transmission of thermal nociceptors compared to other neurons.

On the other hand, approximately 180 compounds belonging to several chemical classes (aldehydes, alcohols, esters, ketones, hydrocarbons, acids) have been separated from the volatile fractions of EVOOs of different quality. Typical flavours and off-flavour compounds that affect the volatile fraction of an oil obtained from olives originate by different mechanisms: positive odours are due to molecules that are produced enzymatically by the so-called lipoxygenase (LOX) pathway. Specifically both C_6 aldehydes, alcohols and their corresponding esters and minor amounts of C_5 carbonyl compounds, alcohols and pentene dimers are responsible for pleasant notes. In contrast, the main defects or off-flavours are due to sugar fermentation (*winey*), amino acid (leucine, isoleucine, and valine) conversion (*fusty*), enzymatic activities of moulds (*musty*) or anaerobic microorganisms (*muddy*), and to auto-oxidative processes (*rancid*).

Volatile molecules can be perceived in very small amounts (micrograms per kilogram or ppb) and these compounds do not have the same contribution to the global aroma of EVOO; in fact, their influence must be evaluated not only on the basis of concentration, but also on their sensory threshold values (Angerosa et al., 2004; Kalua et al., 2007). In addition, antagonism and/or synergism among different molecules can occur, affecting the global flavour of EVOO. Chemical factors of molecules (volatility, hydrophobic character, size, shape, conformational structure), type and position of functional groups appear to affect the odour and taste intensity more than their concentration due to their importance in establishing bonds with receptor proteins (Angerosa et al., 2004).

In general, it is correct to surmise that from healthy olives, picked at the right degree of ripening and properly processed, it is always possible to obtain an EVOO, independent of the olive variety. However, from unhealthy olives or from those harvested off the ground it is inevitable to produce an olive oil characterized by unpleasant flavours and sensory defects. Thus, both natural (olive variety, environmental conditions, degree of ripening and health status of olives) and extrinsic (technological processing by olive farmer/mill worker) factors may profoundly influence olfactory and gustative notes.

Several agronomic and climatic parameters can affect the volatile and phenolic composition of VOOs. The genetic characteristics of the olive cultivar are some of the most important aspects that determine the level of enzymes in fruit (Angerosa et al., 1999) that are involved in synthesis of volatile molecules (LOX pathway) and phenol compounds (biosynthetic pathways via PPO and β-glucosidase) present in VOOs.

Even if enzymatic activity depends on the stage of ripeness (Morales et al., 1996; Aparicio & Morales 1998) agronomic (fertilization, irrigation) and climatic (temperature and rainfall) conditions also play an important role.

2.1 Key points in obtaining a high quality VOO

• Processing of healthy olives:

When the common olive fly (*Bactrocera oleae*) attacks olives (from the beginning of summer to the start of harvesting), damage occurs as a result of larval growth: oils from damaged fruits show changes in both volatile and phenolic compounds that influence negatively the sensory properties and oxidative stability of the product, especially during oil storage (polar phenols have a fundamental role as antioxidants during storage). The bad taste due to these changes caused by the olive fly is well known as a *grubby defect* (Angerosa et al., 1992; Gómez-Caravaca et al. 2008).

In order to obtain a high quality olive oil, it is necessary to process olives that are not overripe. The use of fruits that have partially degraded tissues cause an increase in enzymatic and microrganism activities and oxidative reactions; therefore the produced oil probably will be characterized by an higher free acidity and perceivable sensory defects. When olives are accumulated in piles for many days, the high temperature and humidity inside the mass promotes proliferation of bacteria, yeasts and moulds, producing undesirable fermentation and degradation that give rise to specific volatile molecules responsible for unpleasant odours (i.e. winey, fusty and mouldy).

Winey, the typical pungent sensory note perceptible in oils produced by olives stored in piles or in jute sacks for several days, arises from alcoholic fermentation: *Lactobacillus* and *Acetobacter* have been detected in olives inducing fermentative processes. The main microorganism found in olives depends on the length of storage: at the beginning the enterobacteriaceae genera *Aerobacter* and *Escherichia* prevail, while *Pseudomonas*, *Clostridium* and *Serratia* are predominant after longer periods of time. The activity of these microorganisms results in the presence of low concentrations of biosynthetic volatiles and large amounts of compounds such as the branched alcohols due to degradation of amino acids that lead to the typical undesirable sensory note known as fusty (Angerosa, 2002; Morales et al 2005). The most abundant deuteromycetes found in olives stored at high humidity are several species of the genus *Aspergillus* together with ascomycetes *Penicillium*; these organisms oxidize free fatty acids producing mainly methyl ketones, in contrast to yeasts of the genera *Candida*, *Saccharomyces* and *Pichia* which are able to reduce carbonylic compounds. Enzymes from these microorganisms interfere with the LOX pathway to produce volatile C_8 molecules characterized by very low odour thresholds, and reduce some C_6 compounds. This volatile profile is responsible for the musty defect of EVOO.

• Selection of the most suitable milling conditions

The phenolic content is greatly influenced by this technological step. In general, the use of the more violent crushing systems (i. e. with hammers instead of blades) causes an increase in extraction of phenolic compounds due to more intense tissue breaking; therefore, a more vigorous milling system should be used to process olive varieties that are naturally low in phenolic compounds, and permit enrichment of bitter and pungency intensities. The use of more violent milling systems also produces a significant increase in olive paste temperature and a corresponding decrease of the activity of enzymes that play a key role in the production of volatile compounds responsible for fruity and other green notes (Salas & Sanchez, 1999; Servili et al., 2002).

Concerning the malaxation phase, which consists in a slow kneading of the olive paste, the time-temperature pair should be carefully controlled to obtain a high quality EVOO. The lipoxygenase pathway is triggered by milling of olives and is active during malaxation. The volatile compounds produced are incorporated into the oil phase to confer its characteristic aroma. Specifically, a temperature above 28°C for more than 45 min should be avoided; in fact, these conditions can lead to the deactivation of enzymes that produce both positive volatile compounds and oxidize the phenolic compounds causing changes in oil flavour (Salas & Sanchez 1999; Kalua et al., 2007). The reduced concentration of oxygen in paste, obtained by replacing air with nitrogen in the headspace of malaxer during processing, can inhibit these enzymes and minimize the oxidative degradation of phenolic compounds during processing (Servili et al., 1999; Servili et al., 2003). Malaxation under erroneous conditions is responsible for the unpleasant flavor known as a *"heated defect"* due to the formation of specific volatile compounds (Angerosa et al., 2004).

- The application of different oil separation systems

One of the main disadvantages of discontinuous mill systems is the possible fermentation and/or degradation phenomena of residues of pulp and vegetation waters on filtering diaphragms; these reactions give rise to a defect termed *"pressing mats"*, but also promote winey and fusty defects (Angerosa et al., 2004). It is well known that among continuous systems, discontinuous mill systems with a three-phase decanter need lukewarm water to dilute olive paste in contrast to a two-phase decanter, which has two exits producing oil and pomace and separates the oil phase from the olive paste This latter system has advantages in terms of water reduction and major transfer of phenols from the olive paste to the oil, with a consequent increase in oxidative stability, bitterness and pungency.

The amount of water added determines the dilution of the aqueous phase and lowers the concentration of phenolic substances that are more soluble in vegetable waste water. Consequently, a large amount of antioxidants is lost with the wastewater during processing. In addition to phenolic compounds, some volatile compounds accumulate more in oil from a dual-phase decanter than in oils extracted with three-phase decanters. Therefore, the use of a two-phase decanter promotes greater accumulation of volatile and phenolic compounds that are not lost in the additional water as in a three phase decanter. The higher concentrations of these compounds are related to the high intensities of bitter, pungent, green fruity, freshly cut lawn, almond and tomato perceptions (Angerosa et al., 2000; Angerosa et al., 2004; Kalua et al., 2007).

- Storage of oil under suitable conditions

In unfiltered oil, the low amounts of sugars or proteins that remain for extended times in oil can be fermented or degraded by specific anaerobic microrganisms of the *Clostridium* genus, producing volatile compounds responsible for an unpleasant muddy odour by butyric fermentation. The filtration of newly-produced oil can avoid this phenomenon. It is known (Fregapane et al., 2006; Mendez & Falque, 2007; Lozano-Sanchez et al., 2010) that EVOO has a low amount of water, and for this reason it can be considered as a water-in-oil emulsion (Koidis et al., 2008)

The orientation of phenolic compounds in the oil-water interface and the active surface of water droplets can protect against the oxidation of oil. According to some researchers (Tsimidou et al., 2004; Gómez-Caravaca et al., 2007), the stability of unfiltered samples is significantly higher than that of the corresponding filtered oils. This coincides with a higher total phenolic content in unfiltered oils due to a greater amount of emulsified water. On the other hand, higher water levels are expected to favour enzymatic catalysis, including lipase, lipoxygenase and polyphenol oxidase activities. Thus, a more rapid oxidation of unfiltered oil is expected. Some authors (Montedoro et al., 1993) observed that hydrolytic processes occurr in parallel with oxidation during long term storage.

Lipid oxidation is an inevitable process that begins immediately after oil extraction and leads to a deterioration that becomes increasingly problematic during oil storage. The presence of a rancid defect, typical off-flavour for the fatty matrices, can be avoided or substantially slowed. The most advanced oxidation stages are characterized by the complete disappearance of compounds arising from the LOX cascade and by very high concentrations of saturated and unsaturated aldehydes together with unsaturated hydrocarbons, furans and ketones that contribute mainly to the rancid defect because of their low odour thresholds (Guth & Grosch, 1990; Morales et al., 1997; Bendini et al., 2009). To avoid the rancid perception, it is fundamental to control factors that promote lipid oxidation. These include a decrease in the availability of oxygen, the protection of the oil from light and storage at a temperature of 12-14°C. Before bottling, it is advisable to maintain the oil in stainless steel tanks under an inert gas such as nitrogen equipped with devices that periodically eliminate sediments from the bottom of the tank.

3. Sensory methodology for evaluating the quality of VOO: Basic concepts

A sensory codified methodology for virgin olive oils, known as the "COI Panel test", represents the most valuable approach to evaluate the sensory characteristics of VOO. The use of statistical procedures to analyze data from assessors' evaluation provides results that can be trusted as well as methods usually adopted in scientific fields. The purpose of this international method is to standardize procedures for assessing the organoleptic characteristics of VOO, and to establish the methodology for its classification. This methodology, incorporated into regulations of the European Union since 1991, uses, as an analysis tool, a group of 8-12 persons selected in a controlled manner, who are suitably trained to identify and measure the intensity of positive and negative sensations (EEC Reg. 2568/91).

A collection of methods and standards has been adopted by the International Olive Oil Council (IOOC or COI) for sensory analysis of olive oils. These documents (IOOC/T.20/Doc. 4/rev.1 and IOOC/T.20/Doc.15/rev.2) describe the general and specific terms that tasters use. Part of the vocabulary is common to sensory analysis of all foods (general vocabulary), while a specific vocabulary has been developed *ad hoc* and established by sensory experts of IOOC. In addition, the official method (IOOC/T.20/Doc.5/rev.1 and IOOC/T.20/Doc.14/rev.2) includes precise recording of the correct tasting temperature, as well as the dimensions and colour of the tasting glass and characteristics of the test room.

The panel leader is the person responsible for selecting, training and monitoring tasters to ascertain their level of aptitude according to (IOOC/T.20/Doc.14/rev.2). The number of candidates is generally greater than that needed in order to select people that have a grater sensitivity and discrimination capability. Screening criteria of candidates are founded on sensory capacity, but also on some personal characteristics of candidates. Given this, the panel leader will personally interview a large number of candidates to become familiar with their personality and understand habits, hobbies, and interest in the food field. He uses this information to screen candidates and rejects those who show little interest, are not readily available or who are incapable of expressing themselves clearly.

The determination of the detection threshold of the group of candidates for characteristic attributes is necessary because the "threshold concentration" is a point of reference common to a "normal group" and may be used to form homogeneous panels on the basis of olfactory-gustatory sensitivity.

A selection of tasters is made by the intensity rating method, as described by Gutiérrez Rosales (Gutiérrez Rosales et al., 1984). A series of 12 samples is prepared by diluting a VOO characterized by a very high intensity of a given attribute in an odourless and tasteless medium (refined oil or paraffin). The panel leader sends out the candidate, removes one of the 12 tasting glasses from the series, and places the remaining together; the candidate is called back in the room and is asked to correctly replace the testing glass withdrawn from the series by comparing the intensity of this last with that of the others. The test is carried out for fusty, rancid, winey and bitter attributes to verify the discriminating capacity of the candidate on the entire scale of intensities.

The stage training of assessors is necessary to familiarize tasters with the specific sensory methodology, to heighten individual skill in recognizing, identifying and quantifying the sensory attributes and to improve sensitivity and retention with regards to the various attributes considered, so that the end result is precise and consistent. In addition, they learn to use a profile sheet.

The maintenance of the panel is made through continuous training over all duration of life of the same panel, the check of the sensory acuity of tasters, and exercises that allow the measurement of the panel performance.

Every year, all panels must assess a number of reference samples in order to verify the reliability of the results obtained and to harmonize the perception criteria; they must also update the Member State on their activity and on composition changes of their group.

3.1 Evolution of sensory methodology: From old to new

A method for the organoleptic evaluation of olive oils was introduced in the Regulation (EEC) No 2568/91, Annex XII, that is inspired by the COI/T.20/Doc. no.15, published in 1987. In the profile sheet of EEC Reg. 2568/91, a number of positive attributes and defects were evaluated, giving each a score from 0 to 5 (Figure 1).

Drawing on experience, the International Olive Oil Council has devised a new method of organoleptic assessment of VOOs (Decision Dec-21/95-V/07) that is simpler and more reliable than that in EEC Reg. 2568/91. In particular, the EC Reg. 796/2002 introduced a reduction of the attributes of the old profile sheet, asking tasters to consider only the defects of the oil (fusty, mustiness/humidity, winey/vinegary, muddy sediment, metallic, rancid and others) and only the three most important positive attributes (fruity, pungent and bitter). The most important innovation of EC Reg. 796/2002 is the use of continuous scales, from 0 to 10 cm, for evaluating the intensity of perception of the different attributes (positive and negative), as reported in Figure 2. In this way, tasters are free to evaluate the intensity of each attribute by ticking the linear-scale, without having a prefixed choice (as with the discrete scale of EEC Reg. 2568/91, see Figure 1).

Sensory analysis and its application to olive and virgin olive oil

VIRGIN OLIVE OIL

PROFILE SHEET
GRADING TABLE
OLFACTORY-GUSTATORY-TACTILE NOTES

	0	1	2	3	4	5
Olive fruity (ripe and green)						
Apple						
Other ripe fruit						
Green (leaves, grass)						
Bitter						
Pungent						
Sweet						
Other allowable attribute(s)						
(Specify.....................)						
Sour/Winey/Vinegary/Acid						
Rough						
Metallic						
Mustiness/humidity						
Muddy sediment						
Fusty ("Atrojado")						
Rancid						
Other unallowable attribute(s)						
(Specify.....................)						

DEFECTS	CHARACTERISTICS	OVERALL MARK: POINTS
None	Olive Fruity Olive fruity and fruitness of other fresh fruit	9 8 7
Barely perceptible	Weak fruitness of any tipe	6
Slight perceptible	Rather imperfect fruitness, anomalous odours and tastes	5
Considerable, on the border of acceptability	Clearly imperfect, unpleasant odours and tastes	4
Great and/or serious, clearly perceptible	Totally inadmissible odours and tastes for consumption	3 2 1

1 Barely perceptible
2 Slight perceptible
3 Average
4 Great
5 Extreme

REMARKS...

....

NAME OF ASSESSOR...
LEGEND OF SAMPLE..

DATE...

Fig. 1. Profile sheet for EVOO used for designation of origin (EEC Reg. 2568/91, annex XII).

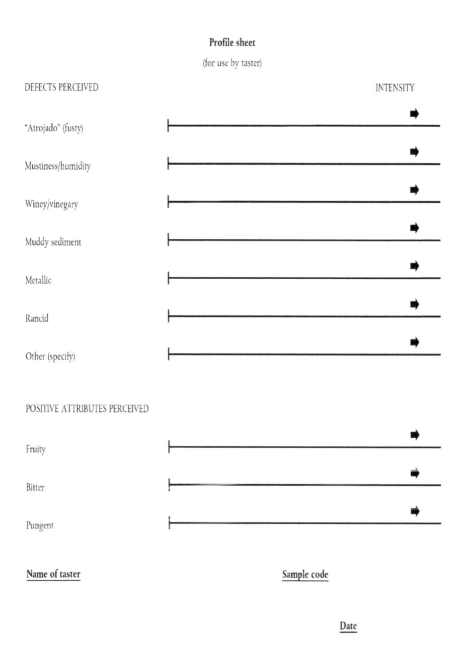

Profile sheet

(for use by taster)

DEFECTS PERCEIVED   INTENSITY

"Atrojado" (fusty)

Mustiness/humidity

Winey/vinegary

Muddy sediment

Metallic

Rancid

Other (specify)

POSITIVE ATTRIBUTES PERCEIVED

Fruity

Bitter

Pungent

Name of taster   Sample code

                                              Date

Fig. 2. Profile sheet for VOO assessment currently adopted by the EU (EC Reg. 796/02).

Each attribute is calculated, and the median value of each is used to classify the oil according to the median of the defect perceived with greatest intensity and the median for "fruity". It is important to remember that the value of the robust variation coefficient for this negative attribute must be no greater than 20%.

The classification of olive oils, according to sensory attributes, has also undergone evolution. According to EC Reg. 796/2002, oils are classified as:

a.   extra virgin olive oil: the median of the defects is 0, and the median for "fruity" is above 0;
b.   virgin olive oil: the median of the defects is above 0, but not above 2.5 and the median for "fruity" is above 0;
c.   ordinary virgin olive oil: the median of the defects is above 2.5, but not above 6.0, or the median of the defects is not above 2.5 and the median for "fruity" is 0;
d.   lampante virgin olive oil: the median of the defects is above 6.0.

Since November 2003, categories c) and d) have been replaced by (c) "lampante olive oil": the median of defects is above 2.5, or the median of the defects is not above 2.5 and the median for "fruity" is 0.

EC Reg. 640/08 introduced a new upper limit of defect for discriminating between virgin and defective oils: in particular, the evaluation of the median defect ('2.5') was replaced by '3.5'. An important innovation of Reg. 640/08 was also the grouping in only one negative attribute of two different defects: fusty and muddy sediment.

A revised method for the organoleptic assessment of VOO was adopted by the IOOC in November 2007 (Decision No DEC-21/95-V/2007, 16 November 2007) and adopted by the European Community with EC Reg. 640/2008. This revision updated the descriptions of the positive and negative attributes of VOO and the method. It also amended the maximum limit for the perception of defects in VOO. The IOOC's revised method for the organoleptic assessment of VOO also specifies the conditions for the optional use, on labels, of certain terms and expressions relating to the organoleptic characteristics of VOO (optional terminology for labelling purposes).

The most recent change is Decision No Dec-23/98-V/2010 of the IOOC, which defined a new method for assessing the organoleptic properties of VOO and to establish its classification on the basis of those characteristics (IOOC/T.20/Doc. No 15/Rev. 3).

3.2 The method for assigning commercial class: The official profile-sheet and expression of results

The organoleptic assessment of VOO is officially regulated in Europe by a Commission Regulation (EC Reg. 640/2008). This regulation describes the procedures for assessing the organoleptic characteristics of VOOs, the method for classification according to sensory characteristics, the specific vocabulary for sensory analysis of VOOs, including positive and negative attributes, and the optional terminology for labelling purposes. The selection, training and monitoring of skilled VOO tasters, the skills and responsibilities of the panel leader, the specific characteristics of the glass for oil tasting and the test room were also considered, according to previous regulations and IOOC documents (IOOC, 2007 and 2010).

The official profile sheet intended for use by tasters, shown in Figure 3 (EC Reg. 640/08), is quite simple and is formed by an upper section for evaluation of the intensity of defects, and

Profile sheet for virgin olive oil

INTENSITY OF PERCEPTION OF DEFECTS

Fusty/muddy sediment | _____ →

Musty-humid-earthy | _____ →

Winey-vinegary — acid-sour | _____ →

Metallic | _____ →

Rancid | _____ →

Other (specify) | _____ →

INTENSITY OF PERCEPTION OF POSITIVE ATTRIBUTES

Fruity | _____ →
                            greenly ☐   ripely ☐

Bitter | _____ →

Pungent | _____ →

Name of taster:

Sample code:

Date:

Comments:

Fig. 3. Profile sheet for VOO assessment currently adopted by the EU (EEC Reg. 640/08).

a lower part for the evaluation of the three most important positive sensory attributes (fruity, bitter, pungent). Tasters have to smell the sample, taste the oil (overall retronasal olfactory, gustatory and tactile sensations) and evaluate the intensity with which they perceive each of the negative and positive attributes on the 10-cm scale. If a taster identifies greenly or ripely as fruity attributes, the correct options must be indicated in the profile sheet. Green fruitiness is a characteristic of the oil which is reminiscent of green olives, dependent on the variety of the olive and coming from green, sound, fresh olives. Ripe fruitness is reminiscent of ripe fruit. If any negative attributes not listed in the upper section of the profile are perceived, the taster records them under the "others" heading, using the descriptors among those in the specific vocabulary for the sensory analysis of olive oils (IOOC/T.20/Doc. 4/rev.1).

The panel leader collects the profile sheets and elaborates the results by a statistical approach. In particular, the medians of the greatest perceived defect and fruity attribute are calculated. According to these two parameters, the oil can be graded in different quality categories. Such values are expressed to one decimal place, and the value of the robust coefficient of variation which defines them shall be no greater than 20%. As already mentioned, the classification of the oil is carried out by comparing the medians of the defects and the fruity attribute with the reference ranges established by EC Reg 640/08 for the different categories:

1. Extra virgin olive oil: the median of the defects is 0 and the median of the fruity attribute is above 0;
2. Virgin olive oil: the median of the defects is above 0, but not more than 3.5, and the median of the fruity attribute is above 0;
3. Lampante olive oil: the median of the defects is above 3.5, or the median of the defects is not more than 3.5 and the median of the fruity attribute is 0.

The panel leader can also state that the oil is characterized by greenly or ripely fruity attributes if at least 50% of the panel agrees.

Actually the most important result for sensory analysis of VOO is to identify the presence of defects instead of evaluating the positive attributes, in agreement with the aim of such an analysis, which is essentially to classify the product in different commercial classes.

3.2.1 Optional terminology for labelling purposes

Upon request, the panel head may certify that an oil complies with the definitions and ranges that correspond to the following adjectives, according to the intensity and perception of attributes:

a. for each of the positive attributes mentioned (*fruity* — whether *green* or *ripe* — *pungent* or *bitter*):
    i.   the term "intense" may be used when the median of the attribute is greater than 6;
    ii. the term "medium" may be used when the median of the attribute is between 3 and 6;
    iii. the term "light" may be used when the median of the attribute is less than 3;
    iv. the attributes in question may be used without the adjectives given in points (i), (ii) and (iii) when the median of the attribute is 3 or more;

b. the term "well balanced" may be used when the oil does not display a lack of balance, which is defined as the smell, taste and feel that the oil has when the median of the *bitter* and/or *pungent* attributes is two points higher than the median of its *fruitiness*;

c. the term "mild oil" may be used when the medians of the *bitter* and *pungent* attributes are 2 or less.

3.3 Method for organoleptic assessment of EVOO to assign designation of origin: Sensory profile and data processing

In 2005, the IOOC issued a document on methods to be used for the organoleptic assessment of EVOO for granting designation of origin (D.O.) status (IOOC/T.20/Doc. no 22). This document declared that the D.O. authority shall select the characteristic descriptors of the designation of origin (10 at the most) from those defined and reported in Table 1, and shall incorporate them into the profile sheet of the method.

Direct or retronasal aromatic olfactory sensations	
Almond	Olfactory sensation reminiscent of fresh almonds
Apple	Olfactory sensation reminiscent of the odour of fresh apples
Artichoke	Olfactory sensation of artichokes
Camomile	Olfactory sensation reminiscent of that of camomile flowers
Citrus fruit	Olfactory sensation reminiscent of that of citrus fruit (lemon,orange, bergamot, mandarin and grapefruit)
Eucalyptus	Olfactory sensation typical of Eucalyptus leaves
Exotic fruit	Olfactory sensation reminiscent of the characteristic odours of exotic fruit (pineapple, banana, passion fruit, mango,
Fig leaf	Olfactory sensation typical of fig leaves
Flowers	Complex olfactory sensation generally reminiscent of the odour of flours, also known as floral
Grass	Olfactory sensation typical of freshly mown grass
Green pepper	Olfactory sensation of green peppercorns
Green	Complex olfactory sensation reminiscent of the typical odour of fruit before it ripens
Greenly fruity	Olfactory sensation typical of oils obtained from olives that have been harvested before or during colour change
Herbs	Olfactory sensation reminiscent of that of herbs
Olive leaf	Olfactory sensation reminiscent of the odour of fresh olive leaves
Pear	Olfactory sensation typical of fresh pears
Pine kernel	Olfactory sensation reminiscent of the odour of fresh pine kernels
Ripely fruity	Olfactory sensation typical of oils obtained from olives that have been harvested when fully ripe
Soft fruit	Olfactory sensation typical of soft fruit: blackberries,raspberries, bilberries, blackcurrants and redcurrants
Sweet pepper	Olfactory sensation reminiscent of fresh sweet red or green peppers
Tomato	Olfactory sensation typical of tomato leaves
Vanilla	Olfactory sensation of natural dried vanilla powder or pods,different from the sensation of vanillin
Walnut	Olfactory sensation typical of shelled walnuts
Gustatory sensations	
Bitter	Characteristic taste of oil obtained from green olives or olives turning colour; it defines the primary taste associated with aqueous solutions of substances like quinine and caffeine
"Sweet"	Complex gustatory-kinaesthetic sensation characteristic of oil obtained from olives that have reached full maturity
Qualitative retronasal sensation	
Retronasal persistence	Length of time that retronasal sensations persist after the sip of olive oil is no longer in the mouth
Tactile or kinaesthetic sensations	
Fluidity	Kinaesthetic characteristics of the rheological properties of the oil, the set of which are capable of stimulating the mechanical receptors located in the mouth during the test
Pungent	Biting tactile sensation characteristic of oils produced at the start of the crop year, primarily from olives that are still unripe

Table 1. List of descriptors for D.O. of EVOO.

The characteristic descriptors are identified according to the round-table method: the panel supervisor leads a discussion based on a series of samples of known origin that display the most important specific characteristics of the VOO undergoing preparatory analysis. When the descriptor recognition stage is completed, the panel supervisor opens discussions with panel members to establish a list of all descriptors that are considered to be most important and characteristic of the designation that is undergoing preparatory analysis.

Validation should take into account the possible natural variations that may occur in the oil from one crop year to the next. When the profile sheet is completed, tasters shall assess the intensity of perception of the descriptors cited in the profile sheet on the 10-cm scale used for commercial grading of oils. The D.O. authority shall fix the maximum and minimum limits of the median for each descriptor included in the profile sheet and shall establish the limits for the robust coefficient of variation of each descriptor. It shall then enter these values in the *IOOC spreadsheet folder-profile* (software) accompanying this method to define the intervals of the characteristic sensory profile of the designation of origin.

Most of the specifications for the designation of origin of oils before 2005 or those that have not undergone revisions after this date, do not refer to the method IOOC just explained, but to the use of a previous procedure (EEC Reg. 2568/1991) for sensory evaluation of the oils. In Figure 1, the profile sheet according to the old regulation for the commercial grading is shown (EEC Reg. 2568/1991). This method provides a partial description of flavour: tasters are requested to define the fruity type, green or ripe, and recognize the presence of attributes such as grass, leaf, apple and other fruits. For each attribute, a discreet score from 0 to 5 is assigned (0: absence of perception; 1: intensity slightly perceptible; 2: intensity light; 3: average intensity; 4: great intensity; 5: extreme intensity), and there are many positive attributes to evaluate in addition to defects. Tasters rate the overall grading by using a 9-point scale: 9 for oils with exceptional sensory characteristics, and 1 for products with the worst qualities. The mean score identifies the category. An oil could be classified as EVOO if it obtains a final score (expressed as an average of the panel's judgement) of 6.5.

In the case of specifications for the designation of origin of some D.O oils, which have not yet been reviewed according to the new IOOC regulation (IOOC, 2005), it is firstly necessary to verify that the sample has the characteristics provided in the extra virgin category using current methods (EC Reg. 640/08), and to subsequently analyze it according to the old profile sheet (EEC Reg. 2568/1991) to verify the presence of characteristic descriptors. The final score for the D.O must be at least 7, but can be even higher.

4. Consumer acceptability of the sensory characteristics of VOO: An overview of literature data

As previously stated, a virgin oil that is not subjected to any subsequent tecnological refining has a sensory profile standardized by a rich/robust/harmonized regolatory environment (Conte & Koprivnjak, 1997) strongly linked to the quality of the starting olives. Any damage to drupes, which can lead to hydrolysis or fermentation, produces molecules that remain in the product and irreversibly affect its quality. There is no way of correcting chemical and/or sensory defects in a virgin product. On the other hand, technological refining results in the loss of the superior quality of "extra virgin/virgin" oil, and the transition to a lower category with weaker sensory attributes. The difference in the overall quality between a virgin and a refined oil, the latter adjusted in both quality and the flavour, is not always correctly perceived by the consumer.

Generally, consumers appreciate what is familiar, what is strongly linked to the territory (tradition/origin) or to which they have a precise expectation (brand, other values) (Caporale et al., 2006, Costell et al., 2010). Furthermore, as demonstrated in a recent large study, people do not understand dietary fat, either the importance of the quality or the quantity needed for health and this generally results in consumers adhering to fat choices they are comfortable with (Diekman & Malcolm,2009). In the case of EVOO, for a correct perception of the overall quality the fruity (green or ripe) and bitter and pungent attributes should be perceived by consumers as "healthy" indicators of quality and genuine taste, linked to the raw oil and its richness in pungent and bitter minor components (phenols) (Carluccio et al, 2003). To achieve this purpose, consumers should be made capable, by research dissemination, to appreciate bitterness (primary taste of oil obtained from green olives or olives turning colour) and pungency (biting tactile sensations characteristic of oils produced at the start of the crop year, primarily from olives that are still unripe) (COI/T.20/Doc. no 22) as healthy substances related attributes.

By law, the virgin oil "ideal" sensory profile is quite simple and easy, the fruity attribute is universally recognized as the primary sensory characteristic, and the bitter and pungent aspects are reported as positive attributes (*CODEX STAN 33-1981*). However, due to the superficial knowledge in terms of fat quality, technology (virgin and refined) and sensory characteristics, consumers do not appear to practice an informed/univocal consumption of EVOO. In this regard, research on consumer behaviour has intensified in recent years, and some of the more salient findings are provided below.

A study in Turkey (Pehlivan & Yilmaz, 2010) comparing olive oils originating from different production systems (continuous, organic, stone pressed, refined) declared that, for a sample of 100 consumers, hedonic values of the refined samples were close to the values of the virgin samples. Similar findings were previously reported by Caporale et al (2006), by which consumers are able to differentiate EVOO on their characteristic sensory attributes, but buying intentions (blind test) of the refined samples were as high as the values for the virgin samples. Again, the sensory attributes of EVOO, even if perceived, did not seem to be drivers to purchase it.

In Italy, Caporale et al. (2006) demonstrated that information about origin creates a favourable hedonic expectation, with regards to specific sensory attributes, such as pungency and bitterness. This means that, if familiar with bitter/pungent oils, consumers can have high and positive expectations of bitter and pungency attributes as distinguishing characteristics of typical olive oils (i.e. *Coratina* cv.). To confirm this physiological opportunity to perceive pungent as a positive attribute can be cited an interesting paper on the unusual pungency of EVOO (Peyrot des Gachons et al., 2011), sensed almost exclusively in the throat, suggesting that it is, therefore, perhaps no coincidence if phenols with potent anti-inflammatory properties (oleocanthale, ibuprofen) also elicit such a localized/specific pungency. In this paper the authors ask what is the functional significance of the pungency to the human upper airways; they suggest that the posterior oral location of toxin and irritant detectors can protect against their intake either by inhalation or ingestion. But if the role of these ion channels, in general, is to protect tissue from harmful compounds, then it is a mystery how one (TRPA1-channel), mediating throat irritation of extra-virgin olive oils, came to be valued as a positive sensory attribute by those who consume them. The authors hypothesize that this pungency, distinguisheing particularly good olive oils in the European Union standards, similarly to other common food irritants (e.g., capsaicin, menthol, and so forth), also important positive components in many cuisines, turns, from a usually negative taste-kinesthetic sensation into positive, because the molecules that elicit it have a body healthy action. This theory requires considerably more investigations to be demonstrated, but is true that many compounds eliciting pungency are also linked to decreased risks of cancer, degenerative and cardiovascular diseases (Boyd et al., 2006; Peng & Li, 2010).

In the case of EVOO, but this is a very general question, the authors suggest that people can transform an inherently unpleasant sensation into a positive one, commonly experienced around the world when consuming pungent EVOO, because it has beneficial health effects (Peyrot des Gachons et al., 2009). If this theory is correct, it means that this kind of pungency colud be easily taught as a positive sensation quality-related, to the unfamiliar consumers.

Infact, it has been reported (Delgado & Guinard, 2011) in the USA, an emergent market, that in a study on 22 samples evaluated in blocks of 5, for the majority of 100 consumers bitterness and pungency were negative drivers of liking.

Descriptive analysis (Delgado & Guinard, 2011) has been proposed as a more effective method to provide a more detailed classification of EVOO; the final method consisted of 22 sensory attributes, some of which were original but infrequent (butter/green tea). But, in the case of EVOO, the challenge for the future does not appear descriptive analysis, which has had the most interesting developments for the characterization/valorization of monovarietal, PDO and PGI (Inarejos-García et al., 2010; Cecchi et al. 2011) with many targeted/robust attributes. Rather it concerns the fact that consumers are actually able to appreciate/perceive its fundamentals of sensory profile (fruity, bitter, pungent) as related to its quality.

Finally, the worldwide problem of two different qualities of EVOO, a high one (expensive) and a "legal" one (less flavour/cheaper), was also highlighted in a means-end chain study (Santosa & Guinard, 2011), explaining that the attributes associated with EVOO generally have high (more flavour, more expensive, smaller size) or, unfortunately, low (cheaper/on sale, big quantity/bulk size, less flavour) levels of product involvement.

5. Conclusion

Sensory analysis of EVOO has been used for classification for more than 20 years. Since 1987, the "COI Panel test" has undergone many revisions, became law in 1991 in Europe and actualy COI/T.20/Doc. no. 15. is the method of analysis accepted by the Codex Alimentarius. Over the years, the profile sheet has undergone simplifications that have restricted selected specific positive (fruity, bitter, pungent) attributes and defects (fusty/muddy sediment, winey-vinegary-acid-sour, metallic, rancid, others).

On the other hand, in 2005 the IOOC issued document COI/T.20/Doc. no 22 that provides specifics about the methods to be used for sensory assessment of EVOO when granting designation of origin (D.O.) status. The method contains a list of 23 direct or retronasal aromatic olfactory sensations, 2 (bitter, sweet) gustatory sensations, 2 tactile or kinesthetic sensations (fluidity/pungent) and a qualitative retronasal persistence. Even taking into account the recent development of sensory analysis, there is no other food that has such a rich/robust/harmonized regulatory environment regulated by the EU, International Olive Oil Council and, as any food, Codex Alimentarius (FAO-OMS).

At present, origin, tradition and habits, more than sensory profile, are purchase drivers for EVOO and the real challenge for the future is improving consumer education in appreciating the foundamental attributes: fruity, together with taste and tactile sensations of phenols, functional and healthy substances naturally present in EVOO, respectively, bitterness and pungency.

Therefore, nowadays, the key to provide the consumer a truly effective EVOO organoleptic knowledge is the worldwide dissemination of the three basic quality-related and "healthy" sensory attributes.

6. Acknowledgment

The authors thank all tasters and the panel leader of the recognized professional panel at the Department of Food Science, University of Bologna.

7. References

Andrewes, P.; Busch, J. L. H. C.; De Joode, T.; Groenewegen, A. & Alexandre, H. (2003). Sensory properties of virgin olive oil polyphenols: identification of deacetoxy ligstroside aglycon as a key contributor to pungency. *Journal of Agriculture and Food Chemistry*, Vol. 51, pp. 1415-1420.

Angerosa, F.; Di Giacinto, L. & Solinas, M. (1992). Influence of Dacus oleae infestation on flavor of oils, extracted from attacked olive fruits, by HPLC and HRGC analyses of volatile compounds. *Grasas y Aceites*, Vol. 43, pp. 134-142.

Angerosa, F.; Basti, C. & Vito, R. (1999). Virgin olive oil volatile compounds from lipoxygenase pathway and characterization of some Italian cultivars. *Journal of Agriculture and Food Chemistry*, Vol. 47, pp. 836-839.

Angerosa, F.; Mostallino, R.; Basti, C.; Vito, R. & Serraiocco, A. (2000). Virgin olive oil differentiation in relation to extraction methodologies. *Journal of the Science of Food and Agriculture*, Vol. 80, pp. 2190-2195

Angerosa F. (2002). Influence of volatile compounds on virgin olive oil quality evaluated by analytical approaches and sensor panels. *European Journal of Lipid Science and Technology*, Vol. 104, pp. 639–660.

Angerosa, F.; Servili, M; Selvaggini, R.; Taticchi, A.; Esposto, S. & Montedoro, G.F. (2004). Volatile compounds in virgin olive oil: occurance and their relationship with the quality. *Journal of Chromatography A*, Vol. 1054, pp. 17-31.

Aparicio, R. & Morales, M. T. (1998). Characterization of olive ripeness by green aroma compounds of virgin olive oil. *Journal of Agriculture and Food Chemistry*, Vol. 46, pp. 1116-1122.

Bendini, A.; Cerretani, L.; Vecchi, S.; Carrasco-Pancorbo, A. & Lercker, G. (2006). Protective effects of extra virgin olive oil phenolics on oxidative stability in the presence or absence of copper ions. *Journal of Agriculture and Food Chemistry*, Vol. 54, pp. 4880-4887.

Bendini, A.; Cerretani, L.; Carrasco-Pancorbo, A.; Gómez-Caravaca, A.M.; Segura-Carretero, A.; Fernández-Gutiérrez, A. & Lercker, G. (2007). Phenolic molecules in virgin olive oils: a survey of their sensory properties, health effects, antioxidant activity and analytical methods. An overview of the last decade. *Molecules*, Vol. 12, pp. 1679-1719.

Bendini, A.; Cerretani, L.; Salvador, M. D.; Fregapane, G. & Lercker G. (2009). Stability of the sensory quality of virgin olive oil during storage: an overview. *Italian Journal of Food Science*, Vol. 21, pp. 389-406.

Boyd, L. A.; McCann, M. J.; Hashim, Y; Bennett, R. N; Gill, C. I; Rowland, I. R. (2006) Assessment of the anti-genotoxic, anti-proliferative, and anti-metastatic potential of crude watercress extract in human colon cancer cells. *Nutr. Cancer* , Vol. 55, pp. 232-241.

Caporale, G.; Policastro, S.; Carlucci, A. & Monteleone, E. (2006). Consumer expectations for sensory properties in virgin olive oils. *Food Quality and Preference*, Vol. 17, pp. 116-125.

Carluccio, M. A.; Siculella, L; Ancora, M. A.; Massaro, M.; Scoditti, E.; Storelli, C.; Visioli, F.; Distante, A.; De Caterina, R. (2003) Olive oil and red wine antioxidant polyphenols inhibit endothelial activation: antiatherogenic properties of Mediterranean diet phytochemicals. *Arterioscler. Thromb. Vasc. Biol., Vol. 23*, pp. 622– 629.

Carrasco-Pancorbo, A.; Cerretani, L.; Bendini, A.; Segura-Carretero, A.; Del Carlo, M.; Gallina Toschi, T.; Lercker, G.; Compagnone, D. & Fernandez-Gutierrez, A. (2005). Evaluation of the antioxidant capacity of individual phenolic compounds in virgin olive oil. *Journal of Agriculture and Food Chemistry*, Vol. 53, pp. 8918-8925.

Cerretani, L.; Salvador, M.D.; Bendini, A. & Fregapane, G. (2008). Relationship between sensory evaluation performed by Italian and Spanish official panels and volatile and phenolic profiles of virgin olive oils. *Chemosensory Perception*, Vol. 1, pp. 258-267.

Cecchi, T.; Passamonti, P.; Alfei, B. & Cecchi, P. (2011) Monovarietal extra virgin olive oils from the Marche region, Italy: Analytical and sensory characterization. *International Journal of Food Properties*, Vol. 14, pp. 483-495.

CODEX STAN 33-1981 Formerly CAC/RS 33-1970. Adopted in 1981. Revisions in 1989, 2003. Amendment in 2009.

Conte, L. & Koprivnjak, O. (1997). Quality control of olive oils in EEC: Origins, evolution and recent trends. *Food Technology and Biotechnology*, Vol. 35, pp. 75-81.

Costell, E.; Tárrega, A. & Bayarri, S. (2010). Food acceptance: the role of consumer perception and attitudes. *Chemosensory Perception*, Vol. 3, pp. 42-50.

Decision No DEC-21/95-V/2007 of 16 November 2007 on the International Olive Oil Council's revised method for the organoleptic assessment of virgin olive oil.

Decision No DEC-23/98-V/2010 of 26 November 2010 on the International Olive Oil Council's revised method for the organoleptic assessment of virgin olive oil.

Delgado, C. & Guinard, J.-X. (2011). How do consumer hedonic ratings for extra virgin olive oil relate to quality ratings by experts and descriptive analysis ratings. *Food Quality and Preference*, Vol. 22, pp. 213-225.

Delgado, C. & Guinard, J.-X. (2011). Sensory properties of Californian and imported extra virgin olive oils, *Journal of Food Science*, Vol. 76, pp. 170-176.

Diekman, C. & Malcolm, K. (2009). Consumer perception and insights on fats and fatty acids: Knowledge on the quality of diet fat. *Annals of Nutrition and Metabolism*, Vol. 54 (Suppl. 1), pp. 25-32.

European Community, Commission Regulation (1991) 2568/91 on the characteristics of olive oil and olive residue oil and on the relevant methods of analysis. *Official Journal of the European Communities*, July 11, L248, 1-83.

European Community, Commission Regulation (2002) 796/2002. Amending Regulation No 2568/91/EEC. *Official Journal of the European Communities*, May 6, L128, 8-28.

European Community, Commission Regulation (2003) 1989/2003. Amending Regulation No 2568/91/EEC. *Official Journal of the European Communities*, November 6, L295, 57-77.

European Community, Commission Regulation (2008) 640/2008. Amending Regulation No 2568/91/EEC. *Official Journal of the European Communities*, July 4, L178, 11-16.

Fregapane, G.; Lavelli, V.; Leon, S.; Kapuralin, J. & Salvador, M.D. (2006). Effect of filtration on virgin olive oil stability during storage. *European Journal of Lipid Science and Technology*, Vol. 108, pp. 134-142.

Gallina Toschi, T.; Cerretani, L.; Bendini, A.; Bonoli-Carbognin, M. & Lercker, G. (2005) Oxidative stability and phenolic content of virgin olive oil: an analytical approach by traditional and high resolution techniques. *Journal of Separation Science*, Vol. 28, pp. 859 – 870.

Gómez-Caravaca, A.M., Cerretani, L., Bendini, A., Segura-Carretero, A., Fernández-Gutiérrez, A. & Lercker, G. (2007). Effect of filtration systems on the phenolic content in virgin olive oil by HPLC DAD-MSD. *American Journal of Food Technology*, Vol. 2, pp. 671-678.

Gómez-Caravaca, A. M.; Cerretani, L.; Bendini, A.; Segura-Carretero, A.; Fernandez-Gutierrez, A.; Del Carlo, M.; Compagnone, D. & Cichelli, A. (2008). Effects of fly attack (Bactrocera oleae) on the phenolic profile and selected chemical parameters of olive oil. *Journal of Agriculture and Food Chemistry*, Vol. 56, pp. 4577–4583.

Guth, H. & Grosch, W. (1990). Deterioration of soya-bean oil: quantification of primary flavour compounds using a stable isotope dilution assay. *Lebensmittel-Wissenschaft und-Technologie*, Vol. 23, pp. 513-522.

Gutiérrez-Rosales, F.; Risco, M.A. & Gutiérrez González-Quijano, R. (1984). Selección de catadores mediante el método de clasificación por intensidad. *Grasas y Aceites*, Vol. 35, pp. 310-314.

Gutiérrez-Rosales, F.; Rios, J. J. & Gomez-Rey, MA. L. (2003). Main polyphenols in the bitter taste of virgin olive oil. Structural confirmation by on-line High-Performance Liquid Chromatography Electrospray Ionization Mass Spectrometry. *Journal of Agriculture and Food Chemistry*, Vol. 51, pp. 6021-6025.

Kalua, C.M.; Allen, M.S.; Bedgood, D.R.; Bishop, A.G.; Prenzler, P.D. & Robards K. (2007). Olive oil volatile compounds, flavour development and quality: A critical review. *Food Chemistry*, Vol. 100, pp. 273-286.

Koidis, A.; Triantafillou, E. & Boskou D. (2008). Endogenous microflora in turbid virgin olive oils and the physicochemical characteristics of these oils. *Eur. J. Lipid Sci. Technol.*, Vol. 110, pp. 164-171.

Inarejos-García; A.M., Santacatterina; M., Salvador, M.D.; Fregapane, G. & Gómez-Alonso, S. (2010). PDO virgin olive oil quality-Minor components and organoleptic evaluation. *Food Research International*, Vol.43, pp. 2138-2146.

International Olive Oil Council (1987) Sensory analysis of olive oil method for the organoleptic assessment of virgin olive oil. IOOC/T.20/Doc. no. 3

International Olive Oil Council (2005) Selection of the characteristic descriptors of the designation of origin. IOOC/T.20/Doc. no 22

International Olive Oil Council (2007). Sensory analysis: general basic vocabulary. IOOC/T.20/Doc. 4/rev.1

International Olive Oil Council (2007). Glass for oil tasting. IOOC/T.20/Doc. 5/rev.1

International Olive Oil Council (2007). Guide for the installation of a test room IOOC/T.20/DOC. 6/rev.1

International Olive Oil Council (2007). Guide for the selection, training and monitoring of skilled virgin olive oil tasters. IOOC/T.20/Doc.14/rev.2

International Olive Oil Council (2007). Specific vocabulary for virgin olive oil. IOOC/T.20/Doc. No 15/rev.2

International Olive Oil Council (2010). Sensory analysis of olive oil. Method for the organoleptic assessment of virgin olive oil. IOOC/T.20/Doc. No 15/Rev. 3.

Lozano-Sanchez, J.; Cerretani, L.; Bendini, A.; Segura-Carretero, A. & Fernandez-Gutierrez, A. (2010). Filtration process of extra virgin olive oil: effect on minor components, oxidative stability and sensorial and physicochemical characteristics. *Trends in Food Science & Technology*, Vol. 21, pp. 201-211

Mateos, R.; Cert, A.; Pérez-Camino, C. M. & García, J. M. (2004). Evaluation of virgin olive oil bitterness by quantification of secoiridoid derivatives. *Journal American Oil Chemist's Society*, Vol. 81, pp. 71-75.

Mendez, A.I. & Falque, E. (2007). Effect of storage time and container type on the quality of extra-virgin olive oil. *Food Control*, Vol. 18, pp. 521-529.

Montedoro, G. F.; Servili, M.; Baldioli, M.; Selvaggini, R.; Miniati, E. & Macchioni, A. (1993). Simple and hydrolysable phenolic compounds in olive oils: Note 3 Spectroscopic characterization of the secoiridoid derivates. *Journal of Agriculture and Food Chemistry*, Vol. 41, pp. 2228-2234.

Morales, M. T.; Calvente, J. J. & Aparicio, R. (1996). Influence of olive ripeness on the concentration of green aroma compounds in virgin olive oil. *Flavour and Fragrance Journal*, Vol. 11, pp. 171-178.

Morales, M. T.; Rios, J. J. & Aparicio R. (1997). Changes in the volatile composition of virgin olive oil during oxidation: flavors and off-flavors. *Journal of Agriculture and Food Chemistry*, Vol. 45, pp. 2666-2673.

Morales, M.T.; Luna, G. & Aparicio, R. (2005). Comparative study of virgin olive oil sensory defects. *Food Chemistry*, Vol. 91, pp. 293-301.

Pehlivan, B. & Yilmaz, E. (2010). Comparison of oils originating from olive fruit by different production systems. *Journal of the American Oil Chemists' Society*, Vol. 87, pp. 865-875.

Peng, J. & Li, Y. J. (2010) The vanilloid receptor TRPV1: role in cardiovascular and gastrointestinal protection. *Eur. J. Pharmacol.* Vol. 627, pp. 1-7.

Peyrot Des Gachons, C.P.; Beauchamp, G.K. & Breslin, P.A.S. (2009). The genetics of bitterness and pungency detection and its impact in phytonutrient evaluation. *Annals of the New York Academy of Sciences*, 1170, pp. 140-144.

Peyrot Des Gachons, C.; Uchida, K.; Bryant, B.; Shima, A.; Sperry, J.B.; Dankulich-Nagrudny, L.; Tominaga, M.; Smith III, A.B.; Beauchamp, G.K. & Breslin, P.A.S. (2011). Unusual pungency from extra-virgin olive oil is attributable to restricted spatial expression of the receptor of oleocanthal. *Journal of Neuroscience*, Vol. 31, pp. 999-1009.

Salas, J. J. & Sanchez, J. (1999). The decrease of virgin olive oil flavor produced by high malaxation temperature is due to inactivation of hydroperoxide lyase. *Journal of Agricultural and Food Chemistry*, Vol. 47, pp. 809-812.

Santosa, M. & Guinard, J.-X. (2011). Means-end chains analysis of extra virgin olive oil purchase and consumption behavior. *Food Quality and Preference*, Vol. 22, pp. 304-316.

Servili, M.; Mariotti, F.; Baldioli, M. & Montedoro, G. F. (1999). Phenolic composition of olive fruit and VOO: distribution in the constitutive parts of fruit and evolution during the oil mechanical extraction process. *Acta Horticulturae*, Vol. 474, pp. 609-619.

Servili, M.; Piacquadio, P.; De Stefano, G.; Taticchi, A. & Sciancalepore V. (2002). Influence of a new crushing technique on the composition of the volatile compounds and related sensory quality of virgin olive oil. *European Journal of Lipid Science and Technology*, Vol. 104, pp. 483-489.

Servili, M.; Selvaggini, R.; Taticchi, A.; Esposito, S. & Montedoro, G. F. (2003). Volatile compounds and phenolic composition of virgin olive oil: optimization of temperature and time of exposure of olive pastes to air contact during the mechanical extraction process. *Journal of Agriculture and Food Chemistry*, Vol. 51, pp. 7980-7988.

Sinesio, F.; Moneta, E. & Esti M. (2005). The dynamic sensory evaluation of bitterness and pungency in virgin olive oil. *Food Quality and Preference*, Vol. 16, pp. 557–564.

Tsimidou, M. Z.; Georgiou, A.; Koidis, A. & Boskou, D. (2004). Loss of stability of "veiled" (cloudy) virgin olive oils in storage. *Food Chemistry*, Vol. 93, pp. 377-383.

7

Innovative Technique Combining Laser Irradiation Effect and Electronic Nose for Determination of Olive Oil Organoleptic Characteristics

K. Pierpauli[1], C. Rinaldi[1,2], M. L. Azcarate[2,3] and A. Lamagna[1,4]
[1]*Comisión Nacional de Energía Atómica*
[2]*Consejo Nacional de Investigaciones Científicasy Técnicas*
[3]*Centro de Investigaciones en Láseres y Aplicaciones CEILAP (CITEDEF-CONICET)*
[4]*Escuela de Ciencia y Tecnología, Buenos Aires,*
Argentina

1. Introduction

1.1 Olive oil

Olive oil has a characteristic flavor that distinguishes it from other edible vegetable oils. Its quality depends on the aroma, taste and colour, which in turn depend on many variables including location.

The International Olive Oil Council (IOOC,2001) Standards and European Commission regulations have defined the quality of olive oil based on parameters derived from spectrophotometric studies that include free fatty acid content, but these methods only give information about the samples' oxidation level. A specific vocabulary has been developed for virgin oil sensory descriptors (IOOC, 1987). The positive attributes are classified as fruty, bitter and pungent and negative attributes as fusty, musty-humid, muddy-sediment, winey-vinegary, metallic and rancid.

Odour is an important parameter determining the sensory quality of olive oils and it is therefore of interest to investigate if volatile compounds contributing to the characteristic odour can be measured.

In the last decades many efforts have been made to study the aromatic fraction of olive oils based mainly on chromatographic determinations (S. de Koning et al 2008, S. Mildner-Szkudlarz, H. H. Jeleń 2008, C. M. Kalua 2007). The presence or absence of particular volatile compounds is a good indicator of olive oil quality.

The aroma of olive oil is attributed to aldehydes, alcohols, esters, hydrocarbons, ketones, furans and probably, other volatile compounds, not yet identified. More than 120 volatile compounds that contribute both positively and negatively to the sensory properties of olive oil have been identified (Aparicio, R., Morales, M .T. & Luna, G. 2006). Table 1 lists some volatile compounds associated with negative attributes determined by Morales et al. in 2005.

Descriptor	Volatile compounds
Mustiness-humidity	1-octen-3-ol
Fusty	Ethyl butanoate, propanoic and butanoic acid
Winey-vinegary	Acetic acid, 3-methyl butanol and ethyl acetate
Rancid	Several saturated and unsaturated aldehydes and acids

Table 1. Volatile compounds associated with negative attributes of olive oils

Odour activity is a measure of the importance of a specific compound for the odour of a sample. It is calculated as the ratio between the concentration of an individual substance in a sample and the threshold concentration of this substance. The minimum concentration of a compound able to give rise to an olfactory response is the compound's odour thereshold value. For this reason, high concentration of volatile compounds is not necessarily the main contribution to odour. For example, Reiners and Grosch reported a concentration of 6770 µg/g of trans-2-hexenal with an odour activity value of 16 whereas 1-penten-3-one with a much lower concentration of 26 µg/g had a higher odour activity value of 36 (C. M. Kalua, 2007).

According to the European Community Regulations (ECR 640/2008,ECR 1989/2008) olive oil can be classified in extra-virgin (high quality), virgin (medium quality) and lampante (lower quality). The first two categories can be bottled and consumed.

The quality and uniqueness of specific extra virgin olive oils is the result of different factors such as cultivar, environment and cultivation practices. The European Community (ECR 2081/1992) allows the Protective Denomination of Origen (PDO) labeling of some European EVOO with the names of the areas where they are produced.

1.2 Analytical techniques

The methods used and / or proposed to evaluate the oxidative deterioration of olive oil based on the determination of volatile compounds are HPLC / GC-MS, analytical methods associated with some headspace extractive techniques. The volatile profile of VOO closely depends on the extraction used (S. Vichi, 2010).

Some of the traditional distillation methods applied in the analysis of plant materials as steam distillation (SD), simultaneous distillation/extraction (SDE) and microwave-assisted extraction (MAE) were used for this purpose (Marriott, Shellie, & Cornwell, 2001).

Among these distillation techniques, SDE appeared to provide the most favourable uptake for mono- and sesquiterpenes, as well as for their oxygenated analogues (Marriott et al., 2001). Hydro distillation (HD) has been applied for the analysis of leaf, fruit and virgin oil volatiles of an Italian olive cultivar (Flamini, Cioni, & Morelli, 2003). With hydro distillation, the volatiles in the steam distillate are strongly diluted in water when collected in cold traps. This can be overcome in simultaneous distillation/extraction (SDE) via solvent extraction of the distillate.

Dynamic headspace techniques have been used to correlate the composition of the olive oil headspace to sensory attributes (Angerosa et al., 1996; Angerosa et al., 2000; Morales et al., 1995; Servili et al., 1995) and or flavors "defects" (Angerosa, Di Giacinto, & Solinas, 1992; Morales, Rios, & Aparicio, 1997).

More recently, the solid-phase microextraction (SPME) technique has been introduced as a sample pre-concentration method prior to chromatographic analysis as an alternative to the dynamic headspace technique. Among other applications, SPME allowed the characterization of virgin olive oils from different olive varieties and geographical production areas (Temime et al., 2006; Vichi et al., 2003a), and the evaluation of varietal and processing effects (Dhifi et al., 2005; Tura et al., 2004). Since the SPME uptakes are strongly dependant on the distribution of analytes among the sample matrix, the gas phase and the fiber coating (Pawliszyn, 1999), some compounds present in virgin olive oil may remain undetected. In the case of other techniques such as SDE, the recovery of analytes is mainly related to their volatility.

These last techniques are complex, expensive and time-consuming. They generally highlight only one or few aspects of the oxidation process, providing only partial information. On the other hand, the olive oil industry needs a rapid assessment of the level of oil oxidation in order to predict its shelf-life. Consumers usually expect manufacturers and retailers to provide products of high quality and seek for quality seals and brands. Therefore, the development of innovative analytical tools for quick and reliable quality checks on extra virgin olive oil is required.

1.3 Electronic nose

Gardner and Barllet (1993) defined the electronic nose as an instrument which comprises an array of electronic chemical sensors of partial specificity and an appropriate pattern-recognition system, capable of recognizing simple or complex odours.

The sensors used in the array of an electronic nose should have the following characteristics: high sensitivity to chemical compounds, low sensitivity to humidity and temperature, medium selectivity, high stability, high reproducibility and reliability; short reaction and recovery time; robustness and durability; easy calibration and data processing and small dimensions (Schaller et al., 1998).

The chemical interaction between the odour compounds and the gas sensors alters the state of the sensors giving rise to electrical signals which are registered by the instrument. Since each sensor is sensitive to all odour components, the signals from the individual sensors determine a pattern which is unique for the gas mixture measured and that is then interpreted by multivariate pattern recognition techniques.

Nowadays, there are different gas sensor technologies available, but only four of them are currently used in commercialized electronic noses: metal oxide semiconductors (MOS); metal oxide semiconductor field effect transistors (MOSFET); conducting organic polymers (CP); piezoelectric crystals (Bulk Acoustic (Wave–BAW), Surface Acoustic (Wave SAW)). Others, such as fiber-optic, electrochemical and bi-metal sensors, are still in the developmental stage.

The processing of the multivariate output data generated by the gas sensor array signals represents another essential part of the electronic nose concept. The statistical techniques used are based on commercial or specially designed software using pattern recognition routines like principal component analysis (PCA), cluster analysis (CA), partial least squares (PLS), linear discriminator analysis (LDA) and artificial neural network (ANN).

The use of an electronic nose for quality evaluation as a means of olfactory sensing is becoming widespread due to its advantages of low cost, reliability and high portability. Electronic noses based on different sensor technologies and using different recognition schemes have been employed for this task.

When samples of olive oil are analyzed with an electronic nose, the standard procedure is to put a fixed quantity in a vial and sense the headspace. The main drawback of this method is that the concentration of some compounds in the headspace may be quite different from their concentration in the liquid phase. For example, the concentration of methanol and ethanol is usually much higher in the vapor phase than in the liquid. However, these substances have been found to be irrelevant in the definition of the olive oil characteristics (S. de Koning, 2008). On the other hand, substances such as hexanal and trans-2-hexanal which are responsible for the organoleptic properties, are more abundant than methanol and ethanol (Cosio et al, 2006, C. Di Natale et al, 2001) in the oil, but are scarcely present in the headspace. It is well known that the odour activity of hexanal and trans-2-hexanal is higher than that of those alcohols because of their low odour thresholds (Morales et al., 2005, J. Reiners, W. Grosch 1998, A. Runcio et al., 2008). Despite the abovementioned drawback, several efforts have been made to use the electronic nose for olive oil quality control (Guadarrama et al., 2000). The combination of electronic nose fingerprinting with multivariate analysis enabled the study of the profile of olive oil in relation to its geographical origin (Ballabio et al., 2006; Cosio et al., 2006). García Gonzalez and Aparicio, 2002, detected the vinegary defect in Spanish VOO with the use of metal oxide sensors. They used an Alpha MOS e-nose equipped with 18 MOS sensors distributed in three chambers, and heated the samples to 34 °C during 10 minutes before testing the headspace. Servili et al., 2009, reported the first study of the use of an Electronic Olfactory System (EOS 835) on-line to control the formation and evolution of the volatile compounds that characterize the most important sensory notes of VOO during the malaxation process.

In 2010, M. J. Lerma-García et al. compared the response of an electronic nose (EOS 507) to classify oils containing the five typical virgin olive oil sensory defects with that of a sensory panel. They demonstrated the usefulness of this tool when combined with panels to perform a fast screening of a large set of samples with the aim of discriminating defective oils. Each sample was incubated at 37°C for 7 minutes before injection.

In the same year, Massacane et al. proposed a method to improve the electronic nose performance for discriminating among different olive oils without changing the properties of the original oil sample. This task is carried out by IR laser vaporization (IRLV) which seems to be a promising technique that modifies only slightly the headspace by volatilizing certain organic compounds or by cracking them. Thus IRLV improves the selectivity of the overall response of the electronic nose. Due to the extremely low sample vaporization that it produces this method can be considered non-destructive as most ablation laser Analytical methods (C. A. Rinaldi and J. C. Ferrero, 2001).

1.4 Laser irradiation effect

Normal vaporization occurs when the vapour pressure in the ambient gas is lower than the saturation pressure of the liquid at the liquid temperature (Xu, X., and D. A. Willis, 2002). As the liquid's temperature increases, so do the saturation pressure and the rate of vaporization.

Laser vaporization (LV) produces a local heating of the irradiated liquid surface and, in consequence, some molecules are driven to the gas phase. This phenomenon can be produced by the use of either pulsed or continuous wave (cw) lasers. For a fixed wavelength, the main difference lies in the amount of energy emitted per unit time, or power. Pulsed lasers produce an increase of the liquid surface temperature without producing a significant change of the bulk volume temperature (Christensen, B., and M. S. Tillack, 2003). These lasers produce only a local heating of the surface allowing a large amount of vapour to be generated in a short time period.

Due to the intrinsic nature of the LV this surface effect is produced immediately after the irradiation time lapse. Therefore, it is common to speak of a "thermal spike" rather than simply "thermal heating", because of the transient nature of the high temperature. The characteristics of this spike are determined by the laser fluence and its pulse length (Taglauer, E, A. W. Czanderna and D. M. Hercules, 1991).

The cw lasers used to vaporize liquids can cause an increase of the bulk sample temperature and can induce chemical reactions, thus, modifying the sample's properties. However, the appropriate choice of the irradiation time lapse and the laser power make them suitable for this application.

In the present work the results of experiments carried out to illustrate the use of the combined techniques of electronic nose and pulsed or continuous wave laser irradiation for olive oil quality determination are reported.

2. E-nose + laser vaporization technique

2.1 Pulsed laser irradiation

In this experiment the Infrared Laser Vaporization, IRLV, properties to improve the e-nose selectivity are investigated. The role of the laser wavelength is additionally analyzed. This is due to the fact that the quality and the quantity of the chemical compounds incorporated the headspace depend on the laser parameters, particularly, the fluence and the pulse length, as it was mentioned in Section 1.4.

Two extra virgin olive oils produced in the same geographical region of Argentina (San Juan) classified as A and B, were tested. Three samples of 15 ml of each oil were subjected to three different analytical methods in order to compare the effects of the laser vaporization.

All analytical methods were carried out under the same temperature and humidity conditions, of 25 °C and 43%, respectively. The samples were introduced in 100 ml T-shaped Pyrex test tubes with screw-caps in air inlet and outlet channels and a CaF_2 window in order to allow the laser beam admission, referred to as vials.

The following analytical methods were undertaken:

Method I: Vial with oil sample A is kept closed during 2 minutes. Immediately afterwards the vial headspace is subjected to 35 seconds sampling with a Cyranose® 320. This procedure is repeated 5 times. The same operation is performed with oil sample B.

Method II: Vial with oil sample A is kept closed for 1min. The sample is subsequently irradiated with Nd:YAG laser pulses of 1064 nm at a repetition rate of 10 Hz during 1 minute. The laser is turned off and the vial headspace is immediately subjected to 35 seconds sampling with a Cyranose® 320. This procedure is repeated 5 times. The same operation is then performed with oil sample B.

The Nd:YAG laser (Continuum, Surelite I) has a pulse length of 5 ns and an output energy of 80 mJ. The laser beam is focused in order to obtain a spot size of 0.037 cm^2 at the surface of the sample so that a fluence of 2.14 J/cm^2 is therefore achieved.

Method III: The vial with oil sample A is kept closed for 1 minute. The sample is subsequently irradiated with a homemade TEA CO_2 laser (D. Petillo, J. Codnia, M. L. Azcárate, 1996) operating at 10.59 μm with a repetition rate of 1 Hz during 1 minute. The laser is turned off and the vial headspace is immediately subjected to 35 seconds sampling with a Cyranose® 320. This procedure is repeated 5 times. The same operation is made with oil sample B.

The TEA CO_2 laser has a pulse length of 100 ns and output energy of 1.45 ± 0.04 J/pulse. The beam is focused in order to obtain a spot size of 0.68 cm^2 at the surface of the sample; a fluence of 2.14 J/cm^2 is therefore achieved. The software provided by the Cyrano® 320TM electronic nose allowed the processing of the raw data given by the 32 sensors responses.

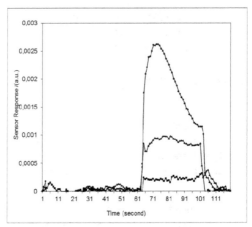

Fig. 1. Sensor response time dependence. (•) Without laser, (▲) Nd:YAG laser, and (•) CO_2 laser. (Massacane et al., 2010) Permission?

Figure 1 shows the signals measured with one sensor corresponding to the samples of oil A subjected to the three analysis methods. We observe that the signal-to-noise ratio (S/R), is considerably increased by the laser vaporization. The highest S/N ratio is obtained when vaporization is performed with the CO_2 laser.

As it is well known, the fluence and the pulse length determine the laser radiation absorption mechanisms and these parameters modify the laser power. In this work the same power was used in both IRLV methods although different total energies were delivered to the sample in each analysis method: 47.5 and 87 J. in Methods II and III, respectively. This energy range produces a negligible temperature increment. Even assuming an ideal oil absorption of 100% of laser energy the sample temperature increment would be about 3°C, as may be calculated from the volume of the sample and its average heat capacity. Therefore, the temperature of the sample remains constant throughout the experiment.

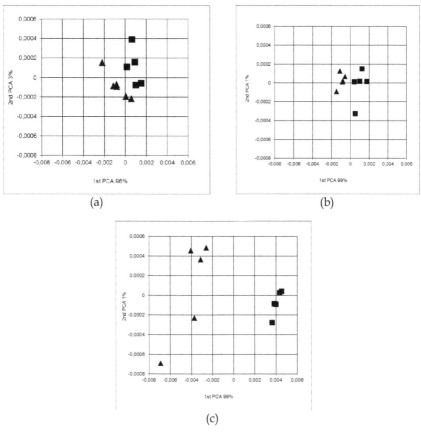

Fig. 2. PCA for olive oils (■:A; ▲: B) performed for experiments (a) without laser, (b) with Nd:YAG laser, and (c) with CO_2 laser. (Massacane et al., 2010) Permission

Not only the S/N ratio is increased by IRLV but also the discrimination of samples corresponding to different oils is dramatically increased. This fact can be verified by the results shown in Figures 2(a), 2(b) and 2(c). Thus the modification of the headspace is made evident by this result.

IR spectra of the liquid oil samples were registered both before and after methods II and III were performed. There were no significant differences between them indicating that IRLV does not produce changes in the liquid oil. On the other hand, IRLV does modify the vapour-liquid equilibrium conditions improving the selectivity of the electronic nose overall response. The only effect of IRLV is to increase the vapor concentration of the olive oil (Massacane et al., 2010)

It has been shown that the electronic nose selectivity is dramatically increased by the use of IRLV and that it is rather insensitive to the recognition pattern employed.

2.2 Continuous wave laser irradiation

A homemade portable electronic nose, Patagonia nose, was used in this study. The instrument comprises three parts: the automatic sampling system, the sensors' chamber with the sensor array, and the software for the e-nose control, data recording and processing. The first two are integrated in the same device and the software can be installed on any notebook (Figure 3)

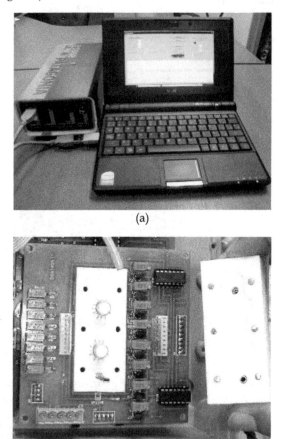

(a)

(b)

Fig. 3. a) Homemade e-nose (Patagonia) b) Sensor chamber

The chamber contains 2 MOS commercial sensors (Silsens MSGS 4000, sensor array). Each sensor has four thin SnO_2 films, one of them is doped with Pd. Each thin film is maintained at the temperature range between 300 and 500 °C during all the measurements.

Two EVOO produced in neighboring geographical regions of Argentina (San Juan) were classified as A and B. About 15 samples of both A and B EVOO were tested. Each sample was divided into three 15 ml samples so that three sets of the 15 samples were obtained to be subjected to three different analytical methods.

All the analytical methods were carried out under the same temperature and humidity conditions: 22 °C and 36 %, respectively. The samples were introduced in 100 ml T-shaped test tubes with screw caps, air inlet and outlet channels and BK7 upper windows to allow the laser beam admission referred to as vials.

The three analytical methods differ in whether the samples are irradiated or not and in the wavelength of the laser used for the irradiation. The vials were kept closed for about 90 seconds to allow the stabilization of the samples:

Method I: The vial headspace is subjected to a 15 seconds sampling.

Method II: After the headspace stabilization takes place, the closed vial is irradiated during one minute with a continuous wave diode laser emitting radiation of 98 mW at 450 nm. Immediately after the laser is turned off, the vial headspace is subjected to a 15 seconds sampling.

Method III: After the headspace stabilization takes place, the closed vial is irradiated during one minute with a continuous wave diode laser emitting radiation of 98 mW at 650 nm. Immediately after the laser is turned off, the vial headspace is subjected to a 15 seconds sampling.

Figure 4 shows a brief scheme of the experimental set up.

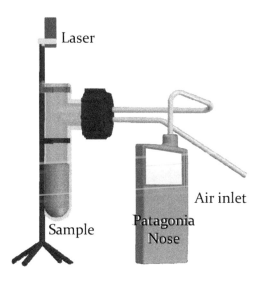

Fig. 4. Experimental set up

We have measured the V-UV spectra of the liquid oil samples before and after being irradiated during different time periods. We have noticed significant differences in the spectra of the liquid samples that had been irradiated during 5 minutes. Figure 5 shows the spectrum of each EVOO used before being irradiated.

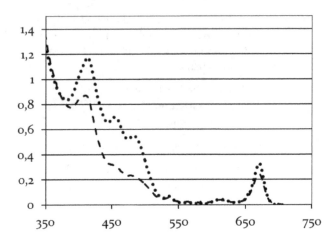

Fig. 5. V-UV Spectra of liquid EVOO A(- - - -) and liquid EVOO B (. . . .) before irradiation

Figure 6 shows the Principal Components Analysis (PCA) of the Patagonia Nose results obtained with Method I.

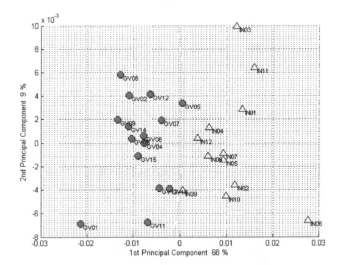

Fig. 6. PCA of EVOO A(GV) and B(IN) with Method I

Figure 7 shows the V-UV spectra registered for each liquid EVOO after the application of the three analytical methods. The experiment was repeated 5 times for each sample. It can be observed that the spectra of the irradiated samples of EVOO A exhibit significant changes with respect to the non-irradiated sample. The largest effect is produced by the irradiation with blue light (450 nm). This more energetic radiation affects the sample composition probably due to a photochemical effect. On the other hand, irradiation with the red light, gives rise to a thermal effect which causes the introduction of more molecules into the gas phase.

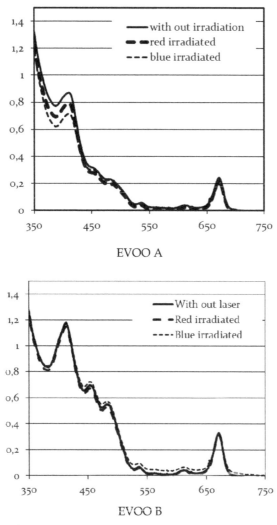

Fig. 7. V-UV spectra of EVOO (A) and (B)- Method I (_____) - Method II (- - - -) –
Method III (- - - -)

The discrimination ability of the three methods mentioned above was analyzed. For the data
processing the absorption and desorption rates were taken into account in addition to the
ratio of the peak value of each sensor response to the base-line value. Since different sensors
have distinct desorption times, for each chemical compound, we have considered the
integral of each time signal to the base line integral ratio.

The usual PCA of the data obtained with Methods II and III was performed. Figures 8 (a)
and 8 (b) show the score plot obtained with both methods. It is evident that better
discrimination is achieved with Method III.

A classification following the discrimination step was designed for Method III using about 70% of the available samples of both A and B oils.

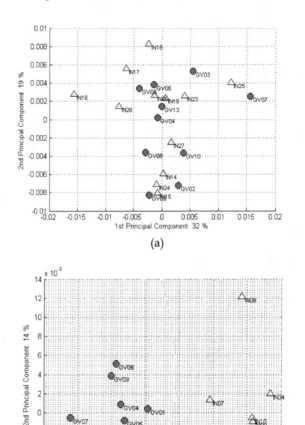

Fig. 8. PCA (a) Method II (b) Method III (discriminating regions are shown).

For validation purposes, 30% of the measured samples were disregarded. Within the resultant restricted space, an unsupervised PCA was performed. It was then possible to establish multivariate confidence regions for the two identified classes: olive oil A and olive oil B. The assignment of a new multivariate measurement x to a given category k occurs when the quadratic discriminant $d_k(x)$ is maximized:

$$d_k(x) = -\frac{1}{2}\ln|\Sigma_k| - \frac{1}{2}m_k(x)$$

$$m_k(x) = (x - \mu_k)^T \Sigma_k^{-1}(x - \mu_k)$$

where $m_k(x)$ is the statical or Mahalanobis distance, μ_k is the mean value of the corresponding class and μ_k is the covariance matrix within each class sample.

3. Conclusions

An easy to implement method to improve electronic nose discrimination ability of a priori similar odours has been presented. This technique has been then applied to the case of virgin olive oils. The way laser vaporization of the samples improves this task has been additionally explored. The behaviour of the gas phase headspace following pulsed and cw laser irradiation with different wavelengths has been analyzed.

The use of pulsed IR lasers increases the sensitivity of the e-nose performance. Furthermore, the use of a CO_2 laser allows a better discrimination than the use of a Nd:YAG laser. When using the CO_2 laser, the signal-to-noise ratio (S/N) is increased by an order of magnitude with respect to the S/N without laser vaporization effect. The IR laser wavelength influences the discrimination capabilities of the method, probably due to the different IR absorption properties of the sample compounds. Further experiments in progress may shed some light on this question.

The use of continuous wave visible diode lasers (methods II and III) produces significant changes in the V-UV spectrum of one of the EVOO, (EVOO A). Irradiation with the diode laser at 450 nm produces larger changes than those produced by the 650 nm diode laser irradiation. The 450 nm laser induces chemical reactions in the liquid oil surface and as a result precludes the discrimination. On the other hand, Laser Vaporization at 650 nm modifies the vapour-liquid equilibrium conditions improving the selectivity of the electronic nose.

It is important to emphasize that although the discrimination obtained with IRLV is larger than that resulting from LV at 650 nm it is insensitive to the recognition pattern used. On the other hand, diode lasers are considerably cheaper than high power TEA CO_2 lasers and they produce very good results considering the benefit-cost ratio.

4. Acknowledgements

The authors acknowledge, Agencia Nacional de Promoción Cientíca y Tecnológica (ANPCyT) and the Comisión Nacional de Energía Atómica (CNEA) and CONICET, for financial support

5. References

Angerosa, F., Camera, L., d'Alessandro, N. & Mellerio, G., (1998). Characterization of seven new hydrocarbon compounds present in the aroma of virgin olive oil. *J. Agric. Food Chem.* 46, 648–653.

Angerosa, F., Mostallino, R., Basti, C. & Vito, R., (2000). Virgin olive oil odour notes: their relationship with volatile compounds from the lipoxygenase pathway and secoiridoid compounds. *Food Chem.* 68, 283–287.

Angerosa, F., (2002). Influence of volatile compounds on virgin olive oil quality evaluated by analytical approaches and sensor panels. Eur. J. Lipid Sci. Technol. 104, 639–660.

Angerosa, F., Basti, C. & Vito, R., (1999). Virgin olive oil volatile compounds from lipoxygenase pathway and characterization of some italian cultivars. J. Agric. Food Chem. 47, 836–839.

Angerosa, F., Di Giacinto, L. & Solinas, M., (1992). Influence of Dacus oleae infestation on flavor of oils, extracted from attacked olive fruits, by HPLC and HRGC analyses of volatile compounds. Grasas Aceites 43, 134–142.

Angerosa, F., Lanza, B. & Marsilio, V., (1996). Biogenesis of "fusty" defect in virgin olive oils. Grasas Aceites 47, 142–150.

Aparicio, R., Morales, M .T. & Luna, G. (2006). Characterisation of 39 varietal virgin olive oils by their volatile compositions. Food Chemistry. 98, 2, 243-252

Ballabio, D., Cosio, M.S., Mannino, S. & Todeschini, R., (2006). A chemometric approach based on a nouvelsimilarity/diversità measure for the characterisation and selection of electronic nose sensors. Anal. Chim. Acta. 578, 170–177.

Christensen, B. & M. S. Tillack, (2003) Survey of mechanisms for liquid droplet ejection from surface exposed to rapid pulsed heating. University of California, UCSDENG- 100.

Cosio, M. S., Ballabio, D., Benedetti, S. & Gigliotti, C. (2006) Geographical origin and authentication of extra virgin olive oils by an electronic nose in combination with artificial neural networks, Anal. Chim. Acta 567, 202-210.

Cosio, M.S., Ballabio, D., Benedetti, S. & Gigliotti, C. (2007) Evaluation of different storage conditions of extra virgin olive oils with an innovative recognition tool built by means of electronic nose and electronic tongue Food Chemistry, 101, 2, 485-491

Dhifi, W., Angerosa, F., Serraiocco, A., Oumar, I., Hamrouni, I., & Marzouk, B. (2005). Virgin olive oil aroma: characterization of some Tunisian cultivars. Food Chem. 93, 697–701.

Di Natale, C., Macagnano, A., Nardis, S., Paolesse, R., Falconi, C., Proietti, E., Siciliano, P., Rella, R., Taurino, A. & D'Amico, A. (2001). Comparison and integration of arrays of quartz resonators and metal-oxide semiconductor chemoresistors in the quality evaluation of olive oils. Sens. and Actuators B, 78 , 303-309.

Esposto, E., Montedoro, J., Selvaggini, R., Rico, I., Taticchi, A., Urbani, S. & Servili, M. (2009). Monitoring of virgin olive oil volatile compounds evolution during olive malaxation by an array of metal oxide sensors. Food Chem. 113, 345 – 350.

European Community Regulation 1989/2003, Off. J. Eur. Community, L295 57-58

European Community Regulation 640/2008, Off. J. Eur. Community, L178 11-16

F. Lacoste (1999) Proceedings of the User Forum of the NOSE network, Ispra, VA,

Flamini, G., Cioni, P. L. & Morelli, I., (2003). Volatiles from leaves, fruits, and virgin oil from olea Europaea. Cv. Olivastra Seggianese from Italy. J. Agric. Food Chem. 51, 1382-1386

Garcia Gonzalez, D. & Aparicio, R. (2002) Detection of Vinegary Defect in Virgin Olive Oils by Metal Oxide Sensors. J. Agric. Food Chem.50, 1809-1814.

Gardner, J.W. & Bartlett, P.N. (1993). Brief history of electronic nose. Sensors and Actuator B. 18, 211–220.

Guadarrama, A., Mendz, M.L.R., Saia, J.A., Ros, J.L. & Olas, J.M. (2000). Array of sensors based on conducting polymers for the quality control of the aroma of the virgin olive oil. Sensors and Actuators B. 69, 276–282.

Kalua, C.M., Allena, M.S., Bedgood, D.R., Jra, Bishopa, A.G., Prenzlera, P.D. & Robards, K. (2007). Olive oil volatile compounds, flavour development and quality: A critical review. *Food Chem.* 100, 273-286.

Koning, S., Kaal, E., Janssen, H., Van Platerink, C., Brinkman & U. A.Th.(2008). Characterization of olive oil volatiles by multi-step direct thermal desorption–comprehensive gas chromatography–time-of-flight mass spectrometry using a programmed temperature vaporizing injector, *J. Chromatogr.,* A. 1186, 228-235.

Lerma-García, M. J., Cerretani, L., Cevoli, C., Simó-Alfonso, E. F., Bendini, A. & Gallina Toschi, T. (2010). Use of Electronic Nose to determine defect percentage in oil. Comparison with sensory panel result. *Sensors and Actuator B* 147, 283-289

Marriot, PJ. Sheillie & R. Cornwell, C. (2001). Gas Cromathografy Technologies for the analysis of essential oils. *J. Chromaogr.* A 936, 1-22

Massacane, A., Vorobioff, J., Pierpauli, K., Boggio, N., Reich, S., Rinaldi, C.A., et al. (2010) Increasing Electronic Nose recognition ability by sample laser irradiation. *Sensor and Actuator B:* Chemical 146, 2, 534-538

Morales, M.T., Berry, A.J., McIntyre, P.S. & Aparicio, R., (1998). Tentative analysis of virgin olive oil aroma by supercritical fluid extraction–high-resolution gas chromatography–mass spectrometry. *J. Chromatogr.* A 819, 267–275.

Morales, M.T., Luna, G. & Aparicio, R. (2005). Comparative study of virgin olive oil sensory defects, *Food Chemisty,* 91, 293-301

Morales, M.T., Rios, J.J., Aparicio, R., (1997). Changes in the volatile composition of virgin olive oil during oxidation: flavours and off-flavours. *J. Agric. Food Chem.* 45, 2666–2673.

Pawliszyn, J., (1999). Quantitative aspects of SPME. In: Pawliszyn, J. (Ed.), *Applications of Solid Phase Microextraction.* 3–21.

Petillo, D., Codnia, J. & Azcárate, M. L. (1996) Construcción de un láser de CO2 TEA, *Anales AFA,* 34-36

Reiners, J. & Grosch, W.(1998) Odorants of virgin olive oils with different flavor profiles. *J. Agric. Food Chem.* 46 (1998) 2754 -2763.

Rinaldi, C.A. & Ferrero, J. C. Analysis of Ca in BaCl2 Matrix using Laser Induced Breakdown Spectroscopy. *SpectroChimica Acta* Part B 56, 1419-1429, 2001

Rousseeuw, P.J. (1987) Silhouettes: a graphical aid to the interpretation and validation of cluster analysis, *J.of Computational and Applied Mathematics,* 20, 53-65.

Runcio, A., Sorgoná, L., Mincione, A., Santacaterina, S. & Poiana, M.(2008) Volatile Compounds of Virgin olive oil obtained from Italian cultivars grown in Calabria. Effect of Processing methods, cultivar, stone removal and antracnose attack. *Food Chemistry,* 106, 735-740

S. Mildner-Szkudlarz & H. H. Jeleń,(2008) The potential of different techniques for volatile compounds analysis coupled with PCA for the detection of the adulteration of olive oil with hazelnut oil, *Food Chem.* 110 , 751-761.

Schaller, E., Bosset, J.O. & Escher, F.(1998). "Electronic noses" and their applications to food. *Lebensmittel Technologies* 31, 305–308.

Servili, M., Conner, J.M., Piggott, J.R., Withers, S.J. & Paterson, A.(1995). Sensory characterisation of virgin olive oil and relationship with headspace composition. *J. Sci. Food Agric.* 67, 61–70.

Taglauer, E A.W. Czanderna & D. M. Hercules (1991). Ion scattering spectroscopy, in Ion Spectroscopies form Surface Analysis, *Plenum Publ. Co., New York*.

Temime, S.B., Campeol, E., Cioni, P.L., Daoud, D. & Zarrouk, M. (2006). Volatile compounds from Chétoui olive oil and variations induced by growing area. *Food Chem.* 99, 315–325

Tura, D., Prenzler, P.D., Bedgood, D.R., Antolovich, M. & Robards, K., (2004). Varietal and processing effects on the volatile profile of Australian olive oils. *Food Chem.* 84, 341–349.

Vichi, S. (2010)Extraction Techniques for the Analysis of Virgin Olive Oil Aroma. *Olives and Olive Oil Health and Disease Prevention* ISBN: 978-0-12-374420-3

Vichi, S., Castellote, A.I., Pizzale, L., Conte, L.S., Buxaderas, S. & Lopez Tamames, E.(2003). Analysis of virgin olive oil volatile compounds by headspace solid-phase microextraction coupled to gas chromatography with mass spectrometric and flame ionization detection. *J. Chromatogr.* A 983, 19–23.

Xu, X., and D. A. Willis, 2002, "Non-equilibrium phase change in metal induced by nanosecond pulsed laser irradiation," *J. Heat Transfer*, 124:293-298.

8

Traceability of Origin and Authenticity of *Olive Oil*

Zohreh Rabiei and Sattar Tahmasebi Enferadi
National Institute of Genetic Engineering and Biotechnology, Tehran, Iran

1. Introduction

Olive (*Olea europaea* L.) is one of the oldest agricultural tree crops worldwide and is an important source of oil with beneficial properties for human health. *Olive oil* is produced solely from the fruit of the olive tree (*Olea europaea* L.) and differs from most of the other vegetable oils in the method of extraction, allowing it to be consumed in crude form, hence conserving its vitamins and other natural healthy high-value compounds.

In comparison to commonly used vegetable oils, the cost of *olive oil* is higher. As such, *olive oil* is more prone against adulteration with other cheaper oils in order to increase profits. Several grades of *olive oil* are marketed which also command different prices. There is also the possibility of the addition of cheaper grades of *olive oil* to better graded ones for the same economic reasons. The presence in highly prized *olive oil*s of lower grade material is sometimes revealed by specific analytical methods.

Mixing low-grade sunflower, canola or other oil with olive's industrial chlorophylls, and flavouring it with beta-carotene has been brought to light and sold as *olive oil*. The FDA does not have the resources to test all the imported *olive oil* for adulteration, and some products are difficult to test. One can always expect adulteration and mislabeling *olive oil* products (Mueller, 1991). Christy et al., (2004) used near-infrared (NIR) spectra in the region of 12,000–4000 cm ($^{-1}$) to detect adulteration of *olive oil* with sunflower oil, corn oil, walnut oil and hazelnut oil. It has reported that adulteration of virgin *olive oil* with hazelnut oil could be detected only at levels of 25% and higher with Fourier transform infrared spectroscopy (Kumar et al., 2011). However, Rabiei (2006) has reported the use of molecular approach in revealing hazelnut-adulterated *olive oil* at level of less than 10% of hazelnut.

Several methods have been proposed for monitoring the adulteration of virgin *olive oil*s with other edible oils. In the last 10 years, technology and knowledge have undergone a great advance in the fight against adulteration; however, in the same way, knowledge of defrauders has also been increased. This enables them to prepare more sophisticated adulterations that make useless the methodologies proposed to detect them. Those oils normally added to virgin *olive oil* can be, either *olive oil*s of lower quality (e.g. olive-pomace *olive oil* or virgin *olive oil* obtained by second centrifugation of the olives), or seed oils (e.g. corn, soybean, palm or sunflower oil, among others) (Peña et al., 2005).

It has long been known that the chemical composition of virgin *olive oil* is influenced by genetic (variety) and environmental (climatological and edaphologic conditions) factors. So the olive production area is greatly responsible for the specific characteristics of *olive oil*.

The analytical analyses have their limits. This has promoted a growing interest towards the application of DNA-based markers since it is independent from environmental conditions. Specific protocols for DNA isolation from *olive oil* have been developed (Breton et al., 2004; Busconi et al., 2003; Consolandi et al., 2008; De la Torre et al., 2004). The application of DNA-based methods requests the knowledge on nucleotide sequences of olive. This information for olive is back to 1994, when the first *Olea europaea* L. sequence has deposited in NCBI. Table 1 provides brief information on olive genomics presented on NCBI from 1994 to July 2011 consist of Nucleotides, ESTs and GSS accessions.

Traceability in food is a recently developed concept of control of the whole chain of food production and marketing that would trace back to each step of the process. In a narrower sense, *genetic traceability* is performed to find out the genetic identity of the plant material from which the transformed products have originated. The recognition of the genetic background underlying food products aims to prove the authenticity of valuable food and to discourage from the adulteration with extraneous material of lower cost and value. Recently, Rotondi et al., (2011) has performed *olive oil* traceability by means of a combination of the chemical and sensory analyses with SSR biomolecular profiles. Her group demonstrated that the genetic (SSR analysis) component and the selected fatty acids (eicosenoic, linoleic, oleic, stearic, palmitic and linolenic), seems to represent a possible tool for inter- and intra-varietal characterisation and for monovarietal traceability.

Year	Olea europaea accessions on NCBI database*			
	Total nucleotide sequences	Nucleotide	EST**	GSS***
1994	3	3		
1995	1	1		
1996	1	1		
1997	13	13		
1998	1	1		
1999	11	11		
2000	39	39		
2001	57	57		
2002	57	57		
2003	88	64	24	
2004	23	23		
2005	213	213		
2006	44	44		
2007	337	335		2
2008	186	186		
2009	4891	55	4,836	
2010	1871	690	1,159	22
07/2011	33			
Total sequences	7,869	1,793	6,019	24

*http://www.ncbi.nlm.nih.gov, **EST: expressed sequence tags, ***GSS: genome survey sequences

Table 1. Olive genomics information present on NCBI database from 1994 to July 2011

2. General description of olive plant

Olive (*Olea europaea* L.) is the main cultivated species belonging to the monophyletic *Oleaceae* family that comprises 30 genera and 600 species, within the clade of Asterids, in which the majority of nuclear and organellar genomic sequences are unknown. The Olea genus includes 30 species and has spread to Europe, Asia, Oceania and Africa (Bracci et al., 2011). The wild olive or oleaster (*Olea europaea* subsp. *europaea* var. *sylvestris*) and the cultivated olive (*Olea europaea* subsp. *europaea* var. *europaea*) are two co-existing forms of the subspecies *europaea* (Green 2002). Other five subspecies constitute the *Olea europaea* complex including *laperrinei*, present in Saharan massifs; *cuspidata*, present from South Africa to southern Egypt and from Arabia to northern India and south-west China; *guanchica* present in the Canary Islands; *maroccana* present in south-western Morocco; and *cerasiformis* present in Madeira (Green 2002).

The Mediterranean form (*Olea europaea*, subspecies *europaea*) includes the wild and cultivated olives is a diploid species (2n = 2x = 46) (Kumar et al., 2011). The origin of the olive tree is lost in time, coinciding and mingling with the expansion of the Mediterranean civilisations which for centuries governed the destiny of mankind and left their imprint on Western culture.

The common olive is an evergreen tree that grows up to ~12m in height with a spread of about 8 m. However, many larger olive trees are found around the world, with huge, spreading trunks. The trees are also tenacious, easily sprouting again even when chopped to the ground. Sometimes it is difficult to recognize which is the primary trunk. The tree can be kept at a height of about 5m with regular pruning. Olives are long-lived, with a life expectancy of greater than 500 years (Kumar et al., 2011).

Most olive-growing areas lie between the latitudes 30° and 45° north and south of the equator, although in Australia some of the recently established commercial olive orchards are nearer to the equator than to the 30° latitude and are producing a good yield; this may be because of their altitude or for other geographic reasons.

The olive fruit is termed a *drupe* botanically, which are green in color at the beginning and generally become blackishpurple when fully ripe. A few varieties are green even when ripe, and some turn a shade of copper brown. Olive fruits consist of a carpel, and the wall of the ovary has both fleshy and dry portions. The skin (exocarp) is free of hairs and contains stomata. The flesh (mesocarp) is the tissue that is eaten, and the pit (endocarp) encloses the seed. Olive cultivars vary considerably in size, shape, oil content and flavor. Raw olive fruits contain an alkaloid that makes them bitter and unpalatable. A few varieties are sweet enough to be eaten after sun-drying (Wiesman, 2009).

Olive cultivars are basically classified into "oil olives" and "table olives," and oil cultivars predominate. Olive cultivars are also classified according to the origin of the cultivar – for example, Spanish, Italian, Greek, Syrian, Moroccan, Israeli, etc. The most popular cultivars are: Picual, Arbequina, Cornicabra, Hojiblanca and Empeltre in Spain; Frantoio, Moraiolo, Leccino, Coratina and Pendolino in Italy; Koroneiki in Greece; Chemlali in Tunisia; Ayvalik in Turkey; Nabali, Suori and Barnea in Israel and The West Bank; Picholin in France; Mission in California; and various varieties in Australia. The table olive cultivars include Manzanilla and Gordal from Spain; "Kalamata" from Greece; "Ascolano" from Italy; and "Barouni" from Tunisia (Jacoboni & Fontanazza, 1981; Weissbein, 2006).

The large number of cultivars, added to the many cases of synonymous and homonymous name, makes particularly difficult the description and classification of olive varieties (Fabbri et al. 2009). Notice that two varieties are synonymous when they have different names but the same profile of fingerprinting, and two varieties are homonyms when they have the same name but different fingerprinting profiles.

The size of olive germplasm is controversial: about 1,250 varieties (or in some other references 1,275 cultivars, Sarri et al., 2006), cultivated in 54 countries and conserved in over 100 collections, were included in the FAO olive germplasm database (Bartolini 2008), also if this number is certainly higher because the lack of information on many local cultivars and ecotypes (Cantini et al. 1999). The most part of these cultivars comes from southern European countries such as Italy (538 varieties), Spain (183), France (88) and Greece (52) (Baldoni & Belaj 2009). Due to this richness of the germplasm, olive is an unusual case among horticultural crops and its biodiversity can represent a rich source of variability for the genetic improvement of this plant.

3. Olive oil

It has been known that climate, soil, variety of tree (cultivar) and time of harvest account for the different organoleptic properties of different *olive oils*. Two factors are influential: where the olives are grown and which harvesting methods are implemented. Certain locations yield more bountiful harvests; consequently their oil is sold for less. Olive trees planted near the sea can produce up to 20 times more fruit than those planted inland, in hilly areas like Tuscany. It is in these land-locked areas that the olive trees' habitat is pushed to the extreme; if the conditions were just a little more severe, the trees would not survive. Extra-virgin oils produced from these trees have higher organoleptic scores.

It is extremely difficult to establish the origins of *olive oil* using DNA technologies. One approach has been **to target yeasts associated with olives and** *olive oil*. Target for characterization was the LTR retrotransposon (Ty element) (Lenoir et al., 1997, as cited in Popping, 2002) using amplified fragment length polymorphism or similar techniques. This method has been more successful for olives, where different yeast strains are associated with olives and *olive oil*. The yeast strains in *olive oil* appear to be associated with the production site (fattoria) where the *olive oil* was produced. And since the number of production sites is limited, the *olive oil* can be traced back to the fattoria.

But this technique is not yet applicable for routine analysis. For the identification of the origin of *olive oil*, a second, non-DNA-based technology has proven very useful. The technology is called **site-specific natural isotope fractionation nuclear magnetic resonance (SNIF-NMR)** (Gonzalez et al., 1999; Martin et al., 1996).

The basis of this technology is that certain elements have naturally occurring stable isotopes (^{16}O and ^{18}O, ^{1}H and ^{2}H, ^{12}C and ^{13}C). The ratios of the different stable isotopes vary from one geographic location to another. These ratios are maintained in the organic material from that region, e.g. plants, animals etc. The SNIF-NMR technology allows measuring these stable isotope ratios at individual positions in a given molecule.

With an appropriate database listing the location and typical stable-isotope distribution, the origin of *olive oil* can be identified (Popping, 2002).

Other techniques such as proton transfer reaction mass spectrometry (PTR-MS), nuclear magnetic resonance spectroscopy (NMR) or high performance liquid chromatography (HPLC) has also been addressed using different methodologies (Luykx & van Ruth, 2008). However, the chemical composition of *olive oil* may differ among seasons and growing areas. Several investigations concerning the origin and authenticity of *olive oil* have shown that chemical analyses per se are not sufficient to assure *olive oil* authenticity or to reveal *olive oil* region (Gimenez et al., 2010). Christopoulou et al., 2004 expressed that no single known parameter could detect the presence of hazelnut and almond oils in *olive oil* which have many chemical characteristics (fatty acid profile, sterol composition, …) similar to *olive oil*.

Several Protected Denomination of Origin (PDO) *olive oil* regions have been established by legislation to ensure both producer's profits and consumer's rights. In this context, it is mainly the identification of the olive cultivar used for the oil production which is of importance for authentication (Luykx & van Ruth, 2008) since the contribution of cultivars is known for each designation (Gimenez et al., 2010). Different PDO labels such as "Oliva Cilento", "Colline Salernitane", and "Penisola Sorrentina", have been granted for the extra-virgin *olive oil* produced in Campania, and some others are in the process of assignment.

4. World olive oil production

According to the report of the International Olive Council (IOC) (the International *Olive oil* Council, IOOC, until 2006), Mediterranean countries account for around 97 percent of the world's olive cultivation, estimated at about 10,000,000 hectares. There are more than 800 million olive trees currently grown throughout the world, of which greater than 90 percent are grown for oil production and the rest for table olives. It is estimated that more than 2,500,000 tons of *olive oil* are produced annually throughout the world.

Since the mid-1990s, Spain has consistently been the largest producer; in the year 2004/05 it produced 826,300 tons of *olive oil* and it is expected a sum of 2,948,000 tons for total world olive oil production in 2010/11. The main producer is still European Union (EU), with 2.1 million tons, of which 1.2 from Spain (-14% from the previous campaign), 336000 tons produced by Greece, 480000 tons by Italy, 67500 tons by Portugal, 65000 tons by Cyprus and 6000 tons by France. Out of EU, IOOC estimated a production of 193500 tons from Syria, 160000 tons from Turkey, 12000 from Tunisia, 150000 from Morocco, 48000 from Algeria, 24900 from Palestine, 19000 from Jordan, 18000 from Australia, 17500 from Argentina, and 15000 from Lybia.

The world *olive oil* consumption (2010-2011) will reach 2.98 million tons, with a 3.65 % increase from the previous campaign 2009-2010 (*IOC website*).
IOOC has estimated that the world export will increase of 5.05% and reach 707000 tons, with EU (438000 tons) as the main commercial power, followed at great distance by Morocco (40000 tons), Syria (50000 tons), Tunisia (90000 tons), and Turkey (38000 tons).

The import for the period between October 1st 2010 and September 30th 2011 is estimated at 648000 tons, with a 2.93% increase from the previous year (*IOC website*).

The pattern of production of *olive oil* during these years shows big fluctuations from one year to the next; however, Spain, Italy and Greece remain the three largest *olive oil* producing countries, dominating the world annual *olive oil* production. This signifies a high level of uncertainty regarding production levels. In the year 2004/05, Spain, Italy and

Greece produced 32, 28 and 13.5 percent of the world's *olive oil*, respectively. However, the recent expansion of the *olive oil* industry and significant contribution to the global *olive oil* market by several other countries, such as Australia and the United States, may lead to stabilization of the market in the near future.

5. *Olive oil* traceability

Food traceability implies the control of the entire chain of food production and marketing, allowing the food to be traced through every step of its production back to its origin. The verification of food traceability is necessary for the prevention of deliberate or accidental mislabeling, which is very important in the assurance of public health. Thus, several regulations provide the basis for the assurance of a high level of protection of human health and consumers' interest in relation to food.

In the case of *olive oils*, the increase in the demand for high-quality *olive oils* has led to the appearance in the market of *olive oils* elaborated with specific characteristics. They include oils of certain regions possessing well-known characteristics, that is, *olive oils* with a denomination of origin, or with specific olive variety composition, that is, coupage or monovarietal *olive oils*. *Olive oils* obtained from one genetic variety of olive or from several different varieties are called monovarietal or coupage, respectively. Monovarietal *olive oils* have certain specific characteristics related to the olive variety from which they are elaborated (Montealegre et al., 2010). However, coupage *olive oils* are obtained from several olive varieties to achieve a special flavor or aroma.

The appearance of denominations and protected indications of origin has promoted the existence of oils labeled according to these criteria. Regulation 2081/92 (2) created the systems known as Protected Designation of Origin (PDO), Protected Geographical Indication (PGI) and the "Traditional Speciality Guaranteed" (TSG) to promote and protect food products (Table 2).

General regimen	Origin	Characteristics	Restriction
Protected Designation of Origin (PDO)	In that region, specific place, or country	Quality essentially or exclusively due to a particular geographical area	Produced, processed and prepared in a given geographical area
Protected Geographical Indication (PGI)	In that region, specific place, or country	Slightly less strict; food reputation of a product from a given region is sufficient	One of the stages of production, processing, or preparation takes place in the area

*Council Regulation (EC) 510/2006, March 20, 2006.

Table 2. General regimen for food and certain other agricultural products based on Regulation 510/2006*

For example, an *olive oil* with a PDO denomination requires meeting precise definition of several parameters such as cultivar, geographical origin, agronomic practice, production technology, and organoleptic qualities (Gimenez et al., 2010), and all of these parameters have to be investigated to study its traceability and to certify its quality. Among the above-mentioned factors, the two first are the most important (Montealegre et al., 2010).

Additionally, a Database of Origin and Registration (DOOR) was created to support these denominations (Montealegre et al., 2010). Based on the report of the *International Olive Council* (http://www.internationaloliveoil.org) gave the world production of *olive oil* in 2008/2009 as 2,669.5 million tons and it consumption for the same period as 2,831.5 million tons. It is quite clear that some of the *olive oil* sold has been mislabeled. *Olive oil* is priced from $13–105 for 500 mL where as canola oil and sunflower oils available from less than $1–10 for 500 mL.

The introduction of certifications of origin and quality for virgin *olive oil* as PDO makes necessary the implementation of traceability procedures. It seems that DNA analysis to be a promising approach to this problem, since it is less influenced by environmental and processing conditions in respect to other methods (i.e.; metabolites).

Any research dealing with *olive oil* traceability is focused on investigating the botanical or geographical origin. However, the concept of geographical traceability, in which the objective is the geographical location of the olive tree, is slightly different from the concept of botanical traceability, in which the olive used for the *olive oil* production is the aim. In both cases, the selection of the markers (compounds with discriminating power) to be studied is complicated because the composition of extra virgin *olive oil*s is the result of complex interactions among olive variety, environmental conditions, fruit ripening, and oil extraction technology (Araghipour et al., 2008).

5.1 Traceability to the botanical origin

The verification of the cultivars employed to produce an *olive oil* sample may contribute to address the oil origin. This fact may have commercial interest in the case of monovarietal *olive oil*s or *olive oil*s with PDO because these high-quality *olive oil*s may be adulterated by other oils of lower quality, using anonymous or less costly cultivars (Breton, 2004).

As the quality of an *olive oil* depends on the olive variety from which it is elaborated, the production of *olive oil*s from certain varieties has increased (Sanz-Cortes et al., 2003). The olive variety selection is based on its adaptation to different climatic conditions and soils. In addition, whereas some cultivars are characteristic of a given zone, others can be found in several countries (Japon-Lujan et al., 2006). As a consequence, one olive variety can be cultivated and nominated in a different way in distinct geographical locations, which makes the differentiation of olive varieties in *olive oil*s quite complex. (Montealegre et al., 2010)

Traditionally, differentiation among olive cultivars has been supported by numerous morphological (study of the form or shape) and pomological (the development, cultivation, and physiological studies of fruit trees) traits. Unfortunately, morphological traits have been difficult to evaluate, are affected by subjective interpretations, and are severely influenced by the environment and plant developmental stage (Sanz-Cortes et al., 2003). Nowadays, several efforts have been focused on the investigation of one or several compounds present in *olive oil*s usable to differentiate olive varieties (Montealegre et al., 2010). Compositional markers (substances that take part of the composition of the olive oils) include major and minor components. Major, (sterols, phenolic compounds, volatile compounds, pigments, hydrocarbons, and tocopherols, fatty acids and triglycerides, components may provide basic information on olive cultivars. Minor components, can provide more useful information and have been more widely used to differentiate the botanical origin of *olive oils* (Montealegre et al., 2010).

5.2 Compositional markers

There are several parameters, major and minor compositional markers, with varied discriminant power used for *olive oil* traceability according to the variety of olive participated in the production of the oil (Arvanitoyannis et al., 2007).

To relate the fatty acid composition of *olive oils* with the cultivar, Mannina et al. (2003) studied *olive oil* in a well- limited geographical region, with no consideration of the pedoclimatic factor (soil characteristics such as temperature and humidity). A relationship between the fatty acid composition and some specific cultivars has been observed (Montealegre et al., 2010).

The *volatile fraction* in *olive oils*, which represents one of the most important qualitative aspects of this oil, consists of a complex mixture of more than 100 compounds, but the most important substances useful for olive cultivar differentiation are the products of the lipoxygenase pathway (LOX). Only a subset of volatile compounds and a combination among them could provide valuable information for olive cultivar differentiation (Montealegre et al., 2010). Three volatile compounds [hexyl acetate, hexanal, and (E)-hex-2-enal] and the total concentration of ketones have nominated to distinguish the olive varieties (Tena et al., 2007) hich consequently could be used for olive oil, as well. However, it has been found that the level of (E)-hex-2-enal in the analyzed samples showed a variability that suggest an influence of genetic factors on the biosynthesis of this compound. In fact, genetic (Tura et al., 2008; Mahjoub-Haddada et al., 2007) and geographic (Mahjoub-Haddada et al., 2007) factors influence the volatile compound production of the olive fruits and affect the differentiation of *olive oils* according to their olive variety. The volatile compound contents allowed differentiation among monovarietal *olive oils* and even identification of the technique used for *olive oil* production (Torres Vaz-Freire et al., 2009).

The color of a virgin *olive oil* is due to the solubilization of the lipophilic chlorophyll and carotenoid pigments present in the fruit. The green-yellowish color is due to various pigments, that is, chlorophylls, pheophytins, and carotenoids (Cichelli & Pertesana, 2004). Chlorophyll a is the major chlorophyll pigment, followed by chlorophyll b. The carotenoid fraction is included lutein, violaxanthin, neoxanthin, β-carotene, β-cryptoxanthin, and luteoxanthin (Montealegre et al., 2010).

Several researchers reported the same qualitative composition in chlorophyll and carotenoid pigments, independent of the olive variety and the time of picking (Giuffrida et al., 2007 ; Roca et al., 2003).

Cerretani et al. (2006) showed that the carotenoid and chlorophyll content determination using UV-vis spectrophotometry was not useful to discriminate oils produced from different olive varieties. Lutein/β-carotene ratio has been reported as a tool to differentiate oils from a single cultivar.

Tocopherols and hydrocarbons are the compositional markers less studied to date to differentiate *olive oils*. An important common aspect is that the content and composition of these markers are highly affected by the environmental conditions, the fruit ripening, and the extraction technology (Montealegre et al., 2010).

6. DNA-based markers

Molecular markers are investigated as a diagnostic tool for food authenticity and traceability of variety/type composition of complex food matrices in an increasing number of projects (Palmieri et al., 2004). DNA-based methods make an important contribution to protect high-quality *olive oils*.

Significant amounts of DNA are present in *olive oil* obtained by cold pressing (Consolandi et al., 2008). However, the filtration process lowers DNA concentrations, which tend to disappear due to nuclease degradation (De la Torre et al., 2004; Muzzalupo et al., 2002). On the other hand, the length of storage after milling of the oil can affect the use of DNA as an analyte for molecular traceability. Pafundo et al., 2010 observed a significant decrease of quality of DNA extracted from olive oil, with a consequent loss of information a month later from olive oil production.

Spaniolas et al. (2008b) has used lambda DNA as a marker to monitor the length of DNA fragments in olive oil during storage time when determined the varietal origin of olive oil. Lambda DNA is a linear molecule of approximately 50 kb, a length that probably resembles that of olive DNA present in olive oil. Based on the fact, PCR-based fingerprinting techniques, which require templates longer than 100 bp, might not be able to successfully amplify the target sequences from olive oil samples low in DNA content and stored for several months, they conduct their analyses. They have deduced the detection of polymorphic markers requiring DNA templates shorter than 100 bp might have a wider range of applications in DNA fingerprinting of olive oil.

In *olive oil*, once the barrier of DNA extraction has been overcome, several markers could be used to identify olive cultivars that made up a certain *olive oil*. (Consolandi et al., 2008)

DNA recovery methods from *olive oil* have been developed by many authors (Busconi et al. 2003; Doveri et al. 2006; Pasqualone et al. 2007; Consolandi et al. 2008). Several commercial kits, providing adapted protocols, were used in different works (Martins-Lopes et al. 2008; Spaniolas et al. 2008a; Ayed et al. 2009; Pafundo et al. 2010). All of these studies confirmed that the DNA of the cultivars is recoverable from extra virgin *olive oil*, but it has low quantity and quality. The first researches, carried out using genomic DNA extracted from drupes. That DNA had a good potential to amplify correctly using RAPDs markers (Cresti et al. 1997). By means of SCAR and AFLP markers, Busconi et al. (2003) were able to show that DNA recovered from *olive oil* had both organellar and nuclear origin. Pafundo et al. (2005) traced the cultivar composition of monovarietal *olive oils* by AFLPs, suggesting that DNA extraction is the most critical step affecting the procedure. Pafundo et al. (2007), performed amplification of DNA isolated from *olive oil* using AFLPs. They have also developed some SCARs to amplify successfully the DNA extracted from *olive oil*. Using SSR analysis, Pasqualone et al. (2007) demonstrated that microsatellites are useful in checking the presence of a specific cultivar in a PDO oil, thus verifying the identity of the product. However, they obtained only the marker profile of the main cultivar in the oil: no signal was detected for the secondary varieties. Montemurro et al. (2008) analyzed ten virgin monovarietal *olive oils* prepared in the laboratory by AFLP markers. They were able to distinguish all the *olive oils* examined, even if only a partial correspondence with the AFLP profile obtained from the leaves was obtained. Martins-Lopes et al. (2008) evaluated the efficiency of RAPD, ISSR and SSR molecular markers for *olive oil* varietal identification and their possible use in certification purposes (Bracci et al., 2011).

Consolandi et al. (2008) reported the development of a semi-automated SNP genotyping assay to verify the origin and the authenticity of extra-virgin *olive oils*. The authors developed a Ligation Detection Reaction (LDR)/Universal Array (UA) platform by using several olive SNPs. They found that 13 accurately chosen SNPs were sufficient to unequivocally discriminate a panel of 49 different cultivars (Bracci et al., 2011).

Doveri et al. (2006) published a cautionary note on the use of DNA markers for provenance testing. Their observations were based on non-concordance between the genetic profiles of *olive oil* and fruit. They suggested that this could be due to the contribution of pollen donors in DNA extracted from the paste obtained by crushing whole fruits. They concluded that care needs to be taken in the interpretation of DNA profiles obtained from DNA extracted from oil for resolving provenance and authenticity issues (Martins-Lopes et al., 2008). It is to note that, the possible presence of additional alleles due to paternal contribution in oils extracted from entire drupes, should be taken into consideration for variety traceability purposes when comparing the amplification profiles of leaves with the corresponding oils for (Alba et al., 2009)

In a recent study, Pafundo et al. (2010) investigated the effect of the storage time on the degradation of the DNA purified from the oil; a negative correlation between storage time and quality–quantity of recovered DNA has been observed. The authors showed that 1 month after the production of the oil the degradation increases making harder traceability goals (Bracci et al., 2011). Table 3 presents a brief report on the application of DNA-based molecular markers for cultivar traceability in *olive oil*.

Molecular marker	Developers	References
RAPD (Random Amplified Polymorphic DNA)	Williams et al. (1990)	Muzzalupo and Perri (2002), Martins-Lopes et al. (2008)
AFLP (Amplified Fragment Length Polymorphism)	Vos et al. (1995)	Busconi et al. 2003, Pafundo et al. 2005
SCAR (Sequence Characterized Amplified Region)	Paran & Michelmore (1993)	De la Torre et al. (2004), Pafundo et al. (2007)
SSRs (Simple Sequence Repeats)	Morgante & Olivieri (1993)	Martins-Lopes et al. (2008), Alba et al. (2009), Rabiei et al. (2010)
ISSR (Inter Simple sequence Repeats)	Zietkiewicz et al. (1994)	Pasqualone et al. (2001), Martins-Lopes et al. (2008)
Chloroplast and mitochondrial (Direct sequencing)	Botstein et al. (1980)	Intrieri et al. (2007)

Table 3. Applications of DNA-based molecular markers for cultivar traceability in *olive oil*

6.1 Random Amplified Polymorphic DNA (RAPDs)

The applicability to the traceability of cultivars in the *olive oil* by means of RAPDs was evaluated by Muzzalupo & Perri (2002). This kind of markers was used, together with other DNA-based markers, in the construction of the first olive linkage maps (De la Rosa et al. 2004; Wu et al. 2004) as well.

6.2 Amplified Fragment Length Polymorphism (AFLPs)

AFLPs have been widely used for DNA fingerprinting of cultivars (Angiolillo et al. 1999; Owen et al. 2005), to analyse the relationships between wild and cultivated olive (Baldoni et al. 2006), for the construction of linkage maps (de la Rosa et al. 2003) and for cultivar traceability of *olive oil* (Busconi et al. 2003; Pafundo et al. 2005).

6.3 Development of sequence-characterized amplified regions (SCARs)

SCARs have been used for cultivar identification (Busconi et al. 2006) and in *olive oil* traceability (De la Torre et al. 2004; Pafundo et al. 2007). Putative associations of several SCAR markers with fruit characteristics (Mekuria et al. 2002) and resistance to pathogenic fungi (Herna´ndez et al. 2001) were found, suggesting the applicability of this type of marker for marker-assisted breeding programs (Bracci et al., 2011).

6.4 Simple Sequence Repeats (SSRs)

Many authors have reported on SSR development in olive and several of them are currently available for DNA analysis (Cipriani et al. 2002; De la Rosa et al. 2002; Rabiei & Tahmasebi Enferadi 2009; Sabino Gil et al. 2006; Sefc et al. 2000). This technique benefits from the use of microsatellites which are short stretches (1-6-bp long) of DNA, tandemly repeated several times. The number of repeats can vary from one individual to another. Besides, they are abundant in eucaryotic genomes. A combination of several SSRs loci allows virtually discrimination of individuals originated through sexual reproduction, especially in out-crossing species, where the level of heterozygosity is high.

In Olea europaea, these markers have been used for different applications such as cultivar discrimination (Sarri et al. 2006; Fendri et al. 2010), study of relationships between wild and cultivated olive tree (Belaj et al. 2007), construction of association maps (De la Rosa et al. 2003), paternity analysis (Mookerjee et al. 2005) and identification of *olive oil* varietal composition (Alba et al. 2009; Ayed et al. 2009; Rabiei et al. 2010). A list of recommended SSR markers and protocols for olive genotyping has been provided with the aim to develop a robust method to track the origin of olive cultivars (Doveri et al. 2008; Baldoni et al. 2009) (table 4).

Series	Number	Institute	References
ssrOeUA-DCA	15	Boku, Vienna, Austria	Sefc et al 2000
IAS-oli	5	CSIC, Córdoba, Spagna	Rallo et al 2000
GAPU	20	Agrobios, Matera, Italia	Carriero et al 2002
UDO	30	Università di Udine, Italia	Cipriani et al 2002
EMO	7	ETSIAM, Córdoba, Spagna	De La Rosa et al 2002
Totale	77		

Table 4. List of microsatellites isolated in olive which their related primers are available in the literature, most of them have been used in case of *olive oil* traceability (Bracci et al., 2011)

For studying the informative potential of the microsatellites, the observed (*Ho*) and expected (*He*) heterozygosities generally are calculated using the software POPGENE ver. 1.31 (Yeh et al., 1999, as cited in Alba et al., 2009). *He* values were estimated using the formula proposed by Nei et al. (1973):

$$H_e = 1 - \sum p_i^2 \tag{1}$$

where *pi* is the frequency of the *ith* allele. The power of discrimination (PD) [21] of microsatellite primer pairs are also calculated as reported by Cipriani et al. (2002), where the allele frequency of the *He* formula is replaced by the genotype frequency.(Alba et al., 2009)

A research carried on the use of SSRs as a tool to identify the genetic background of *olive oil* which was involved the analysis of DNA sequences using a panel of seven simple sequence repeats (SSRs) to provide genotype-specific allelic profiles (Rabiei et al., 2010). The amplified SSR fragments and the DNA profiles from the monovarietal oil corresponded to the profiles from the leaves of the same cultivar. The most reliable SSR in providing correct allele sizing in distinguishing either single cultivar *olive oil* samples or the different ratios of their blends are DCA3, DCA4, DCA16, DCA17, and GAPU101, while DCA9, GAPU59 produced less concordance against data obtained by the genetic analysis of leaf samples. Desalted PCR product has been analyzed on a MegaBACE 500 capillary sequencer (Amersham Biosciences) using Genetic Profiler v2.0 software to estimate allele sizes (figure 1).

Rabiei et al., 2010 concluded PCR product purification and selection of a set of markers with a highly robust amplification pattern is needed to have reproducible results in certify the genetic background of *olive oil*.

6.5 Inter simple sequence repeat (ISSR) polymorphisms

ISSRs are DNA fragments of about 100–3,000 bp located between adjacent, oppositely oriented microsatellite regions. These markers were used with success to distinguish 10 Italian varieties, by analysing genomic DNA extracted from the olive fruit (Pasqualone et al. 2001), and for the study of cultivar traceability in *olive oil* (Pasqualone et al. 2001, Martins-Lopes et al. 2008).

6.6 Chloroplast genome sequencing

A very important results, recently published, in *Olea europaea* L. genomic studies have been the DNA sequencing of the entire plastome of the Italian cultivar 'Frantoio' (Mariotti et al. 2010). This sequence has a length of 155,889 bp and showed an organization and gene order that is conserved among numerous Angiosperms. The olive chloroplast contains 130 genes and 644 repetitive sequences (among which 633 mono-nucleotide SSRs, 6 di-, 3 tetra- 2 penta-nucleotide SSRs were identified) (Bracci et al., 2011)

The annotated sequence was used to evaluate the content of coding genes, the extent, and distribution of repeated and long dispersed sequences and the nucleotide composition pattern. These analyses provided essential information for structural, functional and comparative genomic studies in olive plastids. Furthermore, the alignment of the olive plastome sequence to those of other varieties and species identified 30 new organellar polymorphisms within the cultivated olive. chloroplast DNA polymorphisms has been used as molecular markers to identify cultivars of *Olea europaea* L. (Intrieri et al. 2007).

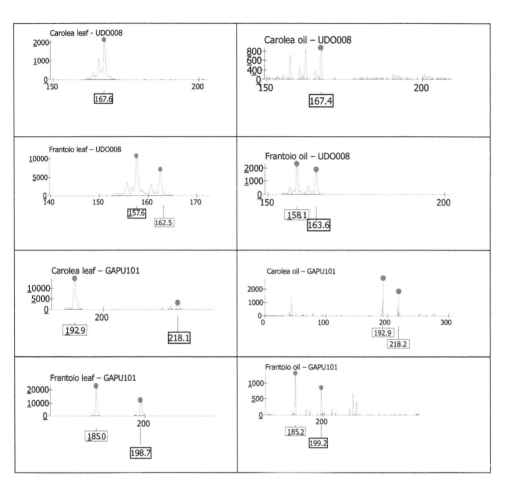

Fig. 1. Electropheregram of PCR products separated by capillary electrophoresis of microsatellite loci UDO008 and GAPU 101, obtained from DNA extracted from Carolea and Frantoio leaves and oils. Allele sizes are below the x axis. The scores on Y-axis are the intensity of amplified allele detection (Rabiei et al., 2010).

6.7 Expressed Sequence Tags (ESTs)

Understanding the function of genes and other parts of the genome is known as functional genomics. In olive, efforts to improve the identification and annotation of genes are prevalently based on EST identification, which are predominantly related to pollen allergens and characteristics of olive fruit (Bracci et al., 2011).

The first nucleotide sequences isolated in 1994 in olive coded for allergenic proteins (Villalba et al. 1994, as cited in Bracci et al., 2011) (Table 1).

6.8 Real time-PCR

The detection of frauds, either due to the mixtures with oils of other species such as hazelnut, or to the certification of PDOs would need quantitative tools. At its best, conventional PCR remains a semi-quantitative technique, and therefore, it is not optimal for authentication purposes when quantification is needed (Gimenez et al.; 2010).

The use of real-time chemistries allows for the detection of PCR amplification during the early phases of the reaction, providing a distinct advantage over detection of amplification at the final phase or end-point of the PCR reaction. qRT-PCR is a useful tool in the development of molecular markers for *olive oil* authentication since it allows inspecting the PCR efficiency. Besides qRT-PCR should be used for the optimisation of the amplicon size and the DNA isolation procedure which are critical aspects in *olive oil* authentication. The potential of cpDNA for *olive oil* authentication should be addressed in the future (Gimenez et al.; 2010).

6.9 DNA barcode

Several sequences from noncoding spacer region between psbA-trnH and partial coding region of matK of plastid genome provided a good discrimination of pure *olive oil* and its admixture by other vegetable oils such as canola and sunflower.

The plastid based molecular DNA technology has a great potential to be used for rapid detection of adulteration easily up to 5% in *olive oil* (Kumar et al., 2011).

7. Paternity analysis

Similar to other woody species, olive is characterized by a long juvenile phase that ranges between 10 and 15 years. This represents a great obstacle to breeding programs and makes the genetic improvement of olive very difficult and expensive. Although seedling-forcing growth protocols have been developed to reduce the length of this phase, the evaluation of the agronomic performance of mature olive plants still requires at least 5 years of experimentation (Santos-Antunes et al. 2005). For this reason, the application of molecular markers both to confirm the parental origins of the progeny and to select early agronomical characteristic-associated markers (Martı́n et al. 2005) can be very useful to reduce the time and cost of the development of new genotypes (Bracci et al., 2011).

With regard the paternity analysis, SSRs are the most suitable to trace the genetic contribution of alleles from the parents to the offspring, being co-dominant and highly polymorphic markers (Mookerjee et al. 2005). The effectiveness of SSRs in the identification of paternity contribution to progeny obtained from olive breeding programs has been demonstrated by several authors (De la Rosa et al. 2004; Diaz et al. 2007). The results demonstrated that SSR analysis is a convenient technique to routinely assess the crosses made in breeding programs and to for check self-incompatibility in olive cultivars (Diaz et al. 2006). These studies have highlighted that no contamination by self-pollen was found, indicating that placing pollination bags well before anthesis is important and that emasculation to avoid selfing is unnecessary (De la Rosa et al. 2004). The analysis also revealed that the main factor affecting the success of crosses seems to be the inter-compatibility among the parental cultivars, since this significantly influences the rate of contamination from external pollen donors. These results indicate that knowledge of cross-compatibility among cultivars is necessary to plan efficient olive breeding crosses (Diaz et al. 2007).

The possibility of associating genetic characteristics and DNA-based molecular markers is very important to select the progeny showing interesting agronomical traits and even specific organoleptic characteristics at the first stages of development which may use as a marker for future *olive oil* identification. However, this technique, called marker-assisted selection (MAS), requires some knowledge on the co-segregation of molecular markers and genetic characteristics in the progeny. (Bracci et al., 2011)

8. Conclusion

For the inefficiency of analytical parameters in showing variability among samples of the same cultivar/blend due to the environmental conditions and pressing technologies, Several DNA-based technologies and traceability analysis has been used to reveal the different origin of lots that have contributed to the olive oil blend. In this regard, DNA-based methods make an important contribution to protect high quality brand names and in turn the consumer

The knowledge of genome nucleotide sequences also could be useful to identify new sequence polymorphisms, which will be very useful in the development of many new cultivar-specific molecular markers (e.g.; Single Nucleotide Polymorphisms, SNPs) and in the implementation of more efficient protocols for tracking and protect *olive oil* origin (in POD *olive oils*).

The greatest challenges one faces while using DNA technology is the low quality and highly degraded DNA recovered from the fatty matrices and the impact of oil extraction processing on the size of the recovered DNA. DNA of low, difficult to determine content and of unknown, variable quality would potentially lead to inconsistent and consequently inconclusive results. Although, the concentration of DNA did not appear to be limiting; rather, successful PCR amplification likely depended on the ability of the DNA extraction method to free DNA from inhibitors of PCR present in the *olive oil*.

It is to be considered if the DNA is damaged, it could be not properly accessible to the DNA polymerase, which stalls at the sites of damage and the reaction may be interrupted; this being able to influence the length and significance of the synthesized amplicons. The use of proteinase K during extraction process has recommended for a better protection of DNA from degradation and increase in DNA yield, as well.

Identification of molecular markers suitable for tracing the genetic identity of olive cultivars from which oil is produced, on the other hand, has a great importance. For making decision, which molecular markers will be more useful in obtaining reliable results through the numerous molecular markers existing in the literature, many of them have been practically examined (including RAPDs, AFLPs, SCARs, SSRs, ISSR, SNPs, …). A combination of molecular markers (RAPD, ISSR, and SSR) to establish a relationship between small-scale-produced monovarietal and commercial *olive oil* samples for certification purposes has been proposed.

Several authors recommended sequences of DNA that show polymorphism at low hierarchical level are therefore suitable for distinguishing between individuals within the same species. They clearly pointed non-coding nuclear DNA sequences could be the best choice. Among those sequences, the microsatellites are likely the most suitable ones. However SNPs that require shorter than 100 bp DNA templates, considered to be successfully used for a wider range of olive oil identification.

In some cases of using microsatellite, the microsatellite profiles obtained with the monovarietal oil-derived DNA were generally consistent with the cultivar used, although some ambiguities were recorded likely due to contamination in monovarietal oils by other cultivars grown in the same block or contaminations occurred at the mill. Moreover, in some cases the lack of matching in leaf and oil profiles has been reported that was due to the presence of embryos in berry seeds that brought the alleles of pollinators. Other cases of mis-amplification were recorded as a missing allele, due either to the preferential amplification of one of the two alleles in oil-derived DNA templates, or to the excess of degradation of the DNA template of the miss allele, that limited the production of a sufficient number of copies of that allele to be detected. In such a case, real-time PCR assay could possibly solve this kind of problems.

To trace out the adulteration in *olive oil* using combined approach of molecular biology and bioinformatics based on unique SNPs present in conserved DNA sequence of plastid genomes of sunflower, canola and olive has been already performed. In general, plastid/chloroplasts are miniature organelles (approx. 5 X 3 μm in size) enclosed in double layer membranes. They are present in abundance (10–100 per cell, and each plastid contains about 100 copies of circular plastid genomes, average size 150 kb) and there is probability that most of the plastid organelles may be left intact due to miniature size when cold pressed to extract the oil from seeds of olive, canola and sunflower.

Moreover, plastid DNA present in extracted oil could be safe from nucleases activities due to double layer membranes and present in large number of copies in comparison to 1–2 nuclear genomic DNA which may be more prone to degradation.

In addition, a new chloroplast marker represents a valuable tool to assess the level of olive intercultivar plastome variation for use in population genetic analysis, phylogenesis, cultivar characterisation and DNA food tracking is recommended.

In summary, molecular biological techniques have become an every-day tool to solve a number of problems and questions in the section of varietal/species identification, fraud, traceability and paternity analysis.

10. References

Alba, V.; Sabetta, W. ; Blanco, A. ; Pasqualone, A. & Montemurro, C. (2009). Microsatellite markers to identify specific alleles in DNA extracted from monovarietal virgin *olive oils*, *Eur Food Res Techno* DOI 10.1007/s00217-009-1062-8

Angiolillo, A. ; Mencuccini, M. & Baldoni, L. (1999). Olive genetic diversity assessed using amplified fragment length polymorphisms, *Theor. Appl. Genet.* Vol.98, pp. 411–421

Araghipour, N.; Colineau, J.; Koot, A.; Akkermans, W.; Moreno Rojas, J. M.; Beauchamp, J.; Wisthaler, A.; Meark, T. D.; Downey, G.; Guillou, C. ; Mannina, L. & van Ruth, S.

(2008). Geographical origin classification of *olive oils* by PTR-MS, *Food Chem.* Vol.108, pp. 374–383

Arvanitoyannis, I. S. & Vlachos, A. (2007). Implementation of physicochemical and sensory analysis in conjunction with multivariate analysis towards assessing *olive oil* authentication/adulteration, *Crit. Rev. Food Sci. Nutr.* Vol.47, pp. 441–498

Ayed, R.B.; Grati-Kamoun, N.; Moreau, F. & Rebaï, A. (2009). Comparative study of microsatellite proWles of DNA from oil and leaves of two Tunisian olive cultivars, *Eur Food Res Technol.* Vol. 229, pp. 757–762

Baldoni, L. & Belaj, A. (2009). *Olive*, In: Vollmann J, Rajean I (eds) Oil crops. Handbook of plant breeding, vol 4. Springer Science Business Media, New York, pp. 397–421. doi 10.1007/978-0-387-77594-4_13

Baldoni, L.; Tosti, N.; Ricciolini, C.; Belaj, A.; Arcioni, S.; Pannelli, G.; Germana, M.A.; Mulas, M. & Porceddu, A. (2006). Genetic structure of wild and cultivated olives in the central Mediterranean basin, *Ann Bot.* Vol.98, pp. 935–942

Bartolini, G. (2008). *Olea databases*, Available at: http://www.oleadb.it

Belaj, A.; Munˉoz-Diez, C.; Baldoni, L.; Porceddu, A.; Barranco, D. & Satovic, Z. (2007). Genetic diversity and population structure of wild olives from the North-western Mediterranean assessed by SSR markers, *Ann Bot.* Vol.100, pp. 449–458

Botstein, D.; White, R.L.; Skolnick, M. & Davis, R.W. (1980). Construction of a genetic map in man using restriction fragment length polymorphisms, *Am J Hum Genet.* Vol.32, pp. 314–331

Bracci, T.; Busconi, M.; Fogher, C. & Sebastiani, L. (2011). Molecular studies in olive (Olea europaea L.): overview on DNA markers applications and recent advances in genome analysis, *Plant Cell Rep.* DOI 10.1007/s00299-010-0991-9

Breton, C.; Claux, D.; Metton, I.; Skorski, G. & Bervillè, A. (2004). Comparative Study of Methods for DNA Preparation from *Olive oil* Samples to Identify Cultivar SSR Alleles in Commercial Oil Samples: Possible Forensic Applications, *J. Agric. Food Chem,*Vol.52, No.3, pp. 531–537

Busconi, M.; Sebastiani, L. & Fogher, C. (2006). Development of SCAR markers for germplasm characterisation in olive tree (Olea europaea L.), *Mol Breed.* Vol.17, pp. 59–68

Busconi, M.;, Foroni, C.; Corradi, M.; Bongiorni, C.; Cattapan, F. & Fogher, C. (2003). DNA extraction from *olive oil* and its use in the identification of the production cultivar, *Food Chem.;* Voh.83, No.1, pp. 127-134

Cantini, C.; Cimato, A. & Sani, G. (1999). Morphological evaluation of olive germplasm present in Tuscany region, *Euphytica.* Vol.109, pp. 173– 181

Carriero, F.; Fontanazza, G.; Cellini, F. & Giorio, G. (2002). Identification of simple sequence repeats (SSRs) in olive (*Olea europaea* L.), *Theor Appl Genet.* Vol.104, pp. 301–307

Cerretani, L.; Bendini, A.; Del Caro, A.; Piga, A.; Vacca, V.; Caboni, M. F. & Gallina Toschi, T. (2006). Preliminary characterisation of virgin *olive oils* obtained from different cultivars in Sardinia, *Eur. Food Res. Technol.* Vol.222, pp. 354–361

Christopoulou, E.; Lazakari, M.; Komaitis, M. & Kaselimis, K. (2004). Effectiveness of determination of fatty acids and triglycerides for the detection of adulteration of *olive oils* with vegetable oils, *Food Chem.* Vol.84, pp. 463–474

Christy, A. A.; Kasemsumran, S.; Du, Y. & Ozaki, Y. (2004). The detection and quantification of adulteration in *olive oil* by near-infrared spectroscopy and chemometrics, *Analytical Sciences*, Vol.20, pp. 935–940

Cichelli, A. & Pertesana, G.P. (2004). High-performance liquid chromatographic analysis of chlorophylls, pheophytins and carotenoides in virgin *olive oils*: chemometric approach to variety classification, *J. Chromatogr. A.* Vol.1046, pp. 141–146

Cipriani, G.; Marrazzo, M.T.; Marconi, R.; Cimato, A. & Testolin, R. (2002). Microsatellite markers isolated in olive (Olea europaea L.) are suitable for individual fingerprinting and reveal polymorphism within ancient cultivars, *Theor. Appl. Genet.* Vo.104, pp. 223-228

Consolandi, C.; Palmieri, L.; Severgnini, M.; Maestri, E.; Marmiroli, N.; Agrimonti, C.; Baldoni, L.; Donini, P.; De Bellis, G.; Castiglioni, G.; (2008). A procedure for *olive oil* traceability and authenticity: DNA extraction, multiplex PCR and LDR-universal array analysis, *Eur. Food Res. Technol.* Vol.227, pp. 1429–1438

Council Regulation (EC) 510/2006, March 20, 2006

Cresti, M.; Linskens, H.F.; Mulcahy, D.L.; Bush, S.; di Stilio, V.; Xu, M.Y.; Vigani, R. & Cimato, A. (1997). Comunicación preliminar sobre la identificación del DNA de las hojas y el aceite de oliva de, *Olea europaea, Olivae*. Vol.69, pp. 36-37

De la Rosa, R.; James, C.M. & Tobutt, K.R. (2004). Using microsatellites for paternity testing in olive progenies, *Horticscience*. Vol.39, No.2, pp. 351–354

De la Rosa, R.; Angiolillo, A.; Rallo, L.; Guerrero, C.; Besnard, G.; Berville´, A.; Martin, A.; Baldoni, L. (2003). A first genetic linkage map of olive (*Olea europaea* L.) cultivars using RAPD and AFLP markers. *Theor. Appl. Genet.* Vol.*106*, pp. 1273-1282

De La Rosa, R.; James, C.M. & Tobutt, K.R. (2002). Isolation and characterization of polymorphic microsatellites in olive (Olea europaea L.) and their transferability to other genera in the Oleaceae, *Molecular Ecology Notes*, Vol.2, pp. 265-267.

De la Torre, F.;, Bautista, R.; Cánovas, F.M. & Claros, G. (2004). Isolation of DNA from *olive oil* and oil sediments: application in oil fingerprinting, *Food. Agric. Envir.*Vol. 2, No.1, pp. 84–9

Diaz, A.; De la Rosa, R.; Rallo, P.; Munoz-Diez, C.; Trujillo, I.; Barranco, D.; Martin, A. & Belaj, A. (2007). Selections of an olive breeding program identified by microsatellite markers, *Crop Science*, Vol.47, No.6, pp. 2317–2322

Diaz, A.; Martin, A.; Rallo, P.; Barranco, D. & De la Rosa, R. (2006). Self-incompatibility of 'Arbequina' and 'Picual' olive assessed by SSR markers, *J Am Soc Hortic Sci.* Vol.131, pp. 250–255

Doveri, S.; Gil, F.S.; Di´az, A.; Reale, S.; Busconi, M.; Da Câmara, M.A.; Marti´n, A.; Fogher, C.; Donini, P. & Lee, D. (2008). Standardization of a set of microsatellite markers for use in cultivar identification studies in olive (Olea europaea L.), *Scientia Horticulturae*. Vol.116, pp. 367–373

Doveri, S.; O'Sullivan, D. M.; & Lee, D. (2006). Non-concordance between genetic profiles of *olive oil* and fruit: A cautionary note to the use of DNA markers for provenance testing, *Journal of Agricultural and Food Chemistry*, Vol.54, No.24, pp. 9221– 9226

Fabbri, A.; Lambardi, M. & Ozden-Tokatli, Y. (2009). *Olive breeding*. In: Mohan Jain S, Priyadarshan PM (eds) Breeding plantation tree crops: tropical species. Springer Science, Business Media LLC, New York, pp. 423–68

Fendri, M.; Trujillo, I.; Trigui, A.; Rodriguez-Garcia, I.M. & De Dios Alche Ramirez, J. (2010). Simple sequence repeat identification and endocarp characterization of olive tree accessions in a Tusinian germplasm collection. *Hortscience*. Vol.45, pp. 1429–1436

Giuffrida, D.; Salvo, F.; Salvo, A.; La Pera, L. & Dugo, G. (2007). Pigments composition in monovarietal virgin olive oils from various Sicilian olive varieties, *Food Chem.* Vol.101, pp. 833–837

Gonzalez, J.; Remaud, G.; Jamin, E.; Naulet, N. & Martin, G.G. (1999). Specific natural isotope profile studied by isotope ratio mass spectrometry (SNIP-IRMS): (13)C/(12)C ratios of fructose, glucose, and sucrose for improved detection of sugar addition to pineapple juices and concentrates, *J. Agric. Food Chem.* Vol.47, pp. 2316_2321

Green, P.S. (2002). A revision of Olea L. (Oleaceae), Kew Bull. Vol. pp. 57:91–140

Giménez, M.J.; Pistón, F.; Martín, A. & Atienza, S.G. (2010). Application of real-time PCR on the development of molecular markers and to evaluate critical aspects for olive oil authentication, *Food Chemistry*, Vol.118, pp. 482–487

Hernández, P.; de la Rosa, R.; Rallo, L.; Martin, A. & Dorado, G. (2001). First evidence of a retrotransposon-like element in olive (*Olea europaea*): Implications in plant variety identification by SCAR marker development, *Theor. Appl. Genet.* Vol.102, pp. 1082-1087

http://www.internationaloliveoil.org

http://www.ncbi.nlm.nih.gov

Intrieri, M. C.; Muleo, R. & Buiatti, M. (2007). Chloroplast DNA polymorphisms as molecular markers to identify cultivars of *Olea europaea* L., *Journal of Horticultural Science and Biotechnology*, Vol.82, No.1, pp. 109–113

Jacoboni, N. & G. Fontanazza. (1981). Cultivar. *REDA, L'Olivo, Roma.* pp. 7-52

Japon-Lujan, R.; Ruiz-Jimenez, J. & Luque de Castro, M. D. (2006). Discrimination and classification of olive tree varieties and cultivation zones by biophenol contents, *J. Agric. Food Chem.* Vol.54, pp. 9706–9712

Kumar, S.; Kahlon, T. & Chaudhary, S. (2011). A rapid screening for adulterants in olive oil using DNA barcodes, *Food Chemistry.* Vol.127, pp. 1335–1341

Luykx, D.M.A.M. & van Ruth, S. M. (2008). An overview of analytical methods for determining the geographical origin of food products, *Food Chemistry.* Vol.107, No.2, pp. 897–911

Mahjoub-Haddada, F.; Manai, H.; Daoud, D.; Fernandez, X.; Lizzani-Cuvelier, L. & Zarrouk, M. (2007). Profiles of volatile compounds from some monovarietal Tunisian virgin olive oils. Comparison with French PDO, *Food Chem.* Vol.103, pp. 467–476

Mannina, L.; Dugo, G.; Salvo, F.; Cicero, L.; Ansanelli, G.; Calcagni, C. & Segre, A. (2003). Study of the cultivar-composition relationship in Sicilian olive oils by GC, NMR, and statistical methods, *J. Agric. Food Chem.* Vol.51, pp. 120–127

Martı´n, A.; Rallo, P.; Dorado, G.; Valpuesta, V.; Botella, M.A. & De la Rosa, R. (2005). *Utilizacio´n de marcadores en la mejora genetic del olivo.* In: Rallo L, Barranco D, Caballero JM, del Rio C, Martı´n A, Tous J, Trujillo I (eds) Variedades del olivo en Espan~a, Junta de Andaluci´a. MAPA and Ediciones Mundi-Prensa, Madrid

Martin, G.G.; Wood, R. & Martin, G.J. (1996). Detection of added beet sugar in concentrated and single strength fruit juices by deuterium nuclear magnetic resonance (SNIFNMR method): collaborative study, *J. AOAC* Vol.79, pp. 917-928

Mariotti, R.; Cultrera, N.G.M.; Munoz-Diez, C.; Baldoni, L. & Rubini, A. (2010). Identification of new polymorphic regions and differentiation of cultivated olives (Olea europaea L.) through platome sequence comparison, *BMC Plant Biol*, Vol.10, pp. 211

Martins-Lopes, P.; Gomes, S.; Santos, E.; & Guedes-Pinto, H. (2008). DNA markers for Portuguese *olive oil* fingerprinting, *Journal of Agricultural and Food Chemistry*, Vol.56, pp. 11786–11791

Mekuria G.T.; Sedgley, M.; Collins, G. & Lavee, S. (2002). Development of a sequence-tagged site for the RAPD marker linked to leaf spot resistance in olive, *J Am Soc Hortic Sci*. Vol.127, pp. 673–676

Montemurro, C.; Pasqualone, A.; Simeone, R.; Sabetta, W. & Blanco, A. (2008). AFLP molecular markers to identify virgin *olive oil*s from single Italian cultivars, *Eur J Food Technol*. Vol.226, pp. 1439–1444

Montealegre, C.; Marina Alegre, M.L. & garci´a-Ruiz, C. (2010). Traceability Markers to the Botanical Origin in *Olive oils*, *J. Agric. Food Chem*. Vol.58, pp. 28–38

Mookerjee, S.; Guerin, J.; Collins, G.; Ford, C. & Sedgley, M. (2005). Paternity analysis using microsatellite markers to identify pollen donors in an olive grove, *Theor Appl Genet*, Vol.111, pp. 1174–1182

Morgante, M. & Olivieri, A.M. (1993). PCR-amplified microsatellites as markers in plant genetics, *Plant J*, Vol.3, pp. 175–182

Mueller, T. (1991). *The trade in adulterated olive oil*. Slippery Business. The New Yorker, 10 August, Letter from Italy: <http://www.newyorker.com/reporting/2007/08/13/070813fa_fact_mueller?curr entPage=al#ixzz0rkvkKfW0>.

Muzzalupo, I. & Perri, E. (2002). Recovery and characterization of DNA from virgin *olive oil*, *Eur Food Res Technol* 214 (6):528–31

Nei, M. (1973). Analysis of Gene Diversity in Subdivided Populations, *Proc Natl Acad Sci USA*. Vol.70, pp. 3321–3323

Owen, R. W.; Haubner, R.; Würtele, G.; Hull, W. E.; Spiegelhalder, B.; & Bartsch, H. (2004). Olives and *olive oil* in cancer prevention, *European Journal of Cancer Prevention*, Vol.13, pp. 319–326

Pafundo, S.; Busconi, M.; Agrimonti, C.; Fogher, C. & Marmiroli, N. (2010). Storage-time effects on *olive oil* DNA assessed by Amplified Fragments Length Polymorphisms, *Food Chemistry*, Vol.123, pp. 787–793

Pafundo, S.; Agrimonti, C.; Maestri, E. & Marmiroli, N. (2007). Applicability of SCAR markers to food genomics: *olive oil* traceability, *J. Agric. Food Chem*. Vol.55, pp. 6052–6059

Pafundo, S.; Agrimonti, C. & Marmiroli, N. (2005). Traceability of Plant Contribution in *Olive oil* by Amplified Fragment Length Polymorphisms, *J. Agric. Food Chem*. Vol.53, pp. 6995-7002

Palmieri, L.; Doveri, S.; Marmiroli, N. & Donini, P. (2004). Molecular marker characterisation of *olive oil* variety composition and SNP development in olive. Proceedings of XII International Plant and Animal Genome Conference, San Diego, CA, pp. 305

Pasqualone, A.; Montemurro, C.; Summo, C.; Sabetta, W.; Caponio, F. & Blanco, A. (2007). Effectiveness of microsatellite DNA markers in checking the identity of Protected

Designation of Origin extra virgin *olive oil*, *J. Agric. Food Chem.* Vol.55, pp. 3857–3862

Pasqualone, A.; Caponio, F. & Blanco, A. (2001). Inter-simple sequence repeat DNA markers for identification of drupes from different *Olea europaea* L. cultivars, *Eur. Food Res. Technol*, Vol.213, pp. 240- 243

Paran, I. & Michelmore, R. (1993). Development of reliable PCR based markers linked to downy mildew resistance genes in lettuce, *Theor Appl Genet.* Vol.85, pp. 985–993

Peña, F.; Cárdenas, S.; Gallego, M.; & Valcárcel, M. (2005). Direct *olive oil* authentication: Detection of adulteration of *olive oil* with hazelnut oil by direct coupling of headspace and mass spectrometry, and multivariate regression techniques, *Journal of Chromatography A.* Vol.1074, Issues 1-2, pp. 215-221

Popping, B. (2002). The application of biotechnological methods in authenticity testing, *Journal of Biotechnology.* Vol.98 pp. 107-12

Rabiei, Z.; Tahmasebi Enferadi, S.; Saidi, A.; Patui, S. & Vannozzi, G.P. (2010). SSRs (Simple Sequence Repeats) amplification: a tool to survey the genetic background of *olive* oils, *Iranian Journal of Biotechnology.* Vol.8, No.1, pp. 24 – 31

Rabiei, Z. & Tahmasebi Enferadi, S. (2009). *Olive oil* fingerprinting using microsatellite markers. 6th Iranian Biotechnology Conference, Tehran, Iran 11 – 13 August 2009

Rabiei, Z.; Messina, R. & Testolin, R. (2006). Molecular Identification of Hazelnut-adulterated *olive oil*. 4th Euro Fed Lipid Congress October 1-4, 2006 Madrid, Spain

Rallo, P.; Dorado, G. & Martín, A. (2000). Development of simple sequence repeats (SSRs) in olive tree (Olea europaea L.), *Theor. Appl. Genet.* Vol.101, pp. 984-989

Roca, M.; Gandul-Rojas, B.; Gallardo-Guerrero, L. & Mı́nguez- Mosquera, M. I. (2003). Pigment parameters determining Spanish virgin *olive oil* authenticity: stability during storage, *J. Am. Oil Chem. Soc.* Vol.80, pp. 1237–1240

Rotondi, A.; Beghe, D.; Fabbri, A. & Ganino, T. (2011). *Olive oil* traceability by means of chemical and sensory analyses: A comparison with SSR biomolecular profiles, *Food Chemistry.* In press. doi:10.1016/j.foodchem.2011.05.122

Sabino Gil, F.; Busconi, M.; Da Caˆmara Machado, A. & Fogher, C. (2006). Development and characterization of microsatellite loci from Olea europaea, *Mol Ecol Notes.* Vol.6, pp. 1275–1277

Sanz-Cortes, F.; Parfitt, D. E.; Romero, C.; Struss, D.; Llacer, G. & Badenes, M. L. (2003). Intraspecific olive diversity assessed with AFLP, *Plant breeding*, Vol.122, pp. 173–177

Sarri, V.; Baldoni, L.; Porceddu, A.; Cultrera, N.G.M.; Contento, A.; Frediani, M.; Belaj, A.; Trujillo, I. & Cionini, P.G. (2006). Microsatellite markers are powerful tools for discriminating among olive cultivars and assigning them to geographically defined populations, *Genome*, Vol. 49, pp. 1606-1615

Sefc, K.M.; Lopes, M.S.; Mendonça, D.; Rodrigues dos Santos, M.; Laimer da Camara Machado, M. & Da Camara Machado, A. (2000). Identification of microsatellite loci in olive (Olea europea) and their characterization in Italian and Iberian olive trees, *Molecular Ecology.* Vol.9, pp. 1171-1173

Spaniolas, S.; Bazakos, C.; Awad, M. & Kalaitzis, p. (2008a). Exploitation of the chloroplast trnL (UAA) intron polymorphisms for the 554 authentication of plant oils by means of a Lab-on-a-chip capillary electrophoresis system, *J Agric Food Chem.* Vol.16, pp. 6886–6891

Spaniolas, S.; Bazakos, C.; Ntourou, T.; Bihmidine, S.; Georgousakis, A. & Kalaitzis, p. (2008b). Use of lambda DNA as a marker to assess DNA stability in olive oil during storage, *Eue food Res Technol*. Vol. 227, 175-179

Tena, N.; Lazzez, A.; Aparicio-Ruiz, R. & Garcı´a-Gonz_alez, D. L. (2007). Volatile compounds characterizing Tunisian Chemlali and Chetouivirgin *olive oils*, *J. Agric. Food Chem. Vol.*55, pp. 7852–7858

Torres VazFreire, L.; Gomes da Silva, M. D. R. & Costa Freitas, A. M. (2009). Comprehensive two-dimensional gas chromatography for fingerprint pattern recognition in *olive oils* produced by two different techniques in Portuguese olive varieties Galega Vulgar, Cobrancosa and Carrasquenha, *Anal. Chim. Acta*. Vol.633, pp. 263–270

Tura, D.; Failla, O.; Bassi, D.; Pedo, S. & Serraiocco, A. (2008) Cultivar influence on virgin olive (Olea europaea L.) oil flavor based on aromatic compounds and sensorial profile, *Sci. Hortic.* Vol.118, pp. 139–148

Vos, P.; Hogers, R.; Bleeker, M.; Reijans, M.; Van de Lee, T.; Hornes, M.; Frijters, A.; Pot, J.; Peleman, J.; Kuiper, M.; & Zabeu, M. (1995). AFLP: a new technique for DNA fingerprinting, *Nucl. Acids Res.* Vol.23, pp. 4407–4414

Weissbein, S.(2006). *Characterization of new olive (Olea europea L.) varieties response to irrigation with saline water in the Ramat Negev area*. Master thesis. Ben-Gurion University of the Negev, Israel

Wiesman, Z. (2009). *Desert Olive oil Cultivation: Advanced Biotechnologies* (first edition), Elsevier Inc.; printed and bounded in United States of America, ISBN: 978-0-12-374257-5

Williams, J.G.; Kubelik, A.R.; Livak, K.J.; Rafalski, J.A. & Tingey, S.V. (1990). DNA polymorphisms amplified by arbitrary primers are useful as genetic markers, *Nucleic Acids Res.* Vol.18, pp. 6531–6535

Zietkiewicz, E.; Rafalski, A. & Labuda, D. (1994). Genome fingerprinting by simple sequence repeat (SSR)-anchored polymerase chain reaction amplification, *Genomics*. Vol.20, pp. 176–183

Wu, S.; Collins, G. & Sedgley, M. (2004). A molecular linkage map of olive (Olea europaea L.) based on RAPD, microsatellite, and SCAR markers, *Genome*. Vol.47, pp. 26–35

9

Quality Evaluation of Olives, Olive Pomace and Olive Oil by Infrared Spectroscopy

Ivonne Delgadillo, António Barros and Alexandra Nunes
Department of Chemistry, QOPNA Research Unit,
University of Aveiro
Portugal

1. Introduction

Olive oil extraction starts by crushing olives and ends by obtaining olive oil, vegetative water and partially de-oiled olive pomace (Petrakis 2006; Di Giovacchino 2000). In the industry it is important to know the amount of oil and water present in both olive fruits and olive pomace. In fact, the amount of oil is the parameter that establishes the price of raw materials and by-products and is critical for the optimization of extraction procedures.

There are publications compiling the various technological aspects of olive oil production, its quality, authenticity, chemical composition and the numerous analytical methodologies used for its characterization (Boskou 2006; Hardwood & Aparicio 1999; Hardwood & Aparicio 2000 Kiritsakis 1998; EC Reg No 2568/1991 and its amendment EC Reg No 1989/2003).

The intrinsic characteristics of the production demand fast decisions based on rapid analytical results. Therefore, conventional analytical determinations of oil, water and acid value should be replaced by short time or real-time/in-line measurements. Rapid characterization of raw material allows the selection of olives according to quality, enabling the production of higher quality oils.

Nowadays, infrared spectroscopy has become widely used as a non-invasive tool for fast analyses with less to no sample pre-preparation. There are numerous publications on the use of infrared spectroscopy for the analysis of oils, some of them will be referred, later in this document.

Baeten et al. (2000) published a paper on infrared and Raman spectroscopies and their potential for olive oil analysis. They described the instrumental techniques, interpretation of the spectra, data treatment and present potential applications.

This chapter reviews various applications of infrared spectroscopy for the analysis of olive oil, presents some results of the authors' work and emphasizes that infrared spectroscopy coupled with proper chemometric tools is an advantageous instrument, to be used in the industry, for olive quality evaluation and olive oil characterization.

2. Overview of the olive oil quality parameters

As mentioned previously the most relevant parameters for olives and olive pomace are fat and water content. Another important parameter is the free fatty acids (FFA) content of the oil of the fruit, which will determine the acid value of the produced olive oil .

Olive oil, after its extraction, classification and quality evaluation should be labeled and priced. Quality evaluation and authenticity are based on organoleptic assessment and chemical characterization according to the European Commission Regulation (EC Reg No 2568/1991 and its later amendments EC Reg No 1989/2003), the Codex Alimentarius Norm (Codex Alimentarius Commission Draft, 2009) and International Olive Oil Council (IOOC) Trade standards (IOC/T.15/NC n° 3/Rev.4, 2009). Usual adulterations of olive oil are the addition of olive residue oil, refined olive oil, low-price vegetable oils, and even mineral oil (Wahrburg et al. 2002; Dobarganes & Marquez-Ruiz 2003).

IOOC standards for olive oil and olive pomace oils contain a set of values and limits for the following parameters: fatty acid composition, *trans* fatty acid content, sterol composition and total sterol content, content of erythrodiol + uvaol, wax content, aliphatic alcohol content, stigmastadiene content, 2-glyceryl monopalmitate, unsaponifiable matter, free acidity, peroxide value, specific absorbance in ultra-violet, moisture and volatile matter, insoluble impurities in light petroleum, flash point, trace metals, α-tocopherol, traces of heavy metals and traces of halogenated solvents (IOC/T.15/NC n° 3/Rev.4, 2009). Olive oil chemical characterization involves complex, time consuming and expensive analytical methodologies, which also destroy the sample.

3. Interpreting and using the infrared spectra

Spectroscopic techniques are neither invasive nor sample destructive and may contribute to rapid quality and authenticity evaluation, with low operating cost. From a physicochemical point of view, infrared spectroscopy is based on the vibrational transitions occurring in the ground electronic state of the molecules. The infrared absorption requires a change of the intrinsic dipole moment with the molecular vibration. The regions of the infrared spectrum which are used for applications in food analysis are: mid-infrared (MID-IR) and near-infrared (NIR).

Mid-infrared spectra present well resolved bands showing absorbances of varying intensity in the range of 4000 to 400 cm^{-1} originating from the fundamental vibrations.

Figure 1 shows a representative olive oil spectrum in the 4000 – 900 cm^{-1} region, where several characteristic bands related to lipid functional groups can be observed. In the 3100 - 2800 cm^{-1} spectral region appear the signals, assigned to C-H stretching mode from methylene and methyl groups of fatty acid and triacylglycerols. The low intensity peak near 3100 cm^{-1} may be explained by the CH=CH elongation and the signals of weak absorption around 2800 cm^{-1} are the result of the presence of secondary oxidation products, such as aldehydes and ketones. At 1800-1700 cm^{-1} the C=O stretching mode is found. The very strong band located at 1743 cm^{-1} can be ascribed to the triacylglycerol n-C=O ester group and a shoulder found around 1710 cm^{-1} is characteristic of the presence of free fatty acids (carboxylic n-C=O). The C-H deformation is detected between 1400 and 900 cm^{-1}, spectroscopic region which is also known as fingerprint region (Tay et al. 2002).

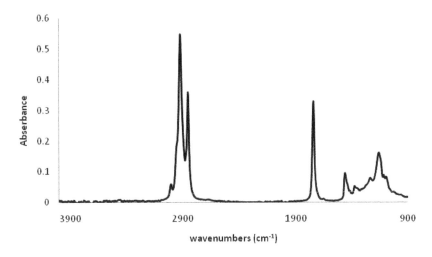

Fig. 1. Typical ATR-Mid-IR spectrum of olive oil.

Near-infrared spectra present less well resolved bands in the range of 14000 to 4000 cm^{-1} corresponding to overtones and combinations of fundamental vibrations.

Figure 2 a) and b) show NIR spectra of olive oil, hammer milled olive and olive pomace. The following main spectroscopic regions can be observed: the region between 9000 – 8000 cm^{-1}, can be ascribed to the second overtone of the C–H stretching vibration of modes of methyl, methylene and ethylene groups of fatty acids and triacylglycerols; the region between 7500 and 6150 cm^{-1} can be attributed to the first overtone of the O-H stretching vibrations; whereas the absorptions located around 6000 – 5700 cm^{-1} correspond to the first overtone of the C-H stretching vibration modes of methyl, methylene and ethylene groups; in next region bands between 5350 and 4550 cm^{-1} result from combinations of fundamentals of the C-H stretching vibration and of bands of water molecules (specially in olives and olive pomace); finally, the 4370 – 4260 cm^{-1} region can be ascribed to the C–H stretching combination of methyl and methylene groups (Galtier et al. 2007; Muick et al. 2004).

Several aspects must be considered when spectroscopic data are used in order to achieve multiple parameter determination, by direct analysis of spectra. A careful calibration framework should be devised, comprising: 1) an adequate sampling strategy, taking in account sampling variability and a suitable physicochemical range set; 2) a robust spectroscopic equipment in order to detect and quantify olive oil parameters in lower amounts, which is particularly important in the industrial in-line process; 3) a proper validation of the results given by infrared spectra and multivariate models; 4) a careful control of the outcome from the instrumental results and chemometric models, by employing control charts to evaluate the performance of the methodology and 5) a plan to address models sustainability through a periodic assessment of models performance, e.g. by performing traditional analysis and comparing to the outcome of the infrared spectra, in order to correct possible deviations. This last aspect is very important due to the nature of the samples (e.g. different harvest periods and samples origin, etc.) and equipment efficacy.

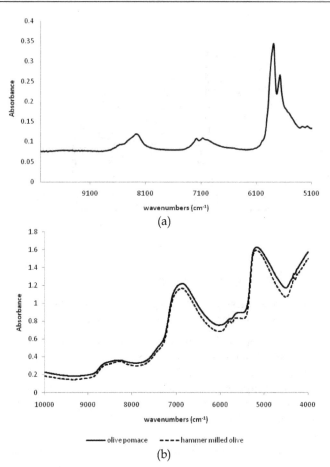

Fig. 2. Typical NIR spectra of (a) olive oil and (b) hammer milled olive and olive pomace.

4. Quantification of chemical parameters in olive oil

To ensure a more reliable control of every step in the extraction process, a good sampling system is necessary. For industrial in-process analysis of olive oil, spectroscopic techniques (mostly, NIR and MID-IR spectroscopy) in tandem with chemometric methods are the corner-stone for quality control. According to Marquez et al. (2005) these techniques have shown a high potential as an alternative to time-consuming and expensive chromatographic or wet chemistry analysis. In fact the application of optical on-line NIR sensor for olive oil characterization is an appealing approach for real-time chemical evaluation, allowing the estimation of acid value, bitter taste (K_{255}) and fatty acids (Marquez et al. 2005). Near infrared spectroscopy has been valuable for the assessment of physicochemical parameters of vegetable oils (Sato 1994; Hourant et al. 2000; Takamura et al. 1995; Franco et al. 2006). Infrared spectroscopy (NIR and MID), has also been applied successfully for olive oil evaluation and geographical origin determination using chemometric approaches (Tapp et al. 2003; Iñón et al. 2003; Galtier et al. 2007). In addition, NIR technique was employed to detect fraudulent addition of other vegetable oils and olive pomace oil, to virgin olive oil (Wesley et al. 1995; Yang & Irudayaraj 2001; Doweny et al. 2002; Vlachos et al. 2006); and for olive oil authentication (Bertran et al. 2000; Downey et al. 2003; Woodcock et al. 2008). The chemical characterization of fatty acids and sterols (Ollivier et al. 2006; Ollivier et al. 2003; Aranda et al. 2004; Leardi and Paganuzzi 1987), is useful for assessing quality and authenticity.

The authors attempted to apply NIR for the quantification of fatty acids, sterols and wax in an industrial scenario (for in-line analysis). 40 chemically characterized olive oil samples, from different origins, were used for this study. Partial Least Squares regression (PLS1) was applied in combination with NIR spectra in the 9000 - 4500 cm^{-1} range. Model validations were carried out using Monte-Carlo Cross-Validation (MCCV) (500 runs to evaluate models robustness); the predictive power of each one of the models were assessed through the computation of 1) Root Mean Square Error of Cross Validation (RMSECV); 2) Root Mean Square Error of Prediction (RMSEP); 3) Coefficient of Determination (R^2) and 4) the cross-validated coefficient of determination (Q^2). At the same time several data pre-treatments were tested in order to find the most suitable ones (predictive ability).

A summary of the results obtained is described in Table 1. Two data pre-treatments were selected: 1) Standard Normal Deviate (SNV) and 2) 2nd derivative computed with the Savistzky-Golay procedure, using a 2nd degree polynomial with 11 points (5+5+1). As it can be seen from the table the obtained models have from reasonable to good predictive power.

Parameter	Range	LV	Q^2	R^2	RMSECV (%)	RMSEP (%)	Pre-treatment
Fatty acids							
16:0	9.0 - 12.1 %	4	0.88	0.82	0.270	0.570	2nd derivative
16:1	0.7 - 1.3 %	5	0.88	0.82	0.050	0.101	2nd derivative
18:0	2.8 - 3.8 %	9	0.81	0.98	0.123	0.216	SNV
18:1	73.3 - 78.9 %	9	0.86	0.98	0.521	1.47	SNV
18:2	5.3 - 9.0 %	8	0.91	0.99	0.263	0.32	SNV
18:3	0.7 - 0.8 %	4	0.64	0.81	0.027	0.038	2nd derivative
24:0	0.0 - 0.1 %	7	0.76	0.80	0.021	0.087	SNV
Sterols							
Campesterol	2.97 - 3.64 %	7	0.82	0.98	0.070	0.131	2nd derivative
Cholesterol	0.07 - 0.27 %	6	0.85	0.96	0.020	0.052	2nd derivative
Stigmastenol	0.23 - 0.33 %	9	0.80	0.96	0.012	0.027	SNV
Stigmasterol	0.52 - 1.21 %	8	0.85	0.96	0.088	0.106	SNV
Sitosterol	94.6 - 98.3 %	7	0.63	0.99	0.111	0.175	2nd derivative
Total Sterols	1434 - 1636 (mg/kg)	9	0.70	0.96	23.90	57.0	SNV
Others							
Wax	59.42 - 240.67 (mg/kg)	5	0.71	0.95	29.20	37.7	2nd derivative

Nomenclature: palmitic acid (16:0), palmitoleic acid (16:1), stearic acid (18:0), oleic acid (18:1), linoleic acid (18:2), linolenic acid (18:3), lignoceric acid (24:0).

Table 1. Fatty acids, sterols and wax determination for olive oil by NIR and PLS1 regression in the spectral range of 9000 – 4500 cm^{-1}.

Several factors may reduce the predictive ability of such modeling techniques: 1) low amount of some parameters present in the olive oil; 2) NIR instrument characteristics (selectivity, specificity, signal to noise ratio, etc.) and 3) sampling distribution. At this stage seven fatty acids can be estimated using NIR, but many more could be quantified. It will be necessary to add more samples with wider ranges of parameters in order to enhance the robustness of the models. Galtier et al. (2007) have managed to quantify 14 fatty acids, squalene and triacilglycerols.

The spectral profiles extracted from infrared spectra using chemometric methods could be in many cases a substitute for the traditional analysis, for olive oil overall characterization.

5. Identification and quantification of olive oil adulterants

Extra virgin olive oil is adulterated with oils of low quality or price. The natural variation due to geographical origin, weather effect during growth and harvesting makes the task of detecting adulteration difficult. Analytical methods applied in the examination of chemical composition include the determination of fatty acid profiles by gas liquid chromatography (Firestone et al. 1988), high-pressure liquid chromatography (Cortesi 1993; Mariani & Fedeli 1993), pyrolysis mass spectrometry (Goodacre et al. 1993), measurement of iodine values and many other determinations. Rapid, non destructive spectroscopic methods such as Raman (Davies et al. 2000), ultraviolet (Calapaj et al. 1993), MID-IR (Lai et al. 1995; Dupuy et al. 1996; Guillen & Cabo 1999; Küpper et al. 2001) and NIR (Wesley et al. 1995; Bewig et al. 1994; Sato 1994; Wesley et al. 1996; Hourant et al. 2000) have all been applied to quantify adulterants in olive oil.

NIR spectroscopy in tandem with PCA and PLS1 regression, as studied in this laboratory, was found to be suitable for the identification and quantification of adulterants (refined olive oil, sunflower oil, maize oil and soya oil) in virgin olive oil. Binary mixtures were prepared with extra virgin olive oil and each one of the selected potential adulterants. Different amounts of refined olive oil and 3 commercial oil samples (sunflower, soya and maize) were mixed with olive oil giving four different data sets. 25 samples were prepared for each data set (binary mixture), containing additions from 5 to 95 mL of adulterant. Additionally, for each adulterant, an independent prediction set of 8 samples was prepared.

NIR spectra from the samples were obtained with a Perkin Elmer Spectrum One NTS FT-NIR spectrometer. The data were recorded in the spectral range between $10000 - 4500$ cm^{-1}, by co-adding 30 scans with a resolution of 8 cm^{-1}. Each sample was acquired five times. PCA allowed the characterization of the sample relationships (scores plans) and the recovery of their sub-spectral profiles (loadings) (Jolliffe 2002). For this analysis, the $6100 - 4500$ cm^{-1} region was selected and each spectrum was SNV corrected. A calibration model was built for each adulterant and was validated with the external and independent prediction data sets.

Oil samples are distributed across PC1 axes according to the olive oil content (Figure 3a). The bands located at 4596, 4668 4704, 5880 and 6024 cm^{-1} are related with the samples with larger amount of olive oil (Figure 3b).

Parameters of the best calibration models built for each adulterant are shown in Table 2. The four calibration models were built in the spectroscopic region of 4536-6108 cm^{-1}.

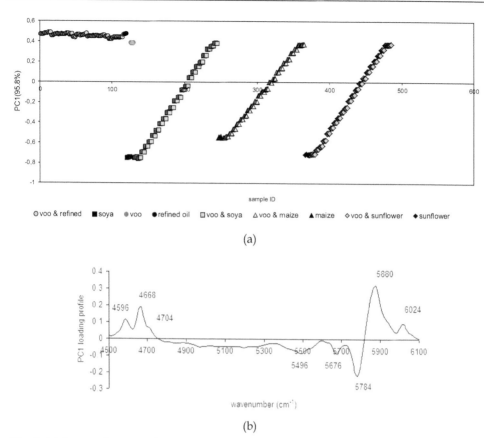

(a)

(b)

Fig. 3. (a) Scores plot of the first principal component (PC1), obtained for the set of virgin olive oil (voo) adulterated with refined oil, sunflower oil, maize oil and soya oil. (b) PC1 loading profile.

Spectral region (cm^{-1})	Pre-treatment	LV	RMSEP(%)	R^2	RMSEC(%)
Model 1: quantification of sunflower oil in virgin olive oil					
4536-6108	SNV	6	0.20	0.99	0.29
Model 2: quantification of maize oil in virgin olive oil					
4536-6108	SNV	5	0.23	0.99	0.34
Model 3: quantification of soya oil in virgin olive oil					
4536-6108	SNV	2	0.38	0.99	0.67
Model 4: quantification of refined olive in virgin olive oil					
4536-6108	1st derivative	7	2.81	0.99	3.74

Table 2. Statistical parameters obtained for the calibration models built for each adulterant.

The coefficient of determination higher than 0.99, and the low root mean squared error of prediction (RMSEP) suggest a good predictive power. PLS1 regression based calibration models were used to predict the percentage of adulterant in the independent data sets. Results presented in Table 3 suggest that NIR spectroscopy in tandem with PLS1 regression is suitable to detect and quantify adulteration with other edible oils (sunflower, soya, maize refined olive oil) in extra virgin olive oil up to 2% (w/w).

Prediction Sample nº	Observed Oil Content in Virgin Olive Oil (%)	Predicted Oil Content in Virgin Olive Oil (%)						
		Model 1 (Sunflower)		Model 2 (Maize)		Model 3 (Soya)		Model 4 (Refined Olive Oil)
P1	0.3	0.1	± 0.1	0.3	± 0.2	0.0	± 0.2	0.4	± 0.4
P2	2.0	1.9	± 0.1	1.8	± 0.3	1.0	± 0.1	1.7	± 1.9
P3	8.0	7.8	± 0.1	7.7	± 0.1	7.5	± 0.1	10.0	± 2.2
P4	16.0	16.1	± 0.1	15.6	± 0.1	15.2	± 0.1	14.9	± 0.7
P5	37.0	36.9	± 0.2	34.5	± 0.1	36.9	± 0.3	36.1	± 2.9
P6	58.0	57.9	± 0.2	56.1	± 0.1	57.7	± 0.1	56.9	± 0.6
P7	71.0	71.1	± 0.1	69.0	± 0.1	70.6	± 0.2	71.2	± 0.8
P8	87.0	86.9	± 0.1	84.9	± 0.2	83.3	± 0.1	82.6	± 1.9

Table 3. Predicted values using the calibration model built for each adulterant.

6. Quality evaluation of the olives at the oil extraction plant

6.1 Determination of oil and water content in olives and olive pomace

Information about olive quality is very important for the olive and olive oil producers as fruits with larger amounts of oil are highly priced. In addition to water content, fat content is also an important parameter for the optimization of the extraction procedures. Olive pomace can be re-extracted in the same industrial facilities or dried and sold in the form of dried olive pomace (O´Brien 2004).

Conventional oil and water analytical determinations could be replaced by real-time methods that avoid mixtures of high quality with low quality fruits. Muick and coworkers (2004) applied NIR and Raman spectroscopy to the determination of oil and water content in olive pomace. Later on, Bendini et al. (2007) were able to determine fat content, moisture and acid value directly in olives, using a Fourier Transform near-infrared (FT-NIR) instrument located in an industrial mill.

Here, the application of NIR in tandem with a multivariate regression method for the quantification of oil and water directly in fresh hammer milled olive and olive pomace samples is discussed (Barros et al. 2009). A total of 159 olive and olive pomace samples were used to build the calibration set. In order to validate the built models (for oil and water), 108 olive and olive pomace samples were used as independent set. In order to build the calibration models for the quantification of oil and water in hammer milled olive and olive pomace using NIR and PLS1 regression, a Monte Carlo Cross-Validation (MCCV) (Xu & Liang 2001) framework was used. This approach was needed for building robust calibration models for real-time industrial application.

The model for water content estimation was built by a preliminary spectrum pre-treatment by computing the 1st derivative, in order to minimize the baseline effect, followed by Standard Normal Deviate (SNV). A PLS1 regression model with 3 LVs (Latent Variables) was needed to obtain predictive power with a Q^2 of 0.96 and a relative RMSECV of 1.1%. The **b** vector plot for the water calibration models and the relationship between measured and predicted water values using a PLS1 are presented, respectively, in Figure 4a and Figure 4b.

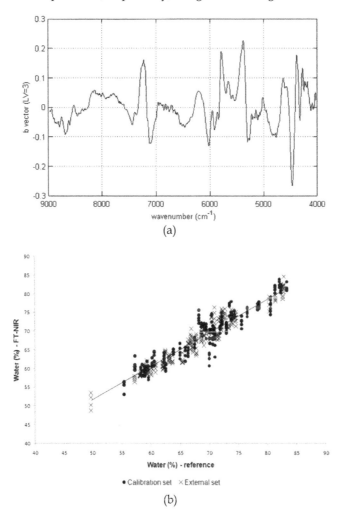

Fig. 4. (a) b vector plot for the water calibration and (b) relationship between measured and predicted water values using a PLS1 model with 3 Latent Variables (Reproduced with permission from Barros et al. 2009 © Springer 2009).

For oil calibration model the spectra pre-treatment was the same as for the water calibration model and, in this case, 3 LVs were needed to obtain predictive power with a Q² of 0.88 and a relative RMSECV of 2.64%. The **b** vector plot for the oil calibration models and the relationship between measured and predicted oil values are shown, respectively, in Figure 5a and Figure 5b.

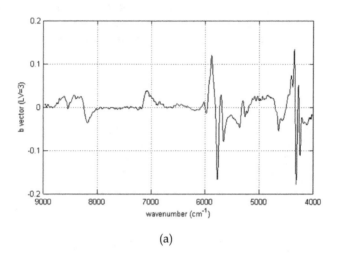

(a)

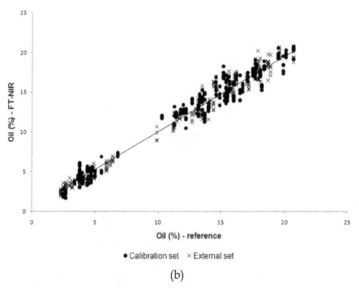

(b)

Fig. 5. (a) b vector plot for the oil calibration and (b) relationship between measured and predicted oil values using a PLS1 model with 3 Latent Variables (Reproduced with permission from Barros et al. 2009 © Springer 2009).

The models showed a good predictive power considering the nature (heterogeneity) of the milled olive fruits and olive pomace samples. NIR technique in tandem with PLS1 regression was found suitable for the quantification of these two important parameters. At industrial scale, the results show that NIR can be used for an extensive screening process of the olive fruits and olive pomace. In fact, when compared to classical approaches of analysis, this methodology is faster, allows larger number of samples in real-time and is environmental sustainable.

6.2 Acid value quantification in olives

Several factors may affect the olive characteristics and consequently its quality (Muick et al. 2004) specially the increase of free fatty acids (FFA) due to the action of lipases (Morelló et al. 2003). Consequently, the classification of olive oils based on their FFA content prior to processing is an important measure to improve and guarantee the production of good quality olive oil.

Previous works (Muick et al. 2003) report the application of Raman spectroscopy to the direct determination of FFA in milled olives. It is not possible to predict FFA content directly in the milled olive by NIR, probably because of the complexity of the matrix: kernel, pulp and skin.

The authors proposed a method for a rapid determination of free fatty acids in olive (Nunes et al. 2009). This procedure combines Soxhlet extraction for 30 minutes with MID-IR spectroscopy coupled to a multivariate regression method (PLS1). The oil extracted from olives (crushed with a hammer mill) was used for infrared analysis and for free fatty acids determination according to UNE 55030 (AENOR 55030:1961) protocol. MID-IR spectra were acquired by ATR in a Golden Gate accessory (one reflection), in the range of 4000 to 600 cm^{-1}. The data set comprising a total of 210 spectra (42 x 5) was imported into an in-house developed procedure for performing PLS1 (Helland 2001; Martens 2001; Wold et al. 2001).

Figure 6a, shows a linear correlation between the actual olive oil acid values and those estimated by the PLS1 model within the considered values range. The corresponding b vector profile shown in Figure 6b clearly identifies the band located at 1710 cm^{-1} (carboxylic n-C=O) as the most important one, related to the quantified olive oil acid value. Moreover, the band located at 1743 cm^{-1} (assigned to triacylglycerols n-C=O ester group), although weaker than the previous one, also contributes to the modeling of the olive oil acidity.

The PLS1 calibration model with a Monte Carlo Cross-Validation approach was built in the spectroscopic region of 1850-1600 cm^{-1} (with SNV pre-treatment, 4 LVs, a RMSECV(%) of 8.7 and a Q^2 of 0.97). It represents an optimized method for calibration of FFA in extracted olive oil and a proposal for an indirect but quick acid value determination in olives and consequently fruit quality. The short extraction time and the spectroscopic determination of FFA from the MID-IR spectra instead of the titration step (with the consequent decreasing use of reagents and analysis time), provide more reliable results and permit a tight sampling control.

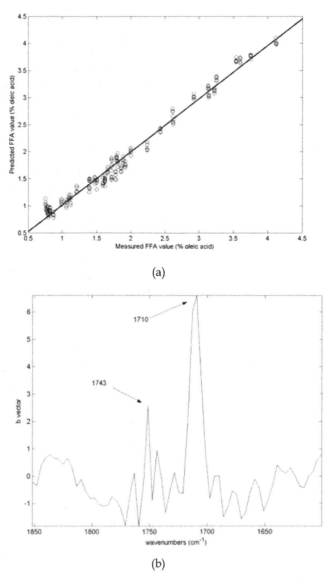

(a)

(b)

Fig. 6. (a) PLS1 regression relationship between actual and predicted value of olive oil acidity from the application of acidity calibration model and (b) the corresponding b vector plot (Reproduced with permission from Nunes et al. 2009 © Springer 2009).

7. Conclusions

The high sensitivity and reproducibility provided by the modern spectrometers allow in-depth studies of food systems, like olives and olive oil. The complexity of these matrices requires chemometric tools to extract both qualitative and quantitative information.

Infrared spectroscopic techniques have a potential in assisting and simplifying olive oil characterization. NIR spectroscopy in tandem with multivariate calibration models could provide a comprehensive chemical characterization of an olive oil sample for waxes, total sterols, sterol composition and free fatty acids composition.

NIR infrared may also contribute to the identification and qualification of adulterants in virgin olive oil (additions of refined olive oil, sunflower oil, maize oil and soya oil with "as low as 2% (w/w)).

Moreover, NIR and MID-IR spectroscopy as tool the advantage that it can be used to quantify oil and water content directly in olive and olive pomace and also to measure FFA directly in olives, allowing a quick quality evaluation that may reduce the processing time and cost.

The spectral profiles extracted from infrared spectra using chemometric methods could in many cases be a substitute for chromatographic and wet chemistry analysis, for olive oil overall characterization. Therefore, spectrometers of this type can be an important tool in modern olive oil analytical laboratory since they have so many advantages such as sensitivity, versatility (several type of analysis with only one equipment), real-time/in-line measurements, minimal sample preparation, relatively low cost implementation and high throughput.

8. Acknowledgment

Authors wish to acknowledge the support from QOPNA unit (project PEst-C/QUI/UI0062/2011). Alexandra Nunes thanks Fundação para a Ciência e a Tecnologia for her Post-PhD grant.

9. References

Aranda F., Gómez-Alonso S., Rivera del Álamo R.M., Salvador M.D. & Fregapane G. (2004). Triglyceride, total and 2-position fatty acid composition of Cornicabra virgin olive oil: Comparison with other Spanish cultivars. *Food Chem.* 86, (August 2004), pp. 485-492, ISSN 0308-8146

Asociación Española de Normalización y Racionalización (AENOR), Norma UNE 55030 (AENOR, Madrid, 1961)

Baeten V., Aparicio R., Marigheto N.A. & Wilson R.H. (1999). Olive oil analysis by infrared and Raman spectroscopy: methodologies and applications. In *Handbook of Olive Oil - Analysis and Properties*, J. L. Harwood & R. Aparicio (eds.). Aspen Publication, ISBN 0-8342-1633-7, London

Barros A., Nunes A., Martins J. & Delgadillo I. (2009). Determination of oil and water in olive and olive pomace by NIR and multivariate analysis. *Sens. & Instrumen. Food Qual.* 3, (July 2009), pp. 180–186

Bendini A., Cerretani L., Virgilio F.D., Belloni P., Lercker G. & Toschi T.G. (2007). In-process monitoring in industrial olive mill by means of FT-NIR. *Eur. J. Lipid Sci. Technol.* 109, (May 2007), pp. 498-504, ISSN 1438-9312

Bertran E., Blanco M., Coello J., Iturriaga H., Maspoch S. & Montolin I. (2000). Near infrared spectrometry and pattern recognition as screening methods for the authentication of virgin olive oils of very close geographical origins. *J. Near Infrared Spectrosc.* 8(1), pp. 45-52

Bewig K. M., Clarke A. D., Roberts C. & Unklesbay N. (1994). Discriminant Analysis of Vegetable Oils by Near-Infrared Reflectance Spectroscopy. *J. Am Oil Chem. Soc.* 71, pp. 195-200, ISSN 0003-021X

Boskou, D. (Ed.) (2006). Olive Oil, Chemistry and Technology, AOACS Press, ISBN 189399788X, Champaign: AOAC Press

Calapaj R., Chiricosta S. & Saija G. (1993). Evaluation of gas chromatographic and spectrophotometric analytical results to check the presence of seed oils in olive oil samples. *Riv. Ital. Sostanze Grasse* 70, pp. 585-594

Codex Alimentarius Commission (2009). Amended draft revised standard for olive oil and olive pomace oils. Report of the eighteenth session of the Codex Committee on fats and oils

Cortesi N. (1993). La cromatografía líquida de alta definición (HPLC) en los análisis del aceite de oliva. *OliVae* 45, pp. 40-42

Davies A. N., McIntyre P. & Morgan E. (2000). Study of the use of molecular spectroscopy for the authentication of extra virgin olive oils. part I: Fourier transform Raman spectroscopy. *Appl. Spectrosc.* 54, (December 2000), pp. 1864-1867, ISSN0003-3530

Di Giovacchino, L. (2000). Technological Aspects. In *Handbook of Olive Oil - Analysis and Properties*, J. L. Harwood & R. Aparicio (eds.), pp 17-56, Aspen Publication, ISBN 0-8342-1633-7, London

Dobarganes C. & Marquez-Ruiz G. (2003). Oxidized fats in foods. *Curr. Opin. Clin. Nutr. Metab. Care* 6, (March 2003), pp. 157-163, ISSN 1473-6519.

Downey G., McIntyre P. & Davies A. N. (2002). Detecting and quantifying sunflower oil adulteration in extra virgin olive oils from Eastern Mediterranean by visible and near-infrared spectroscopy. *J. Agric. Food Chem.* 50, (September 2002), pp. 5520-5525

Downey G., McIntyre P. & Davies A. N. (2003). Geographic classification of extra virgin olive oils from the eastern Mediterranean by chemometric analysis of visible and near-infrared spectroscopic data. *Appl. Spectrosc.* 57, (February 2003), pp. 158-163, ISSN 1943-3530

Dupuy N., Duponchel L., Huvenne J. P., Sombret P. & Legrand P. (1996). Classification of edible fats and oils by principal component analysis of Fourier transform infrared spectra. *Food Chem.* 57, (October 1996), pp. 245-251, ISSN 0308-8146

European Commission Regulation (EC) No 2568/1991 of 11 July 1991 on the characteristics of olive oil and olive pomace oil and the relevant method of analysis. *Off. J. European Union* L248, 5 September 1991

European Commission Regulation (EC) No 1989/2003 of 6 November 2003 amending regulation (EEC) No 2658/1991 on the characteristics of olive oil and olive pomace oil and the relevant method of analysis. *Off. J. European Union* L295, 13 November 2003

Firestone D., Carson K. L. & Reina R. J. (1988). Update on control of olive oil adulteration and misbranding in the United States. *J. Am. Oil Chem. Soc.* 65, (May 1988), pp. 788-792, ISSN 0003-021X

Franco D., Núñez M.J., Pinelo M. & Sineiro J. (2006). Applicability of NIR spectroscopy to determine oil and other physicochemical parameters in Rosa mosqueta and Chilean hazelnut. *Eur. Food Res. Technol.* 222, (February 2006), pp. 443-450

Galtier O., Dupuy N., Le Dréau Y.,Ollivier D., Pinatel C., Kister J. & Artaud J. (2007). Geographical origins compositions of virgin olive oils determined by chemometric analysis of NIR spectra. *Anal. Chim. Acta* 595, (July 2007), pp. 136-144

Goodacre R., Kell D.B. & Bianchi G. (1993). Rapid assessment of the adulteration of virgin olive oils by other seed oils using pyrolysis mass spectrometry and artificial neural networks. *J. Sci. Food Agric.* 63, (October 2003), pp. 297-307

Guillen M. D. & Cabo N. (1999). Usefulness of the Frequencies of some Fourier Transform Infrared Spectroscopic Bands in Evaluating the Composition of Edible Oil Mixtures. *Fett/Lipid* 101, (February 1999), pp. 71-76

Harwood J. & Aparicio R. (Eds.). (2000) *Handbook of Olive Oil - Analysis and Properties*, Aspen Publication, Aspen Publication, ISBN 0-8342-1633-7, London

Helland I.S. (2001). Some theoretical aspects of partial least squares regression. *Chemom. Intell. Lab. Syst.* 58, (October 2001), pp. 97-107, ISSN 0169-7439

Hourant P., Baeten V., Morales M. T., Meurens M. & R. Aparicio (2000). Oil and fat classification by selected bands of near-infrared spectroscopy. *Appl. Spectrosc.* 54, (August 2000), pp. 1168-1174

International Olive Oil Council Trade standards for olive oil and olive pomace oils. COI/T.15/NC nº3, 2009

Iñón F. A., Garrigues J.M., Garrigues S., Molina A. & Guardiã M. (2003). Selection of calibration set samples in determination of olive oil acidity by partial least squares–attenuated total reflectance–Fourier transform infrared spectroscopy. *Anal. Chim. Acta.* 489, (August 2003), pp. 59-75, ISSN 0003-2670

Jolliffe I.T. (2002). Principal component analysis. (2nd Ed.), ISBN 0-387-95442-2, New York: Springer

Kiritsakis A.K. (Ed.) (1998). *Olive Oil From the Tree to the Table*, Food and Nutrition Press, Inc., Trumbull Connecticut, USA

Küpper L., Heise H. M., Lampen P., Davies A. N. & McIntyre P. (2001). Authentication and quantification of extra virgin olive oils by attenuated total reflectance infrared spectroscopy using silver halide fiber probes and partial least-squares calibration. *Appl. Spectrosc.* 55, (May 2001), pp. 563-570

Lai Y. W., Kelmsley E. K. & Wilson R. H. (1995). Quantitative analysis of potential adulterants of extra virgin olive oil using infrared spectroscopy. *Food Chem.* 53, pp. 95-98, ISSN 0308-8146

Leardi R. & Paganuzzi V. (1987). Caratterizzazione dell'origine di oli di oliva extravergini mediante metodi chemiometrici applicati alla frazione sterolica. *Riv. Ital. Sostanze Grasse* 64, pp.131-136

Mariani C. & Fedeli E. (1993). La gas cromatografia nell'analisi dell'olio di oliva. *OliVae* 45, (November 1993), pp. 34-39

Marquez A. J., Díaz A. M. & Reguera M. I. P. (2005). Using optical NIR sensor for on-line virgin olive oils characterization. *Sens. & Act. B: Chemical.* 107, (May 2005), pp. 64-68, ISSN 0925-4005

Martens H. (2001). Reliable and relevant modelling of real world data: a personal account of the development of PLS regression. *Chemom. Intell. Lab. Syst.* 58, (October 2001), pp. 85-95, ISSN 0169-7439

Morelló J.R., Motilva M.J., Ramo T. & Romero, M.P. (2003). Effect of freeze injuries in olive fruit on virgen olive oil composition. *Food Chem.* 81, (June 2003), pp. 547-553, ISSN 0308-8146

Muick B., Lendl B., Molina-Díaz A. & Ayora-Cañada M. J. (2003). Direct reagent-free determination of free fatty acid content in olive oil and olives by Fourier transform Raman spectrometry. *Anal. Chim. Acta.* 487, (July 2003), pp. 211-220, ISSN 0003-2670

Muick B., Lendl B., Molina-Díaz A., Pérez-Villarejo L. & Ayora-Cañada M. J. (2004). Determination of oil and water content in olive pomace using infrared and Raman spectroscopy - A Comparative study. *Anal. Bioanal. Chem.* 379, (May 2004), pp. 35–41

Nunes A., Barros A., Martins J. & Delgadillo I. (2009). Estimation of olive oil acidity using FT-IR and partial least squares regression. *Sens. & Instrumen. Food Qual.* 3, (June 2009), pp. 187–191

O'Brien R.D. (2004). Fats and Oils. In: *Formulating and Processing for Applications*, 2nd ed., CRS Press, London, New York, Washington, DC

Ollivier D., Artaud J., Pinatel C., Durbec J.P. & Guérère M. (2003). Triacylglycerol and fatty acid compositions of French olive oils. Characterization by chemometrics. *J. Agric. Food Chem.* 51, (September 2003), pp. 5723-5731

Ollivier D., Artaud J., Pinatel C., Durbec J.-P. & Guérère M. (2006). Diferentiation of French virgin olive oil RDOs by sensory characteristics, fatty acid and triacylglycerol compositions and chemometrics. *Food Chem.* 97, (August 2006), pp. 382-393, ISSN 0308-8146

Petrakis C. (2006). Olive oil extraction, In *Olive Oil, Chemistry and Technology*, D. Boskou (Ed.), AOACS Press, ISBN 189399788X, Champaign, Ilinois

Sato T. (1994). Application of principal-component analysis on near-infrared spectroscopic data of vegetable oils for their classification. *J. Am. Oil Chem. Soc.* 71, (March 1994), pp. 293-298, ISSN 0003-021X

Takamura H., Hyakumoto N., Endo N. & Matoba T. (1995). Determination of lipid oxidation in edible oils by near infrared spectroscopy. *J. Near Infrared Spectrosc.* 3 (4), pp. 219-225

Tapp H.S., Defernez M. & Kemsly E.K. (2003). FTIR spectroscopy and multivariate analysis can distinguish the geographic origin of extra virgin olive oils. *J. Agric. Food Chem.* 51, (October 2003), pp. 6110-6115

Tay A., Singh R.K., Krisshnan S.S. & Gore J.P. (2002). Authentication of olive oil adulterated with vegetable oils using Fourier transform infrared spectroscopy. *Lebensm–Wiss U-Technol.* 35, (March 2002), 99-103

Vlachos N., Skopelitis Y., Psaroudaki M., Konstantinidou V., Chatzilazarou A. & Tegou E. (2006). Applications of Fourier transform-infrared spectroscopy to edible oils. *Anal. Chim. Acta.* 74, (July 2006), pp. 459–465, ISSN 0003-2670

Wahrburg U., Kratz M. & Cullen M. (2002). Mediterranean diet, olive oil and health. *Eur. J. Lipid Sci. Technol.* 104, (October 2002), pp. 698-705, ISSN 1438-9312

Wesley I. J., Barnes R. J. & McGill A. E. J. (1995). Measurement of adulteration of olive oils by near-infrared spectroscopy. *J. Am. Oil Chem. Soc.* 72, (March 1995), pp. 289-292, ISSN 0003-021X

Wesley I. J., Pacheco F. & McGill A. E. J. (1996). Identification of adulterants in olive oils. *J. Am. Oil Chem. Soc.* 73, (April 1996), pp. 515- 518, ISSN 0003-021X

Wold S., Sjöström M. & Eriksson L. (2001). PLS-regression: a basic tool of chemometrics. *Chemom. Intell. Lab. Syst.* 58, (October 2001), pp. 109-130, ISSN 0169-7439

Woodcock T., Downey G. & O'Donnell C.P. (2008). Confirmation of declared provenance of European extra virgin olive oil samples by NIR spectroscopy. *J. Agric. Food Chem.* 56, (November 2008), pp. 11520-11525

Xu Q.-S. & Liang Y.-Z. (2001). Monte Carlo cross validation. *Chemom. Intell. Lab. Syst.* 56, (April 2001), pp. 1-11, ISSN 0169-7439

Yang H. & Irudayaraj J. (2001). Comparison of near-infrared, Fourier transform-infrared, and Fourier transform-Raman methods for determining olive pomace oil adulteration in extra virgin olive oil. *J. Am. Oil Chem. Soc.* 78(9), (March 2000), pp. 889-895, ISSN 0003-021X

10

Cultivation of Olives in Australia

Rodney J. Mailer
Australian Oils Research
Australia

1. Introduction

Australia, by European standards, is a very young country with the first European settlers arriving as recently as 1788. Olives were not native to Australia but it took only a short time before the species was introduced. The first introduction of an olive tree to Australia was in 1800 (Spennemann, 2000), 12 years after the country was settled. Other importations have been recorded into New South Wales (NSW) including a tree planted by John Macarthur, one of Australia's pioneers and a man considered to be the father of the Australian sheep wool industry. A remaining olive tree still stands at Elizabeth Farm where he lived.

Despite the early start in the new settlement in NSW, little development occurred in that state over subsequent years. As the colony moved to other areas in Australia, olive production was spurred on by European immigrants particularly in the states of South Australia and Victoria. The NSW Department of Agriculture was formed in 1890 with an agenda to introduce new and useful species and study orchard farming and animal husbandry. The Department established experimental farms at sites throughout NSW including Wollongbar and Hawkesbury which became sites for evaluating olive production. In 1891 several Department of Agriculture research stations established schools and experimental farms including one at Wagga Wagga in Southern NSW, which included olive growing.

One of the most significant early developments for the olive industry was through the efforts of Sir Samuel Davenport (1818 – 1906), one of the early settlers of Australia, who became a landowner and parliamentarian in South Australia. His father was an agent of the "South Australia Company" in England and purchased land in South Australia. Samuel and his wife Margaret went to Australia in 1843 and ventured into mixed farming, almonds and vines. He tried sheep-farming and in 1860 he bought land near Port Augusta, SA, and turned to ranching horses and cattle. Davenport strongly promoted agriculture in South Australia and between 1864 and 1872 he published a number of papers, some concerning the cultivation of olives and manufacture of olive oil (en.wikipedia.org). In 1891 Davenport provided the NSW Department of Agriculture and other parts of the colony with olive cuttings from four cultivars, Verdale, Pigale, Blanquette and Bouquettier, from the south of France which were trialled for fruit production at the experimental farms.

In 1894, the farm at Wagga Wagga established orchards for evaluation of various fruits including plums, pears, persimmons and others. It was decided to establish a complete collection of olive cultivars within that orchard (Wagga Wagga Advertiser, 14 June 1894 from Spennemann 2000). Spennemann reports (2000) that by 1895, 8 acres of olives had been sown in Wagga Wagga "which now had the finest collection of cultivars in Australia" including many from California. By the turn of the century approximately 60 cultivars were present in the Wagga Wagga collection.

In future years significant studies were carried out on oil production and fruit pickling based on cultivars including *cvv*: Amelau, Blanquette, Bouquettier, Boutillan, Corregiola, Cucco, Dr Fiaschi, Gros Redondou, Macrocarpa, Nevadillo Blanco, Pigalle and Pleureur. Small scale commercial production and sales occurred after 1900 with the sale of olive oil and the distribution of olive cuttings for orchard development.

New cultivars continued to be introduced and the grove at Wagga Wagga expanded over subsequent years with several lines brought from Spain in 1932. Despite the excellent collection which had been established at Wagga Wagga, in 1959 it was decided to remove many of the trees due to low demand for the product. Although one of each of the cultivars was to be retained, subsequent loses through trees dying or being removed resulted in confusion about tree identification.

Fig. 1. One of over 100 trees and 60 cultivars planted at the Wagga Experimental Farm in 1891.

There was resurgence in interest in olive production in 1995 with the formation of the Australian Olive Association. At that time, Charles Sturt University, which had taken over ownership of the olive collection, made an attempt to resurrect the grove. The trees were severely pruned back from the massive size to which they had grown. A project funded by Rural Industries and Research Organisation (RIRDC) (Mailer & May, 2002) analysed DNA from leaves of the individual trees using RAPD analysis to attempt to identify the collection. This study was successful in naming many of the trees but for others there were no matches and identification was not possible. Some of the trees by this time had been named by areas in which the cuttings had been taken, such as Pera Bore or Hawkesbury Agricultural College, although logically, they were of European origin. At the same time, research on yield, oil content and oil quality was being carried out.

Based on this research, together with data from the original maps and planting diagrams, the Wagga Wagga orchard became the source of cuttings for some of Australia's leading nurseries. Many trees were propagated and sold to new growers. Despite the best attempts to ensure correct identification, many of these new trees were misidentified and created problems for new orchardists in future years.

Amelon	Dr Fiiaschi	Pecholine
Arecrizza	Frantojo	Pecholine de St Chamis
Atro Violacca	Gros Redoneaux	Pendulina
Attica	Hardy's Mammoth	Pera Bore
Attro Rubens	Hawkesbury Agric. College	Pigalle
Barouni	Large Fruited	Polymorpha
Belle d'Espagne	Lucca	Praecox
Big Spanish	Manzanillo No.14	Regalaise de Languedoc
Blanquette	Manzanillo No.2.	Regalis
Borregiola	Marcocarpa	Rubra
Bouchine	Nevadillo Blanco	Saloma
Bouquettier	O de Gras	Sevillano
Boutillon	Oblitza	Tarascoa
Columella	Oblonga	Verdale
Corregiolla	Oje Blanco Doncel	
Cucco	Olive de Gras	

Table 1. Olive Cultivars included in the historic Wagga Wagga Olive Grove. NB. *Names and spelling of cultivars are from the Spennemann report (1997). Some names are descriptive (e.g. large fruited) or the source of cuttings (e.g. Pera Bore).*

Despite an early start, for over 100 years olive production showed only minor indications of becoming a substantial crop in Australia. Olive oil production remained only a boutique industry with the bulk of olive products being imported, almost entirely from Spain, Italy and Greece. There were several feasibility studies carried out which indicated a potential for an olive industry. These included a report published by Farnell Hobman (1995), a Senior Research Officer with the South Australian Department of Primary Industries, on the economic feasibility of olive growing. This reported stimulated further interest.

Olives today are planted throughout Australia, from the most southern point of Western Australia to the northern tropical areas of Queensland (Fig. 2.). The trees have been found to be capable of surviving in a wide range of environments from hot tropical regions to the cold areas of Tasmania. Over many years, birds have spread seeds across the land around many of these established orchards and numerous feral trees now grow throughout olive production areas, reinforcing the suitability of the Australian environment to grow olive trees. Studies to select for new cultivars from these wild trees (Sedgley, 2000) failed to establish any outstanding new cultivars. These wild trees are now considered a pest to native flora and in some States have been declared noxious weeds.

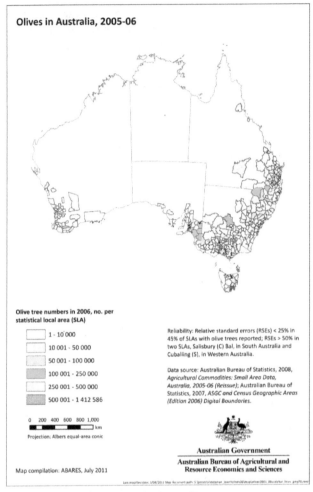

Fig. 2. Olive growing regions and intensity in Australia.

Today, the Australian olive industry is a modern production system for excellent quality oil. High yields have been achieved with low production costs. It is estimated that in the late 1990s, Australia had only 2,000 hectares of traditional olive groves, producing about 400 tonnes of oil. By 2008, Australia produced approximately 12,000 tonnes of oil. By 2013 it is expected that this production will have doubled. Most of this new oil production comes from 30,000 hectares of modern olive groves planted since 2000. There have been significant improvements in mechanical harvesting to achieve high levels of efficiency and economy which is comparable with any in the world. In traditional olive growing regions mechanical harvesting using trunk shakers was once considered as the best and most reliable method for reducing labour costs over the past decade. Today, continuous straddle harvesting machines are used which have been adapted or developed for Australian conditions with great success. These are currently used for more than 75% of Australian production.

Australia produces mostly extra virgin olive oil. The natural diversity of the Australian environment along with the selection of the most productive cultivars, harvested and processed under optimal conditions, is responsible for the exciting range of high quality olive oil products from Australia.

2. Formal development of an Australian Olive Association

The first national symposium on olive growing was held at the Roseworthy Campus of the Adelaide University in 1994, with strong interest spurred on by the economic feasibility report by Farnell Hobman (1995). The symposium was attended by over 100 participants. A decision was made to form an "olive industry group". Over the next two years this olive group drafted a constitution which was to become the Australian Olive Association (AOA).

The AOA committee had identified several issues which were critical to the development of a new industry (Rowe and Parsons 2005). These included:

The lack of any Australian or State quality standard for olives
A lack of knowledge about cultivars suited to the large range of environments
Strong optimism about growing olives in Australia
A network needed to be established for the free transfer of information.

The constitution was adopted by the committee in Mildura in May, 1995. Of the 100 participants at that meeting, 65 became members of the new AOA. The committee adopted a number of objectives:

a.   To promote interest in olive growing and processing
b.   To foster cooperation between regional groups
c.   To facilitate research
d.   Encourage education and information
e.   Develop and distribute superior genetic olive material
f.   Market research and promotion
g.   Quality assurance

Following the formation of the AOA, several State industry organisations were then formed. The first publication of the AOA magazine, the Olive Press, was printed in 1995. By the end of 1995 regional grower groups had been established in Queensland (Qld), New South Wales (NSW), South Australia (SA) and Tasmania (Tas).

The International Olive Oil Council (IOC) provided assistance to the developing industry in Australia. The IOC funded olive experts to attend workshops held in Wagga Wagga NSW and Roseworthy, SA and provided information to further encourage the industry in Australia. This included the development of Australia's first organoleptic panel, in South Australia in 1997. The AOA and State Departments also held industry and grower workshops throughout the country on oil quality and production.

The AOA developed a five year strategic plan for the industry in 1997. This plan described the AOA as an "umbrella organisation" with a national industry structure (Rowe & Parsons 2005) overseeing State grower groups.   In 1999 the Association was well established with the creation of 27 Regional Olive Associations and 1000 members.

The AOA made a commitment to establish an Australian standard. Existing international standards were based on oil produced under limited environmental variation, particularly in Mediterranean climates, and failed to recognise the natural variation in Australian olive oil. The AOA, together with the Australian Government and international organisations has been able to illustrate the high quality of Australian oil and the limitations of existing standards. The Australian standards were approved by Standards Australia in June 2011.

Today the Australian Olive Association Limited is recognised as the Peak Industry Body in Australia for olive growers. In its own words "*The Australian Olive Association (AOA) is the national body representing the Australian olive industry. Formed in 1995 as a result of a burgeoning industry that recognised the need for collective action in developing and supporting the industry, it now represents over 800 people nationwide. Members are involved in all aspects of the Australian olive industry, from grower (small and large), processors to end-user and associated business partners and service industries*" (http://www.australianolives.com.au).

The Australian Olive Association exists to:

- set and maintain quality standards for Australian-produced olive products
- provide quality research & development to create and maintain a sustainable integrated olive industry
- implement an ongoing consumer awareness programme to promote the benefits, and create a preference for Australian olive products
- provide a focal organisation which facilitates progressive direction for stakeholders in the olive industry.

The AOA holds an annual conference (Expo) within Australia to address the latest technology and research. The National Extra Virgin Olive Oil and the Australian Table Olive Competitions are held concurrently with the Expo to highlight the quality of Australian olive products.

3. The code of practice

One of the major outcomes of the AOA has been a Code of Practice. This was developed to ensure to consumers that signatories to this Code have undergone rigorous procedures to certify that the contents of a bottle of olive oil being sold is indeed Australian extra virgin olive oil. Signatories to the Code of Practice are listed on the Australian Extra Virgin website. To conform to the code of practice, producers must apply for registration and have their oil tested to ensure it meets specific criteria.

Australian extra virgin olive oil must have the following characteristics:

- be produced only from olives grown in Australia
- have a free fatty acid content of not more than 0.8 grams per 100 grams (as oleic acid)
- have a peroxide value of less than 20 (mEq peroxide oxygen per kg of oil)
- not exceed the following extinction coefficients for ultra-violet absorbency tests:
  - An absorbance value at 270nm of no greater than 0.22
  - An absorbance value at 232nm of no greater than 2.5
  - A ΔK value of no greater than 0.01

- have been assessed organoleptically by a person or persons accredited by the Australian Olive Association Ltd or in accordance with processes determined by the Australian Olive Association Ltd as having positive attributes such as fruitiness and not having any defects including fusty, muddy, musty, rancid or winey characteristics.

The chemical analyses for these purposes shall be undertaken by a person or organisation accredited by the Australian Olive Association. A sample from each batch identified on labels needs to meet the above tests before the claim that the oil is Australian extra virgin olive oil can legitimately be made. Inclusion of a 'best before' date on a label shall not be more than the equivalent of 30 days for every hour of the oil in Rancimat® at 110°C. Where the oil is a blend and the constituent oils have been tested separately the 'best before' date shall be that for the lowest scoring constituent.

In 2010 there were 230 Australian producers signed up to the Code of Practice including grocery retailers who use the Code as an internal standard for extra virgin olive oils (http://www.australianolives.com.au/).

4. Interaction with the International Olive Council

The Australian olive industry has learnt and benefited from input from the IOC and continues to work with their members. Both chemists and producers utilise the IOC website and advice from the IOC technical experts. Australia has two chemical laboratories and sensory laboratory which continue to participate in the IOC proficiency program. The laboratories utilise IOC methods of analysis and generally follow the limits of IOC standards.

The IOC initially provided funding for representatives to attend IOC meetings and during the 1990s the Australian Government Analytical Laboratories (AGAL), Sydney, gained IOC accreditation. Unfortunately, this provided no support for the industry due to a lack of contact between the two organisations.

In 1996 the NSW Government laboratory at Wagga Wagga pursued accreditation for the laboratory and in 2005, with funds from Horticulture Australia Limited, the organoleptic laboratory was also accredited (Mailer, 2005a). In 2007 the Modern Olives laboratory in Victoria also gained IOC accreditation. The sensory panels and chemical laboratories provide the industry with a resource to monitor quality and to meet the stringent requirements of the IOC and international standards.

During the period through 1995 - 2000, the IOC helped raise awareness of the health benefits of olive oil. Partly, as a result of the Olive Council's interaction, olive oil imports increased in Australia from 7 million litres in 1978 to 15 million litres in 1992 and to 30 million litres by 2000. The IOC continues to play a role in the Australian industry. Accredited Australian chemists are invited to attend chemists meetings in Madrid and the laboratories are invited to participate in proficiency programs and ring tests in the development of new methods.

5. Codex alimentarius

During the early years of the olive industry, although Australian growers were producing high quality olive oil, it was recognised that there were minor differences in the chemical profile of oil from olives grown across the range of Australian environment. These oils had a spectrum of flavours and qualities not apparent in olives grown in the limited environmental fluctuations of the Mediterranean climate. The first workshops in Wagga Wagga in 1996, identified linolenic acid levels from 0.6 – 1.8% whereas the IOC standard for olive oil was <1.0%. Many studies have shown that fatty acid profiles are strongly influenced by environment, particularly the temperature during fruit development. Although insignificant in value, and no problem in terms of nutritional quality, this factor needed to be considered within international standards.

Further studies at WWAI in subsequent years showed other chemical parameters to sometimes vary from the existing and restrictive limits of the IOC standards (Mailer, 2007). Some cultivars being grown in Australia, particularly cv Barnea, from Israel, contained up to 5% campesterol whereas the limit imposed by IOC was <4.0%. These limits, developed as a means to detect adulteration, have no bearing on the nutritional value in olive oil. These components became a limitation for exporters of Australian olive oil but it became apparent that not only did Australian oils fail to meet these standards but many countries from the Southern Hemisphere (Argentina, Australia, Chile and New Zealand) also failed.

Through Food Standards Australia and New Zealand (FSANZ) Australia has attempted to have these limitations in the standards amended so that the standards represent olive oil grown throughout the world. Over several years Australia has sent representatives to Codex meetings to discuss these issues. IOC have responded vigorously, defending the standards on the basis that changes will encourage fraud, despite the fact that genuine oils, such as those growing in countries in the Southern Hemisphere, fail these limits. Due to the lack of agreement Codex have failed to come to a conclusion and particularly, an accepted level for linolenic acid in virgin olive oil has not been agreed upon.

6. An Australian standard

The lack of an Australian standard left Australia exposed to a lack of control of imported product as well as limitations for the domestic industry. Until recently no testing was done within Australia on imported olive oil. Several recent studies of imported olive oil products taken from supermarket shelves has illustrated that a significant proportion of it does not meet the international trade standards of IOC and Codex Alimentarius.   Similar recent studies have shown that many imported oils into the USA do not meet European and international standards (Frankel et al 2011).

Surveys of Australian olive cultivars have been used to determine compliance with international standards (Mailer et al, 2002). Reports show the effects of olive cultivars, the influence of harvest timing and the changes to quality as a result of site and seasonal growing conditions.

From the first inception of the Australian Olive Association, it has been recognised that Australia must have its own standards for olive oil. The standards are required to set guidelines for Australian producers to ensure the oil was authentic and of acceptable quality. It was also critical to allow authorities to determine if the imported and local product meets the quality levels demanded by the industry and the consumers. The standard was created with consultation within the industry including producers, marketing and exporters. It needed to address issues of authenticity, to detect any efforts to blend or mislabel the product. It needed also to be able to detect oil which had been heated and/or refined or if the oil had exceeded its potential use by date.

The standard was established with the support of Australian Standards organisation (www.standards.org.au). A wide spectrum of representatives from the industry contributed. A final draft was made available to the public for comment in early 2011. The draft drew both praise and criticism from all aspects of the industry both domestically and internationally. Ultimately it had strong support and was accepted with the final standard approved in July 2011 for release during 2011.

The new standard is similar in many ways to that of the IOC. The standard allows for a higher level of linolenic acid and campesterol, reflecting the actual properties of the Australian product. It has also included some new tests developed by the German DGF which allow traders to identify fresh oil from old oils or oil which has undergone heating, such as in refining. The standard is available from Standards Australia.

7. Consumption and production

7.1 Development of a boutique industry

In the early years of Australian settlement, there was not a strong demand for olive oil. Olives were grown for personal use or for a small boutique industry. The major edible oils used continue to be refined sunflower, cottonseed and canola oil. It was not until the late 20th Century that the olive industry began to grow. Australia had an increasingly cosmopolitan population including a large portion of Greek and Italian migrant workers who increased the demand for olive oil production. Despite this, the industry continued for many years as a boutique industry with small farms of only a few trees in which people produced their own oil or sold small quantities to others. Olive oil was imported from Spain, Greece and Italy for many years and by the 1990s the value of the imported olive oil products was in excess of $100 million dollars per annum.

7.2 Australian consumption

Outside the Mediterranean region, Australia is currently the largest consumer of olive oil per capita, consuming about 32,000 tonnes of olive oil in 2008. The demand for olive oil continues to grow, creating a good opportunity for the domestic market. The increased demand is highlighted by the increase in total imports of olive products in the last five years (Table 2).

7.3 Australian production

Australia currently has about 10 million olive trees spread across approximately 30,000 ha. Although the initial plantings of olives in Australia included a large number of cultivars, today about 90 percent of Australian olive oil is produced from 10 major cultivars including Arbequina, Barnea, Coratina, Corregiola, Frantoio, Koroneiki, Leccino, Manzanillo, Pendolino and Picual. These cultivars have been found to be agronomically suitable while at the same time producing a good range of oil types. Barnea, a cultivar from Israel, is a recent addition to the other predominantly European cultivars but is today the most prolific.

Commercial production increased rapidly from the mid 1990s onwards, designed using state of the art equipment and methodologies. From an almost non-existent crop prior to 1990, olive oil production in Australia reached 12,000 tonnes in 2008. Due to the modern technologies used, that production is almost 100% extra virgin olive oil with no facilities or requirements for solvent extraction and only limited refining capacity for the oil. Only rarely do harvest conditions produce poor quality fruit which requires refining. These refined oils would generally be marketed as alternative products. Hence, Australian olive oils in supermarkets are all extra virgin olive oil. Additionally, around 10 per cent of Australian groves have organic certification, representing an increase of 60 per cent since 2006 (Australian Olive Association).

Most of Australia's olives are grown in the east, south and west of the country. Although South Australia was originally the largest producer, with 39% of total production in 2003, Victoria has become the leader with 48% of the production in 2009. New South Wales, Queensland, Western Australia and South Australia share the other half of production. The main harvest time is May to July although Queensland tends to harvest first due to climate, although harvest time vary across the states.

Australia's share of the world's extra virgin olive oil production has grown from only 0.31 percent in 2006, to 3 per cent of the world market with a 2008 harvest of 12,000 tonnes. By 2014, production is expected to reach 25,000 tonnes.

| | Production | | EVOO | | Table Olives | |
Year	Table Olives	Olive Oil	Imports (tonnes)	Exports (tonnes)	Imports (tonnes)	Exports (tonnes)
2001		500	27,680	385	11,545	74
2002		750	28,987	300	12,618	199
2003		1,500	28,447	278	14,483	138
2004	2,000	2,500	32,657	501	13,711	265
2005	2,700	5,000	29,062	1,652	15,143	215
2006	3,200	8,650	34,511	2,988	15,608	230
2007	2,500	9,250	43,404	2,502	16,364	207
2008	2,200	12,000	23,952	4,169	17,542	239
2009	3,000	15,000	31,169	6,960	16,210	366

Table 2. Extra virgin olive oil and table olive production, imports and exports (www.australianoliveoil.com)

The majority of olives grown are for oil production. Much of the production is from a few large producers although there are a large number of small producers spread throughout the growing regions of Australia. Despite the rapid increase in production, Australians are continuing to increase their consumption of olive oil and imports have been maintained at around 20-30,000 tonnes per annum. It would seem however, that there is some import replacement with imports of 43,000 tonnes in 2007 being reduced to around 31,000 tonnes in 2009.

Fig. 3. a. Harvesting at night at Boundary Bend and b. aerial photograph of olive harvesting at Boundary Bend (photo courtesy of Boundary Bend)

7.4 Australian imports / exports of extra virgin olive oil

Australia imports in excess of 31,000 tonnes of olive oil per annum. Despite that, an increasing percentage of olive oil is being exported. In 2004, 501 tonnes, or 20 per cent of Australian production was exported while in 2009, 6,959 tonnes, or around 46 per cent, was destined for the export market, an average annual increase of 85 per cent. The value of exports in 2009 was $37.8m (Source ABS).

The top five countries buying Australian extra virgin olive oil have been the United States, England, China, Singapore and Japan. The Australian customers are changing over time, with the United States and Italy now being major destinations of Australian olive oils. Exports to China are also increasing albeit from a very low base.

The export figures from Table 2 indicate that there is a demand and an opportunity for Australian olive producers to continue to sell olive oil overseas. Despite that, a significant level of import replacement is a long term goal for the Australian olive industry and is on track to being achieved. The amount of Australian produced olive oil that is consumed domestically is now one quarter of the sum total of olive oil that is imported. The increased percentage of Australian olive oil that is being consumed domestically has occurred in the context of fairly static import volumes over the last 5 years. Further increases in market share for Australian extra virgin olive oil in the domestic market will require further investment in consumer education.

7.5 Table olives

Data about table olive production in Australia is less well known than for olive oil. Although olives for oil production have been increasing rapidly, table olives have not had such success with production figures increasing from 2000 tonnes in 2004 to only 3000 tonnes in 2009 (Table 2). Although there are many boutique operations, a few operations have the capacity to process hundreds of tonnes of olives. Table olives are appearing more on domestic and export markets but large-scale production is still limited by the costs of harvesting. Despite the limited increase in production of table olives, the demand for them continues to increase. Imports of table olives have increased from 12,000 tonnes in 2001 to 16,000 tonnes in 2009 (Table 2.). Exports of Australian table olives have remained steady over the past few years, with around $800,000 worth of table olives having been exported in 2007.

By far the greater amount of research on olives has been directed toward the production of high quality olive oil. However, in addition to this research on oil and applications in Australia, some work has also been carried out on table olives (Kailis & Harris, 2004). The Australian table olive industry and trade currently has no nationally accepted guidelines for ensuring the quality and safety of processed table olives and the Kailis report was prepared for olive growers and processors from both national and international viewpoints.

Percentage of production for Australian States

Year	Olive Oil (tonnes)	NSW	Qld	SA	Vic	Tas	WA
2001	500	-	-	-	-	-	-
2002	750	-	-	-	-	-	-
2003	1,500	11	12	39	28	1	9
2004	2,500	12	8	16	47	1	16
2005	5,000	12.1	5	16.2	40.1	0.4	26.2
2006	8,650	8.3	4	18.2	53.9	0.2	15.4
2007	9,250	8	4.1	14.6	49.2	0.2	23.9
2008	12,000	7.7	4.1	19.5	53	0.2	15.5
2009	15,000	9.0	4.3	14.5	48.0	0.2	24

Table 3. Percentage of olive oil produced per State.

8. Quality

Australian olive oil quality is generally of high quality using modern technology for growing, harvesting, processing and packaging. The majority of the crop is mechanically harvested and transported to processing facilities within a few hours. Everything from the machines used to harvest the fruit through to the temperature controlled stainless steel storage vessels are built on new technology. Oil extractors are generally centrifugal machines which are kept hygienically clean and housed in temperature controlled facilities.

The oil produced is almost entirely extra virgin olive oil and is ensured through the Australian Olive Association's "Code of Practice". The code requires olive growers to have their oil tested to ensure EVOO quality at the time of bottling. It also requires that the oil remains within specifications throughout the oils "best before" date to provide the consumer with confidence that the product meets the label qualifications. Only fruit that may have been damaged through frost, insect or disease generally fails EVOO quality. In these cases the oil is refined and redirected toward alternative uses.

Australia maintains two IOC accredited laboratories and sensory panels which advise the industry on oil quality. There is a continuous educational program through workshops and conferences to inform producers and consumers to help them understand defects and attributes in olive oil. The AOA presents several industry awards to olive oil producers at the annual AOA Expo. In addition, many regional growers groups have their own olive competitions judged by trained sensory personnel. All of the olive competitions demand that the oil passes basic chemical requirements.

8.1 Fatty acids

For commercial samples, the majority of oil analysis is carried out by the two Australian IOC accredited laboratories. This allows the laboratories to keep accurate records of the quality of oil being produced in Australia from year to year.

Free fatty acid value (FFA) of olive oil is a general indicator of how sound the olive fruit was at harvest and how carefully it was processed into oil. Table 4 shows a typical range for free fatty acids (FFA) and peroxide value (PV). The range shows that the majority of oils are well within acceptable limits with the median value of 0.18 FFA and 8 mEq oxygen/kg. Occasionally, due to fruit damage or climatic factors, oils may be outside of acceptable standards. However, less than 3.3% of FFA samples and 2.2% of PV samples failed to meet the IOC limits in 2006.

2006	Free Fatty Acid	Peroxide Value
Minimum	0.05	0
Maximum	3.48	48
Average	0.26	9
Median	0.18	8
No of samples	585	501

Table 4. Typical range of free fatty acids and peroxide value in Australian olive oil based on commercial samples in 2006.

The quality of Australian extra virgin olive oil has improved over the last decade. A summary of the FFA's of olive oils submitted to the Australian National EVOO Competition since 1997 (Table 5) shows how oil quality has improved. Between 1997 and 2002, only 34%

of the oils were less than 0.19% free fatty acids. In the following six years, from 2003 to 2009, 62% were less than 0.19% FFA.

1997-2002		2003-2009	
<0.19	34%	<0.19	62%
0.20-0.29	33%	0.20-0.29	26%
0.30-0.39	14%	0.30-0.3	8%
>0.40	19%	>0.40	4%

Table 5. Average free fatty acid levels of Australian olive over two periods (AOA)

The variable Australian climate and differences in temperature during fruit development has a strong influence on fatty acid profiles (FAP) as shown for oil analyses carried out in 2006 (Table 6). The profile of the fatty acids covers the full IOC range for acceptable limits but exceeds that range in several instances. Although the range is not indicative of nutritive value, the issues of compliance to international standards are significant.

Sample	C16:0	C16:1	C17:0	C17:1	C18:0	C18:1	C18:2	C18:3	C20:0	C20:1	C22:0
IOC limits	7.5- 20.0	0.3-3.5	≤0.3	≤0.3	0.5 - 5.0	55.0 - 83.0	3.5 - 21.0	≤1.0	≤0.6	≤0.4	≤0.2
Average	12.3	0.9	0.1	0.1	2.1	74.1	8.9	0.7	0.3	0.3	0.1
Min	7.4	0.4	0.0	0.0	1.1	55.7	2.7	0.4	0.2	0.1	0.0
Max	18.3	1.9	0.4	0.6	4.0	84.9	23.4	1.5	0.5	0.5	0.2

Table 6. Average, minimum and maximum limits for FAP of Australian olive oil in 2006 (n=468).

The range of fatty acids is further demonstrated with the analysis of samples from New Zealand, a cooler climate to that of the Australian olive producing areas. The FAP of 56 randomly selected samples in 2006 (Table 7) show that the oleic acid level often (23%) exceeds the IOC values which suggest these oils are nutritionally superior to those with high levels of saturated fat. However, these oils would officially fail the IOC standard. Many samples are lower (17%) than the IOC standard for palmitic (saturated) acid.

Fatty Acids	C16:0	C16:1	C17:0	C17:1	C18:0	C18:1	C18:2	C18:3	C20:0	C20:1	C22:0
Max	12.4	0.9	0.06	0.11	2.59	**85.5**	7.6	0.9	0.4	0.4	0.2
Min	**6.4**	0.3	0.03	0.06	1.12	78.2	3.0	0.5	0.2	0.2	0.1

Table 7. Fatty acid profile of 56 randomly selected New Zealand oils from 2006.

8.2 Phytosterols

There is also a significant range in the phytosterol content and profile in Australian olive oil (Table 8). In particular, the level of campesterol often exceeds 4.0%, generally due to the production level of cv Barnea which is higher in campesterol than other cultivars. Due to the suitability of this cultivar to the Australian climate and its high production rate, this cultivar will continue to be a significant portion of the Australian crop. As for other parameters, these components may exceed the international limits.

Sterols (%)	Cholesterol	Brassicasterol	24-Methylene-cholesterol	Campesterol	Campestanol	Stigmasterol	D-7- Avenasterol	D-7- Stigmastenol	D-7-Campesterol
Average	0.08	0.00	0.09	3.61	0.16	0.63	0.50	0.19	0.13
Min.	0.03	0.00	0.02	2.27	0.10	0.34	0.22	0.00	0.00
Max.	0.16	0.02	0.48	4.89	0.25	1.41	1.00	0.52	0.59
Median	0.07	0.00	0.07	3.49	0.15	0.56	0.47	0.19	0.06

	β-Sitosterol	D-5- Avenasterol	D-5,23-Stigmastadienol	Clerosterol	Sitostanol	D-5,24-Stigmastadienol	Apparent β sitosterol	Diols	Total Sterols (mg/kg)
Average	85.08	7.34	0.01	0.58	0.93	0.68	94.62	1.11	1537.8
Min.	79.45	5.21	0.00	0.20	0.28	0.21	93.83	0.64	1131.7
Max.	88.24	13.66	0.13	0.93	2.51	1.27	96.38	3.09	2153.8
Median	85.75	6.81	0.00	0.60	0.48	0.58	94.56	1.06	1520.9

Table 8. Phytosterols profile in Australian olive oil showing the range and the average and median values for each component.

9. Research

9.1 Funding

A new agricultural industry requires significant research and development support to optimise the industry. Such was the case with the awakening of the olive industry in Australia. It created a need for Australian research scientists to develop an understanding of the agronomy and the chemistry of the crop, essential for producing the highest yield with the best quality. The research effort has been supported strongly by the Australian Olive Association and financial support from some of the larger producers. Much of the financial support has come from the Federal Governments "Rural Industries Research and

Development Council" (www.rirdc.gov.au) which has consistently supported projects the olive industry considered to be of significant value.

Although most Australian agricultural industries pay a levy to the Federal Government to support research, olives have always been considered a new crop and have been exempt from a levy. However, in 2011, through support from the Australian Olive Association, the industry has agreed to contribute to a crop levy. This guarantees ongoing funding for the research and development of this industry in the future.

9.2 Cultivar selection

Determining which cultivars to grow was an early requirement for growers. At the early stages of development one of Australia's best resources was the historic olive orchard at Wagga Wagga. This orchard, with over 50 cultivars and trees which, in some cases, were over 100 years old, provided an ideal resource for study. Such was one of the first research projects funded by RIRDC (Ayton et al., 2001) in which oil content, oil quality and initial attempts to identify cultivars by DNA were carried out. The range of trees, some of which were grafting experiments and others with varying levels of irrigation, appeared to be an ideal study. Although the trees were producing reasonable crops due to poor maintenance for such a long period, and the variable conditions under which each of the trees were grown, the use of the data was limited.

There remained considerable confusion about cultivars being grown in Australia and if they were true to type. After many years, maps of the grove had been altered and many trees removed. Using RAPD DNA methods to discriminate between the cultivars (Mailer & May, 2002), dendrograms were constructed showing the relationship of the cultivars to each other. Although some trees were identified, it was not possible to obtain reference standards for many of the cultivars and they remained unknown. Errors in this labelling on the map became evident as shown by the dendrogram of trees labelled as *cv* Manzanillo in Fig 4. The comparison of trees, reportedly to be the same cultivar, was clearly different, based on DNA patterns and seed morphology.

There was little data on the performance of any olive variety for optimal yields and quality under Australian conditions and the industry has relied on information from the Northern hemisphere, particularly from Mediterranean sources. Performance characteristics of cultivars are the basis on which a selection is made for a particular use or physical situation. The National Olive Variety Assessment (NOVA) project was established to help resolve the confusion in olive variety identity and to evaluate the performance, in different climatic regions of Australia, of the majority of known commercial olive varieties. (Sweeney, 2005). The establishment of a national varietal grove at Roseworthy provided an opportunity for growers to evaluate different cultivars, grown at that site.

At the same time studies were being undertaken on wild olive trees which had become established in the Adelaide Hills to attempt to identify feral olives which may be better adapted to the Australian conditions (Sedgley, 2000). Despite these investigations, the Australian industry has been established on common European cultivars and some more recently introduced including *cv* Barnea from Israel.

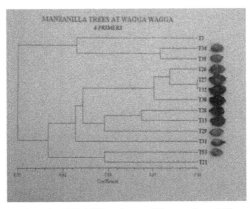

Fig. 4. Dendrogram created with 10 RAPD primers illustrating differences in seed morphology of 11 trees identified by the historic map as *cv* Manzanillo (Mailer unpublished).

9.3 Environment

Perhaps the main issue for Australian oils was the very variable environmental conditions under which the crop was being grown. Oil quantity and quality rely heavily on crop management, moisture availability, harvest timing, processing methodology and storage. As a result of these differences Australian oils showed a wide range in the fatty acid profiles (Mailer, 2005b). The diversity in other quality characteristics and sensory analysis were also significant.

In the initial stages of the development of this industry, there was little understanding of the intricacies of these crop management practices and the resultant crop yield, chemical quality and sensory attributes such as taste, colour and odour. For these reasons, several studies have been undertaken to look at the relationship between oil quality and harvest timing, irrigation treatments, yield and sensory characteristics (Mailer 2007).

As a result of the low rainfall and unpredictable nature of the Australian environment, almost all Australian olive groves are irrigated. Irrigation provides more predictable yields and harvest timing unlike dryland groves. Due to the importance of irrigation, research has focussed on water requirements, particularly in the stages from planting to commercial harvesting (De Barro, 2005). This research has been aimed at increasing the understanding of olive water use and requirements in the period from planting to early fruit bearing. As most Australian olive groves are irrigated several studies have focussed on variation in maturity, yield, oil quality and sensory attributes under variable moisture availability (Ayton et al., 2007) and with different harvest times.

Environmental effects on oil production have created unexpected issues for Australian producers. The variation in quality and sensory has created a new spectrum of oils with unique qualities and sometimes more variable attributes than has been produced in Mediterranean climates. This sometimes results in the oil being outside the limited range of existing international standards (Mailer, 2007).

9.4 Harvesting and processing

Along with the determination of the best cultivars, the methods of harvesting and processing have been evaluated. Many types of harvesters and extractors, generally from European manufacturers, were being used throughout the industry. Hand harvesting and many types of vibrating rakes, tree shakers and accessories were unsuitable for large scale production. Larger producers investigated straddle harvesters (Fig. 3) which underwent several modifications to make them suitable. These are used widely today.

Processing also went through stages. Some producers tried mechanical mat presses or stone mills (Mailer & Ayton, 2004) in the early stages but these were never used for large scale production The majority of processors have installed modern two or three phase centrifugal extraction mills.

Fig. 5. One of the early harvesting methods adopted in an Australian olive grove.

9.5 Quality analysis

As the industry developed, a need for quality evaluation increased. Using the resources of the International Olive Oil Council, Australian laboratories were able to set up methods to determine oil quality. Many of the existing methods were time consuming, reasonably difficult to carry out and expensive. This prompted the investigation of more rapid methods such as the near infra red spectrometry (Mailer, 2004), a rapid screening analytical tool whilst more intensive, wet chemistry methods were maintained as checks where necessary.

Minor compounds in olive oil were recognised as the basis of the sensory attributes, nutritional value and stability, or antioxidant capacity, of the oil. Additionally, minor compounds are used to ensure authenticity in Australian research, particularly the sterol profile. As Australian oil has a wide spectrum for each of the sterol components, which may lie outside the limits of the IOC standards, this has been an important focus for Australian scientists (Guillaume et al., 2010). Environmental effects and irrigation on polyphenols both showed a significant effect (Mailer et al 2007). The influence of frost on these compounds and the resulting changes in sensory and chemical characteristics has also been investigated (Guillaume et al., 2009). Frost is one of the most important weather related hazards for the Australian olive industry and has caused significant economic losses during the past decade. Its impact on oil quality has been significant in 2006 with more than 20 per cent of Australian oil of that year being affected to some degree. Early frosts will normally affect the fruit leading to significant changes in the chemical and organoleptic characteristics of the oils.

9.6 Shelf life

Although oil may be acceptable when it is processed, maintaining the quality after processing became a major consideration. Two studies carried out in 2008 by the Australian Olive Association on supermarket oils (AOA Report – unpublished) included 22 oils initially and later, a further 33 oils of random brands. The reports revealed that many oils would not pass IOC tests most likely because of poor storage or old age, although some oils were clearly adulterated. The AOA and RIRDC have funded long term storage studies under extremes of temperature, light and oxygen exposure to determine potential shelf life and develop an understanding of methods used to advise marketing on potential shelf life.

Shelf life depends heavily on the type of material the oil is stored in. Although most experts would recommend the use of glass or stainless steel, often oil is stored temporarily in plastic bottles or collapsible bags. Studies on the effects of the different types of containers used for transport and sometimes for long term storage have been published (Mailer & Graham, 2009). The study reinforces that the best storage conditions for olive oil is in opaque, impervious and inert containers, stored at cool temperatures. Metallised flexible bags used for short term transport may provide reasonable protection. Storage in clear plastic, particularly in the light and at elevated temperatures, is unacceptable and results in loss of extra-virgin olive oil quality within a short period. Re-use of these containers is highly undesirable and would cause more rapid degradation.

9.7 Pest and disease

Australia has been free of many cosmopolitan olive pests due to its isolation but the rapid expansion of the olive industry in all mainland states has led to increased problems with pests and diseases. The control of these problems became a focus for all growers. A report on sustainable Pest & Disease Management in Australia Olive Production (Spooner-Hart, 2005) describes sustainable management, monitoring pest and beneficial species in groves and identified a number of previously unreported pests and diseases. Further publications have included a field guide to olive pests, diseases and disorders in Australia (Spooner-Hart et al., 2007) designed as a quick reference to take into the grove and use to identify pests and diseases and the damage they cause.

9.8 Waste management

Dealing with by-products of olive oil processing is another important issue in modern agriculture. A study on recycling of solid waste from the olive oil extraction process (Tan & Markham, 2008) and a subsequent report outlines methods for developing an environmentally sustainable system to manage solid waste from the 2- and 3-phase olive oil mill extraction processes. The expanding Australian olive industry over recent years, with significant increase in fruit production, has resulted in vast quantities of solid and liquid wastes generated to the detriment of the environment. The industry is been faced with the challenge to manage these wastes in order to achieve sustainable production under a clean environment. The research work provided the industry with a tool to recycle processed oil mill waste to improve the health of the crop and the status of the soil.

9.9 Reviews

The revival and development of the olive industry stimulated wide areas of research over a relatively short period. The quest for information has been intense. This has led to the publication of several reviews being carried out, particularly regarding the potential for olive production in Australia. These include studies by McEvoy et al. (1999) in which the market for the development of an olive industry in Australia was examined based on analysis of: trends in international production and trade; consumer segments and product characteristics; whether Australia could compete with imported olive products.

Another review contains detailed steps required to establish an olive grove in Australia and is a comprehensive survey of the Australian Industry (Meyers Strategy Group, 2010). It was developed as a method of establishing how Australia could compete in a rapidly growing olive industry worldwide.

10. References

Ayton, J., Mailer R.J. and Robards, K. 2001. Changes in oil content and composition of developing olives in a selection of Australian cultivars. *Australian Journal of Experimental Agricultural*. 41: 815-821

Ayton, J., R. J. Mailer, A. Haigh, D. Tronson, D. Conlan. 2007. Quality and oxidative stability of Australian olive oil according to harvest date and irrigation. *Journal of Food Lipids* 14:138-156

De Barro J. 2005. From Planting to Harvest - A Study of Water Requirements of Olives, from planting to first commercial harvest. Report No. 05-039, 1 Jun 2005. ISBN: 1-74151-138-0. Web: http://www.rirdc.gov.au.

Frankel, E.N., Mailer, R.J., Wang, S. C., Shoemaker, C.F., Guinard, J-X., Flynn, D., and Sturzenberger, N. 2011. Evaluation of extra virgin olive oil sold in California. UC Davis Olive Centre. www.olivecenter.ucdavis.edu.

Guillaume, C., Ravetti, L. And Gwyn, S. 2009 Characterisation of Phenolic Compounds in Oils Produced from Frosted Olives. Report 09-058. 5 May ISBN: 1-74151-860-1. Web: http://www.rirdc.gov.au.

Guillaume, C; Ravetti, L and Johnson J. Sterols in Australian olive oils: The effects of technological and biological factors. RIRDC Report: 14 Oct 2010. ISBN: 978-1-74254-140-2. http://www.rirdc.gov.au.

Hobman F. 1995. An economic study into irrigated olive growing and oil processing in Southern Australia. A report for RIRDC. Research Paper No 95/5. https://rirdc.infoservices.com.au/.

Kailis, S. G. And Harris, D. 2004. Establish protocols and guidelines for table olive processing in Australia. Report No. 04-136. 1 Oct 2004. ISBN: 1-74151-044-9. Web: http://www.rirdc.gov.au.

Mailer, R.J. 2004. Rapid evaluation of olive oil by NIR reflectance spectroscopy. Journal American Oil Chemists' Society. 81(9):823-827

Mailer, R.J. 2005a. Establishment of an olive oil sensory panel. A Report prepared for Horticulture Australia Limited. Report FR02054.
http://www.horticulture.com.au

Mailer, R.J., 2005b. Variation in oil quality and fatty acid composition in Australian olive oil. Australian Journal of Experimental Agriculture. 45:115-119)

Mailer, R.J. 2007. The natural chemistry of Australian extra virgin olive oil. RIRDC Publication No. 06/132, Project DAN239A. https://rirdc.infoservices.com.au/.

Mailer, R.J., Ayton, J. 2004. Comparison of olive oil (Olea europaea) quality extracted by stone mill and hammermill. New Zealand Journal of Horticultural Science. 32(3):325-330.

Mailer, R.J. Ayton, J. and Conlan, D. 2002. Comparison and evaluation of the quality of thirty eight commercial Australian and New Zealand olive oils. *Advances in Horticultural Sciences.* (16)3-4: 259-256).

Mailer R.J., Ayton, J. and Conlan D. 2007. Influence of harvest timing on olive (Olea europaea) oil accumulation and fruit characteristics in Australian conditions Journal of Food, Agriculture & Environment. Vol 5. (3 & 4): 58-63).

Mailer, R. J., and Graham, K. 2009. Effect of storage containers on olive oil quality. Report No. 09-160. 26 Oct 2009. ISBN: 1-74151-957-8. Web: http://www.rirdc.gov.au.

Mailer, R.J. and May, C.E. 2002. Variability and interrelationships of olive trees and cultivars using RAPD analysis. *Advances in Horticultural Sciences.* (16)3-4: 192-197

McEvoy, D., Gomez, E., McCarrol, A and Sevil, J. 1999. The olive industry: a marketing study report. No. 99-086, 1 Jan 1999. ISBN: 0-642-57993-8. Web:
http://www.rirdc.gov.au.

Meyers Strategy Group Pty. Ltd. 2001. Regional Australia olive oil processing plants. RIRDC Report No. 00-187. 1 Feb 2001. ISBN: 0-642-58218-1. Web:
http://www.rirdc.gov.au.

Nair (Tan), N.G. and Markham, J. 2008. Recycling solid waste from the olive oil extraction process. Report Code: 08-165, 28 Oct 2008. ISBN: 1-74151-754-0. Web:
http://www.rirdc.gov.au.

Rowe, I and Parsons, L. 2005 Ten years of the AOA. The Olive Press - Winter Edition. Australian Olive Association newsletter. pp 9-10.

Sedgley, M. 2000. Wild olive selection for quality oil production. A report for the Rural Industries Research and Development Corporation. RIRDC Publication No 00/116 RDC Project No UA-41A. https://rirdc.infoservices.com.au/items/00-116.

Spennemann, D H R. 1997. The spread of Olives (Olea sp.) on Waggas Wagga Campus. I. Biology and History. Charles Sturt University. Johnstone Centre Report No. 100.

Spennemann, DHR. 2000. Centenary of Olive Processing at Charles Sturt University. Charles Sturt University, Faculty of Science and Agriculture. ISBN 1 86467 070 3

Spooner-Hart, R. 2005. Sustainable Pest & Disease Management in Australia Olive Production. Report No. 05-080, 1 Jun 2005. ISBN: 1-74151-143-7.
http://www.rirdc.gov.au.

Spooner-Hart, R., Tesoriero, L. and Hall B. 2007. Field guide to olive pests, diseases and disorders in Australia. Report No. 07-153, 1 Oct 2007. ISBN: 1-74151-549-1. http://www.rirdc.gov.au.

Sweeney, S. 2005. National olive variety assessment - A report for RIRDC. RIRDC Publication No 05/155. Project No. SAR-47A. Rural Industries Research and Development Corporation. 2010. http://www.rirdc.gov.au.

Consumer Preferences for Olive-Oil Attributes: A Review of the Empirical Literature Using a Conjoint Approach

José Felipe Jiménez-Guerrero[1], Juan Carlos Gázquez-Abad[1],
Juan Antonio Mondéjar-Jiménez[2] and Rubén Huertas-García[3]
[1]University of Almería
[2]University of Castile La Mancha
[3]University of Barcelona
Spain

1. Introduction

1.1 Olive oil: Some general aspects

During the last decade, olive oil consumption has experienced a major breakthrough in the world, not only in producing countries but also among those who are not. Undoubtedly, this growth in consumption is a consequence of the consolidation of a cultural phenomenon established between the main producing countries (Spain, Italy and Greece), owing to the so-called Mediterranean diet[1]; a food concept that provides important health benefits and of which olive oil is one of the main components. The recent recognition of United Nations Educational, Scientific and Cultural Organization (UNESCO) — it has declared to the Mediterranean diet 'the intangible cultural heritage of humanity'— offers promising perspectives for the Mediterranean diet in the coming years.

The major producer of olive oil in the world is the European Union (EU), which produces 80 per cent and consumes 70 per cent of the world's total olive oil production (European Commission, 2010). Italy and Spain are the major producers and can influence the prices of olive oil (Blery and Sfetsiou, 2008). Greece takes third place in world production after Spain and Italy (Sandalidou and Baourakis, 2002) and first place in olive oil consumption

---

[1] The Mediterranean Diet is a way of eating based on the traditional foods (and drinks) of the countries surrounding the Mediterranean Sea. The principal aspects of this diet include high olive oil consumption, high consumption of pulses unrefined cereals, fruits and vegetables, as well as moderate consumption of dairy products (mostly as cheese and yogurt), moderate to high consumption of fish, low consumption of meat and meat products, and moderate wine consumption. Olive oil is particularly characteristic of the Mediterranean diet. It contains a very high level of mono-unsaturated fats, most notably oleic acid, which epidemiological studies suggest may be linked to a reduction in the risk of coronary heart disease. There is also evidence that the antioxidants in olive oil improve cholesterol regulation and 'Low-density lipoprotein' (LDL) cholesterol reduction, and that it has other anti-inflammatory and anti-hypertensive effects.

(Hellonet, 2006). Olive farming provides an important source of employment in many rural areas of the Mediterranean, including many marginal areas where it is either a principal employer or an important part-time employer which can be combined with other activities, such as tourism. Olive farming is also an important part of local rural culture and heritage in many areas, and is maintained and 'valorized' through labelling schemes in some cases. Olive production is an important economic sector in many rural areas of the Mediterranean. In some areas, it is the principal economic activity and the basis for other sectors (Beaufoy 2002:11). The greatest concentration of oil production is found in two Spanish provinces, Jaén and Córdoba in Andalusia, which between them account for over one-third of EU output. Olive farming has both positive and negative environmental effects. As Beaufoy's (2002:30) report indicates, such effects depend on several factors, including prevailing environmental conditions in and around the plantation (soil type, slope, rainfall, adjacent land uses, presence of water bodies, etc.) and farm management, pest control, irrigation and the type of land (and previous land cover) on which new plantations are established. In particular, Beaufoy (2002) identified the following as the main categories of actual and potential environmental effects associated with the management of olive plantations: soil, water, air, biodiversity (flora and fauna), and landscape.

Among the EU non-producing countries of olive oil, Germany and the UK are the main consumers (de la Viesca *et al.*, 2005), although the US is the most important market outside the Mediterranean basin (Zampounis, 2006). In the US, interest in and consumption of olive oil has been growing exponentially over the last 20 years (Delgado and Guinard, 2011). Indeed, the US ranks fourth in olive oil consumption after Italy, Spain and Greece. US consumption went from 88,000 tons in 1990 to 260,000 tons in 2009; an increase of 228% (International Olive Council, 2008). Something similar is happening in China, where the demand for olive oil is expected to increase significantly in the next few years (Soons, 2004). According to this author, Chinese tourism to Mediterranean countries will affect the general awareness of the healthy Mediterranean kitchen and its use of olive oil in a positive way.

The increasing preference for olive oil worldwide denotes a change in consumer behaviour, either by strengthening the role of it in their diet or by incorporating it in a novel way. The set of tangible and intangible attributes that consumers believe to particularly meet their needs, is a concept of product marketing. From this point of view, the concept is intended to reflect two fundamentally different approaches: a) considering the product itself as a sum of characteristics or physical attributes; or b) considering the needs of the consumer, where the buyer's choice rests not with the product, but with the service they expect to receive or the problems it can solve.

During the purchase process, consumers form their preferences based on the best combination of attributes, evaluating the brands that are part of their evoked sets, or are considered important in terms of attributes such as price, country of origin, quality or design, among others. Olive oil, like any other commodity, is not immune to this stage of the buying process, despite the uniqueness of its attributes that determine the degree of preference for the consumer. In this chapter, we aim to describe what attributes assume greater importance, and therefore are preferred, by the consumer. In order to do so, a review of the previous literature focusing on this stage of olive oil consumer-buying behaviour is developed.

The rest of the chapter is structured as follows. Section 2 discusses the importance of culture as a factor in the formation of consumer preferences with respect to olive oil. In Section 3, we discuss consumer preferences for olive oil from the literature review, with reference to the methodology on conjoint analysis. Finally, we present the findings of the work.

2. Consumer behaviour in purchasing food: The role of culture in the consumption of olive oil

The study of consumer behaviour and marketing discipline has focused on analysing how individuals make decisions to spend their resources in categories related to consumption (Schiffman and Kanuk, 2001). The act of purchase is considered as an activity aimed at solving a problem (Howard and Sheth, 1986). Typically, the consumer is faced with a multitude of decisions to make, whose complexity varies depending on product and purchase situation (Lambin, 1995). Consequently, understanding consumer behaviour requires assessing how people made and make their purchasing and consumption decisions (Blackwell et al., 2001), considering that a decision is the result of selecting a choice from two or more alternative possibilities (Schiffman and Kanuk, 2001).

In the context of food products, Steenkamp (1997) proposes a conceptual model of consumer behaviour in which four stages in the purchase decision process are identified: (1) problem recognition; (2) information search; (3) evaluation of alternatives; and (4) choice. In addition, there are three groups of factors that influence this process: a) properties of foods; b) Individual-related factors (e.g., biological, psychological and demographic); and c) environmental factors (i.e., economic, cultural factors and marketing aspects (see Figure 1).

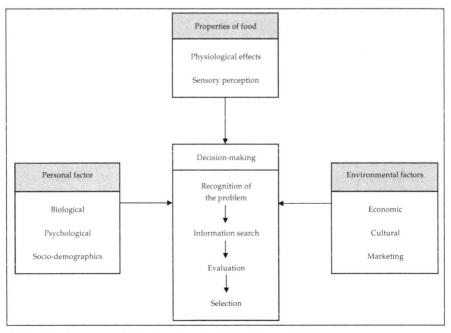

Fig. 1. Model of consumer behaviour for food (Steenkamp, 1997)

As Figure 1 shows, culture is a key concept for understanding consumer behaviour. According to Solomon *et al.* (1999), this is a consequence of culture representing the collective memory of society and the prism through which such society develops its perception. Culture includes a complex of values, ideas, attitudes and other meaningful symbols that allow humans to communicate, interpret and evaluate as members of a society (Blackwell *et al.*, 2001). According to Schiffman and Kanuk (2001), it is the sum of learned beliefs, values and customs helping to determine the behaviour of members of a given society as a consumer. Undoubtedly, culture is a key element not only because it affects all stages of consumer choice, but also because it exerts a major influence on the reasons why people of different cultures buy and consume products (Blackwell *et al.*, 2001). In this manner, culture helps to explain the behavioural differences between them. Specifically, during the evaluation stage, culture mainly influences the way in which consumers assign a greater value to certain attributes of the product over others. As Solomon and colleagues (1999) indicate, a consumer culture determines the priorities of certain products and the success or failure thereof.

When consumers buy a product, they expect it to perform their need. But these needs are different between cultures. This is, for instance, the case of olive oil. There are big differences between olive oil producing, Mediterranean countries and non-producing countries. For the former, olive oil can be considered as a traditional food product. In this respect, the literature shows some important associations between the consumption of such traditional products and cultural aspects such as values, beliefs and life-style orientations (Vanhonacker et al., 2010). In this respect, as noted by Govers and Schoormans (2005), some studies have tested how consumers prefer products or brands with a particular symbolic meaning, compatible with the image they wish to convey of themselves. In some cases, this is intended to resemble the kind of people who normally use the product (Heath and Scott, 1998). Thus, traditional food consumers are generally not caught up in modern ways of life (Guerrero et al., 2009), where time pressure, business and convenience orientation dominate. Housewives are usually portrayed in the literature as typical consumers. In addition, traditional food consumers are also portrayed as liking the familiar; one expression of this preference being the consumption of familiar dishes (Dagevos, 2005). According to this author, these consumers have fairly conservative food habits, maintaining culinary customs across generations. In addition, they are concerned about their health. In this context, olive oil plays an important role.

There exist several studies analysing the role of culture and food habits in the behaviour of consumers regarding such products. Thus, Nielsen *et al.'s* (1998) cross-cultural study showed that there were large differences in the perceptions of virgin olive oils across UK, Denmark and France. Olive oil users from all three countries agreed on the health benefits of virgin olive oil, which led to the feeling of good health and a long life. Therefore, both hedonistic and sensory aspects of virgin olive oil appeared the most varied between countries.

In the UK, García *et al.* (2002) used focus groups and conjoint analysis to analyse the product attribute trade-offs that consumers make when choosing olive-oil products. One of this paper's main findings was that British consumers continue to regard olive oil as a set of individual attributes (e.g., size, taste and health) instead of a product that is perceived as encapsulating all these attributes. They found that price was one of the most influential factors on consumers' preferences for basic olive oil, followed by size of container. The main role of price was recently pointed out by Dekhili and d'Hauteville (2009), who found that price was the most important choice attribute in both producing (e.g., Tunisia) and non-producing (e.g., France) countries. In this respect, although fair prices can be charged for olive oil compared with other vegetable oils, there is a limit to the price many consumers are willing to pay (Mili, 2006). The use of price as a choice criterion for consumers is a consequence of the variety of olive-oil brands. Nevertheless, it is true that aspects such as colour, packaging and product labelling are helping producers to differentiate their brands from those of competing suppliers in the distribution chain for olive oil (Van der Lans et al., 2001).

Moreover, aspects related to the origin of olive oil are becoming more important in consumers' choice behaviour. Dekhili and d'Hauteville (2009) showed that the region of origin was relevant in explaining consumer behaviour. Such a regional image has three components: (i) local agronomic conditions (soils, climate); (ii) traditional human know-how; and (iii) raw product characteristics (variety). Thus, these authors found important differences between France and Tunisia in giving credence to the role of an olive-oil-specific regional image. In particular, these authors found quite significant differences regarding the relative weights of the attributes of this image valued in each country. In this respect, there is a growing segment of consumers who prefer quality food with certification of origin (both Protected Designation of Origin [PDO] and Protected Geographical Indication [PGI]). Dekhili et al. (2011) found that these 'official cues' are more important for those consumers belonging to non-producing olive oil countries (e.g., France), whereas in producing countries (e.g., Tunisia) consumers tend to choose olive oil based on origin and 'sensory cues' (e.g., colour and appearance). For instance, in Spain there are 32 PDOs for olive oil; that is, Andalusia, in southern Spain, the geographical area with the highest number of certifications of origin. Sanz and Macías (2005) confirmed the strategic role of Spanish olive oil PDOs. Thus, these authors found that such PDOs, effectively, add greater value to local production systems and so enhance the final quality and market differentiation of a specific-origin olive oil. In this respect, Scarpa et al's. (2005) study in the context of three products (table grapes, oranges and olive oil), confirmed the importance of PDOs. According to these authors, the role of PDOs was stronger for olive oil compared to the other two categories analysed.

Thompson et al. (1994) used Ajzen & Fishbein's (1980) *theory of reasoned decision* (TRA) successfully, as a mean of identifying the major issues influencing olive oil choice in the UK. These authors found that attitudes were strongly related to the user or non-user of olive oil. In this respect, the most significant attitude related to the flavour-improving attributes of olive oil (e.g., improving the taste of salads and cooked meals).

In Mediterranean countries, Saba & Di Natale (1998) surveyed 909 Italians in order to assess their attitudes towards fats and food choice. The researchers also used Ajzen & Fishbein's (1980) TRA, combined with a measure of 'habit', as a theoretical framework. The findings suggested that in Italy, culture and food habits might predict intention to consume fats and oils better than TRA. Saba *et al.* (2000) recently re-confirmed this attitudinal TRA model in the Italian context.

Another interesting aspect related to culture and purchase habits is the place of purchase. Delgado and Guinard's (2011) study of US consumers, reported that the majority of them bought olive oil (extra-virgin) primarily at the supermarket (68%), specialty stores (50%) and farmers' markets (43%), in contrast with the ways in which Mediterranean consumers most frequently buy their olive oil. Thus, Fotopoulos and Krystallis (2001) reported that 41% of Cretan consumers buy olive oil at the supermarket, while 38% buy in bulk directly from the producer or farm, and 21% make oil from their own olive orchards. Similar figures can be associated with other Mediterranean producing countries such as Italy or Spain. This habit is a consequence of the consumer experience, of belonging to producing countries and the role of olive oil in their intrinsic cultures. Thus, in buying olive oil at supermarkets or hypermarkets, consumers are not exposed to the sensory properties of the product, as they are at farmers' markets or direct from the producers or farms, and so their decisions are based on extrinsic factors such as packaging materials, bottle material and label design (Delgado and Guinard, 2011). This is the case of US consumers.

This is probably the reason why US consumers prefer Italian oil more than Spanish oil. In our opinion, Italian oil possesses a lower quality than Spanish olive oil. However, Italian oil's marketing strategy, from a general point of view, is stronger than the strategies used by Spanish producers. In contrast, when consumers buy oil in bulk directly from the producer, as is the case with consumers belonging to producing countries, they experience the properties of the oil and can make purchasing decisions based on sensory factors. This is the case for consumers living in Mediterranean countries. Even, nowadays these 'experienced consumers' show a greater interest in organic olive oil, given the increasing interest of consumers in ecologically clean products for health and environmental reasons (Gavruchenko et al., 2003). Consumers' need for safer, good quality food has increased over the last years and thus, healthiness and nutritional value are the basic reasons given by consumers for purchasing organic olive oil. In this way, more consumers are willing to pay a higher price, since they take into account organic olive oil's contribution (Sandalidou et al., 2002:405). Nevertheless, Sandalidou et al. (2003) pointed out that there is a large number of people who still do not know what an organic product is. For this reason, these authors suggest that the systematic provision of information, mainly through advertising, is necessary, in order to enhance consumers' awareness of organic olive oil's features and nutritional content.

3. Analysis of consumer preferences of olive oil

What motivates consumers to prefer and purchase olive oil is not clear (Delgado and Guinard, 2011:214). As has been indicated before, some authors highlight, as the main motivators behind consumption, the oil's region of origin, focusing on the influence of PDO designation and the degree to which an oil typifies the characteristics of that particular region. This is especially true for those consumers who are experienced, local or familiar with a particular region of origin, whereas these factors do not seem to affect urban, less knowledgeable and less experienced consumers (Caporale et al., 2006).

Other authors focus on olive oil's health benefits and flavour (including its use to enhance the taste of recipes) as main motivators for olive oil consumption. Thus, olive oil is promoted as beneficial for health, and industrial strategies and advertising are often based on health claims (Duff, 1998) although, nowadays, EU regulation has imposed the use of 'nutritient profiles', which are already in use in the USA and Canada, and which are under development (Blery and Sfetsiou, 2008:1151). However, many critics argue that this aspect would mean that products such as olive oil should not carry health claims. At present, there is no harmonized legislation at EU level to ensure the scientific accuracy and appropriateness of such claims. Nevertheless, homogeneous regulation is expected to set clear parameters across Europe for health claims, and they will be allowed only if they are substantiated scientifically (Tamsin *et al.*, 2005). Given the increasing number of countries being integrated into the EU in the last few years, this seems to be even more important for the success of European olive oil production.

New olive oil consumers seem to be more interested in olive oil for two main reasons: health benefits and flavour (Santosa, 2010). Olive oil is claimed to be beneficial for health, as it is rich in vitamin E and it does not contain preservatives (Blery and Sfetsiou, 2008). Among health benefits, lowering the risk of coronary disease, preventing certain kinds of cancer and reducing inflammation have been highlighted (Medeiros and Hampton, 2007). For these reasons, Duff (1998) pointed out that the preference for olive oil is a result of health reasons because the replacement of saturated fats by olive oil results in a lowering of the rate of heart disease. Nevertheless, it is true that there other cheaper seed oils being used as substitutes (Bernabéu et al., 2009). In this respect, olive oil has a high price, although it depends on its origin and its quality (Bourdieu, 1984). For instance, virgin and extra-virgin olive oils are more expensive than standard olive oil. With regard to flavour, Santosa and Guinard (2011) recently reported that this is an important aspect in both the consumption and purchase motivations for olive oil, especially for extra-virgin olive oil, where sensory characteristics are even more important. According to Thompson *et al.* (1994), this is also a consequence of improving the taste of salads and meals.

3.1 Study of consumer preferences of olive oil through conjoint analysis

The measurement of attitudes/preferences using a multi-attribute methodology, especially a conjoint Analysis of Multivariate technique, is most appropriate.

In fact, this methodology has become an important tool to assess the preferences that a consumer assigns to the various attributes of a specific product/brand (Ruiz and Munuera, 1993). Hair *et al.* (2009) define conjoint analysis as:

'a multivariate technique used specifically to understand how respondents develop preferences about products or services, and is based on the simple premise that consumers assess the value of a product/service idea (real or hypothetical) combining separate amounts of value provided by each attribute. The utility, which is the conceptual basis for measuring this value, is a subjective preference unique to each individual which includes all the features of a product or service, both tangible and intangible, and as such, is the measure of overall preference'.

The most direct application of conjoint analysis is as a tool to find the weight or importance that different levels or categories of product attributes play on the formation of consumer preferences (Múgica, 1989). Therefore, conjoint analysis seeks to establish the relative importance of attributes and levels, inferring the utility (satisfaction) that consumers express when they are presented with a series of product concepts that vary in a systematic way (Walley *et al.*, 1999). The application of this methodology in the field of food has, until recently, been quite limited (Van der Pol and Ryan, 1996), starting in the 1990s when it began to generate a relevant scientific production. This confirms the suitability of this methodology to improve knowledge about consumer behaviour when purchasing food. Thus, when reviewing the literature, it appears that it has only been in recent years that there has been further development in this field of research, ranging from the wine, to the meat, dairy, fruit and vegetable industries. In the case of olive oil, although the literature is not extensive, there are several studies that have examined consumer preferences in deferent countries, as shown in Table 1.

Despite the tradition existing in the consumption of olive oil in the main producing countries, studies carried out to analyse consumer preferences in these countries have been scarce. Probably, this is a consequence of the olive oil's difficulties for differentiating itself in order to better meet the needs of consumers.

One of the first studies that examined consumer preferences is Fotopoulos and Krystallis (2001) that analysed Greek consumer preferences based on two attributes: price and character of protected designation of origin (PDO). Van der Lans *et al.* (2001), in turn, focus their analysis of preferences on extra-virgin olive oil, a variety that is characterized by acidity (oleic acid) to a maximum of 0.8 g per 100 g. According to these authors, the selected attributes were price, colour, origin and appearance, like the unit sample in two Italian regions.

Garcia *et al.* (2002) provide the first work that analyses consumer preferences of olive oil in a non-producing country such as the UK. Following this study, other countries have been used (e.g., Japan (Mtimet *et al.*, 2008) and Canada (Menapace *et al.*, 2011)).

Therefore, in analysing consumer behaviour for olive oil, there are various areas of analysis that take us beyond the attributes that have been considered in each study. In this respect, preference will be conditioned by the variety of olive oil covered by the study, and analysed even if the consumer resides in a country producing this product or not.

A first result that emerges from the literature review is that the extrinsic attributes of olive oil (e.g., price, origin or variety) are the most important when consumers face the act of purchase. Instead, intrinsic attributes, such as colour or flavour, are relegated to second place, with the exception of Mtimet *et al.* (2008), who analysed the Japanese consumer, for whom colour comes first.

Focusing specifically on the extrinsic attributes, a second interesting result refers to the importance of price when buying olive oil. Indeed, in five of the nine studies analysed this is the attribute with the highest relative importance. It should also be noted that this is true for consumers belonging to both producer and non-producer countries, and is also irrespective of the variety of olive oil.

In addition, our analysis shows that origin of oil is also an extrinsic attribute of interest to the consumer. Thus, in a majority of the papers analysed, it is the first or second attribute in order of importance, either in the consumer's conceptualization of country of origin, the region of origin or as part of a Protected Geographical Indication.

With regard to the variety of olive oil, another interesting finding is that,for the case of extra-virgin olive oil, price is always the most important attribute.

Finally, with regard to the sampling unit of analysis, it can be observed that among consumers of non-producing countries, the origin of oil is not the primary attribute on which they make their purchasing decision. Indeed, in two of the three studies analysed, the origin of the oil is the second attribute in importance. In the third study, this factor was not considered. This conclusion is very different among consumers of producer countries, probably as a consequence of the fact that they are more familiar with the product.

	Type	Country	Producing	Attribute	Rank (relative importance) (%)
Fotopoulos and Krystallis (2001)	Olive	Greece	Yes	PDO Price	1st - (55.51) 2nd- (44.49)
van der Lans, van Ittersum, De Cicco and Loseby (2001)	Extra-virgin olive	Italy	Yes	Price Colour Origin Appearance	1st- (n.d.) 2nd- (n.d.) 3rd- (n.d.) 4th- (n.d.)
García, Aragonés and Poole (2002)	Olive	UK	No	Price Size Packaging	1st- (37.40) 2nd- (33.48) 3rd- (29.12)
Scarpa and Del Giudice (2004)	Extra-virgin olive	Italy	Yes	Price Quality Certification Origin Appearance	1st- (44.44) 2nd- (28.60) 3rd- (25.41) 4th- (1.54)
Krystallis and Ness (2005)	Olive	Greece	Yes	Origin Organic label Health info Quality certifications (HACCP, ISO) PDO label Price Glass bottle	1st- (21.71) 2nd- (19.07) 3rd- (16.96) 4th- (11.11) 5th- (9.58) 6th- (8.10) 7th- (7.17) 8th- (6.29)
Mtimet, Kashiwagi, Zaibet and Masakazu (2008)	Olive	Japan	No	Colour Origin Price Olive oil type Taste	1st- (30.14) 2nd- (29.06) 3rd- (20.50) 4th- (10.34) 5th- (9.94)
Bernabeu, Olmeda, Diaz y Olivas (2009)	Olive	Spain	Yes	Oil type Origin Price Production System	1st- (41.09) 2nd- (33.35) 3rd- (25.35) 4th- (0.20)
Chan-Haldbrent, Zhllima, Sisior and Imami (2010)	Olive	Albania	Yes	Price Olive oil type Origin Taste Place purchase	1st- (34.70) 2nd- (22.16) 3rd- (20.96) 4th- (18.66) 5th- (3.52)
Menapace, Colson, Grebitus and Facendola (2011)	Extra-virgin olive	Canada	No	Price Origin Production system Geographic identification Appearance Colour	1st- (36.88) 2nd- (26.54) 3rd- (23.72) 4th- (12.49) 5th- (0.35) 6th- (0.00)

Table 1. Analyses for consumer preferences of olive oil.

4. Conclusions and recommendations

From a marketing point of view, consumers' purchasing behaviour is affected by the presence of heterogenous preferences that are derived from their own needs. This is especially important when the consumer is faced with new products or innovations. With regard to olive oil, an agrofood product that is one of the main components of the well-known Mediterranean diet, this heterogeneity is still greater, given the lower consumer knowledge in relation to such a product. Nevertheless, it is not exclusive of markets in the first stages of adopting olive oil, but is also, to a lesser extent, present in countries with a stronger tradition of olive oil consumption. Thus, the majority of studies report a stronger role of extrinsic aspects over intrinsic attributes. In this context, price is the major factor affecting consumer behaviour. Nevertheless, when the consumer is familiar with the product, the country of origin is the most influential aspect in determining consumer purchases.

The results obtained here lead the authors to offer a number of implications for marketing management in the olive oil sector. Given the strong presence of olive oil in international markets, as well as production systems strongly linked to a specific area of origin, promoting familiarity or experience of the products could provide firms with an effective source of competitive advantage, because of the positive consequences for the product image.

In particular, for those markets with less knowledge of olive oil (e.g., the US or China) communication actions will be critical for increasing consumers' familiarity and consequently their knowledge about these products. Thus the objective of commercial communication should consider not only different perceptions of the product cues, but also the differences in product familiarity. In particular, for countries in which product knowledge is greater and consumers have greater knowledge of the intrinsic characteristics of the product, communication campaigns should aim to reinforce consumers' image of the product. In contrast, in countries where the product knowledge is more limited, communication should focus on raising the consumers' level of familiarity and knowledge. This will raise the consumer confidence and, by extension, their purchase intention (Laroche *et al.*, 1996). In general terms, such campaigns should be divided into four consecutive stages:

1. *Campaign to raise awareness among producers and exporters of olive oil.* This first stage should focus on informing producers/exporters about the positive consequences that a favourable image of the area of origin can have for each firm in particular, and for the sector in general. It is necessary to improve training in marketing and communication, for producing and commercial firms, so that they adopt a market orientation approach. For instance, this is the case for Spanish olive oil. Such campaigns are necessary if Spanish olive oil wants to improve its international market penetration, in relation to Italian olive oil.

2. *Educational campaign aimed at consumers.* The main objective at this stage should be to inform the consumer about the area of origin, the olive oil's area characteristics, the production techniques, and so forth. In short, the objective should be to communicate to the consumer the sector's enormous experience in the cultivation of olive oil, and the quality and safety processes used to obtain the end product. This stage of the campaign should also include communication actions aimed at prescribers (e.g., restaurant owners, food specialists) so that they recommend the product to the end consumers (current and potential consumers).

3. *Campaign aimed at distinguishing the most important product cues in the formation of the image in each target market.* The advertising messages should be adapted to each market as a function of its preferences, and the tools and communications media chosen should be consistent with the messages. The messages should transmit the idea that olive oil from the area of origin has attributes that are the most important for that particular consumer.

4. *Campaign to increase consumption and maintain loyalty to the brand.* After achieving the positioning, the consumer should be reminded of the benefits of consuming olive oil from the specific area of origin. At this stage, the communication campaigns must consolidate the product image and be aimed at different segments. Likewise, in the distribution channel, the communication should be flexible and adapted to the particular needs of each channel agent.

In developing all these communication campaigns, public institutions and governments can play an important role. Given the importance of olive oil sector in their socio-economic context, this is especially relevant for the case of Mediterranean olive oil producing countries. Therefore, public institutions are advised to collaborate with their own countries' olive oil sector.

5. References

Ajzen, I. & Fishbein, M. (1980). *Understanding and predicting social behaviour*, Englewood Cliff, NJ: Prentice Hall

Bernabéu, R.; Olmeda, M.; Díaz, M. & Olivas, R. (2009). Oportunidades comerciales para el aceite de olive de Castilla-La Mancha. *Grasas y Aceites*, Vol.60 No.5, pp. 525-533.

Blackwell, R.D.; Miniard, P.W. & Engel, J. (2001). *Consumer behaviour, 9th ed.*, Hartcourt Collage Publishers, Fort Worth, TX.

Blery, E. & Sfetsiou, E. (2008). Marketing olive oil in Greece. British Food Journal, Vol. 110, No.11, pp. 1150-1162.

Bourdieu, P. (1984). *Distinction: A Social Critique of the Judgment of Taste*, Routledge & Kegan, Paul, London.

Caporale, G.; Policastro, S.; Carlucci, A. & Monteleone, E. (2006). Consumer expectations for sensory properties in virgen olive oils, *Food Quality and Preferences*, 17, pp. 116-125.

Chan-Halbrendt, C.; Zhllima, E.; Sisior, G.; Imami, D. & Leonetti, L. (2010). Consumer preferences for olive oil in Tirana, Albania. *International Food and Agribusiness Management Review*, Vol.13, No. 3, pp. 55-74.

Dagevos, H. (2005). Consumers as four-faced creatures: looking at food consumption from the perspective of contemporary consumers, *Appetite*, Vol. 45, pp. 32-9.

de la Viesca, R.; Fernández, E.; Fernández, S. & Salvador, J. (2005). Situation of European SMEs in the olive oil and table olive area. Survey, *Grasas y Aceites*, Vol. 56, No.3, pp. 209-219.

Dekhili, S. & d'Hauteville, F. (2009). Effect of the region of origin on the perceived quality of olive oil: an experimental approach using a control group, *Food Quality and Preference*, Vol. 20, pp. 525-532.

Dekhili, S.; Siriex, L. & Cohen, E. (2011). How consumers choose olive oil: the importance of origin cues, *Food Quality and Preference*, doi:10.1016/j.foodqual.2011.06.005

Delgado, C. & Guinard, J-X. (2011). How do consumer hedonic rating for extra virgen olive oil relate to quality ratings by experts and descriptive analysis ratings?, Food Quality and Preference, Vol. 22, pp. 213-225.

European Commission (2010). Statistical and economic information report 2009, available at: http://bookshop.europa.eu/en/agriculture-in-the-eu-pbKFAC10001/downloads/KF-AC-10-001-EN-C/KFAC10001ENC_002.pdf?FileName=KFAC10001ENC_002.pdf&SKU=KFAC100 01ENC_PDF&CatalogueNumber=KF-AC-10-001-EN-C (accessed August 2, 2011).

Fotopoulos, C. & Krystallis, A. (2001). Are quality labels a real marketing advantage? A conjoint application on Greek PDO protected olive oil. *Journal of International Food & Agribusiness Marketing*, Vol.12, No.1, pp. 1-22.

García, M.; Aragonés, Z. & Poole, N. (2002). A repositioning strategy for olive oil in the UK market. *Agribusiness*, Vol.18, No.2, pp. 163-180.

Gavruchenko T.; Baltas G.; Chatzitheodoridis F.; & Hadjidakis S. (2003). Comparative marketing strategies for organic olive oil: The case of Greece and Holland, in A. Nikolaidis *et al* (eds) *The Market for Organic Products in the Mediterranean Region*, Cahiers Options Mediterraneennes, Vol 61, pp. 247-255

Govers, P. & Schoormans, J. (2005). Product personality and its influence on consumer preference. *Journal of Consumer Marketing*, Vol. 22, No.4, pp. 189-197.

Guerrero, L.; Guardia, M.; Xicola, J.; Verbeke, W.; Vanhonacker, F.; Zakowska, S.; Sajdakowska, M.; Sulmont-Rosse´, C.; Issanchou, S.; Contel, M.; Scalvedi, L.; Granli, B. & Hersleth, M. (2009). Consumer-driven definition of TFP and innovation in traditional foods. A qualitative cross-cultural study, *Appetite*, Vol. 52, pp. 345-54

Hair, J.F.; Anderson, R.E.; Tatham, R.L. & Black, W.C. (2009). *Multivariate data analysis, 7th ed.*, Prentice Hall.

Heath, A.P. & Scott, D. (1998). The self-concept and image congruence hypothesis: an empirical evaluation in motor vehicle market. *European Journal of Marketing*, Vol.32, No.11/12, pp. 1110-1123.

Hellonet (2006). The Greek olive oil, available at: http://hellonet.teithe.gr/GR/aboutgreece_gr.htm (accessed August 2, 2011).

Howard, J. A. & Sheth, J.N. (1969). *The Theory of Buyer Behavior*, Nueva York, John Wiley & Sons.

International Olive Council (2008). *Olive products market report summary*, Market Commentary.

Krystallis, A. & Ness, M. (2005). Consumer preferences for quality foods from a South European perspective: A conjoint analysis implementation on greek olive oil. *International Food and Agribusiness Management Review*, Vol.8, No.2, pp. 62-91.

Lambin, J.J. (1995). *Marketing Estratégico*. Mc Graw Hill.

Laroche, M.; Kim, Ch.; & Zhou, L. (1996). Brand familiarity and confidence as determinants of purchase intention: an empirical test in a multiple brand context, *Journal of Business Research*, Vol. 37, pp. 115-120.

Lilien, G.L. & Kotler, P. (1990): *Toma de decisiones en mercadotecnia. Un enfoque a la construcción de modelos*. CECSA.

Medeiros, D. & Hampton, M. (2007). Olive oil and health benefits, in R.E.C. Wildman (Ed.), *Handbook of nutraceuticals and functional foods*, CRC Press, Boca Raton, FL.

Menapace, L.; Colson, G.; Grebitus, C & Facendola, M. (2011). Consumers' preferences for geographical origin labels: evidence from the Canadian olive oil market. *European Review of Agricultural Economics*, Vol.38, No.2, pp 193-212

Mili, S. (2006). Olive oil marketing in non-traditional markets: prospects and strategies, *New Medit*, Vol. 5, No.1, pp. 27-37.

Mtimet, N.; Kashiwagi, A.K.; Zaibet, L. & Masakazu, N. (2008). Exploring Japanese olive oil consumer behavior. *International Congress from European Association of Agricultural Economists*, August 26-29, Ghent, Belgium, Paper No. 44447.

Múgica, J.M. (1989). Los modelos multiatributo en marketing: el análisis conjunto. *IP-MARK*, Vol.324, pp. 63-71.

Nielsen, N., Bech-Larsen, T. & Grunert, K. (1998). Consumer purchase motives and product perceptions: a laddering study on vegetable oil in three countries. *Food Quality and Preference*, Vol.9, No.6, pp. 455-466

Ruiz, S. & Munuera, J.L. (1993). Las preferencias del consumidor: Estudio de su composición a través del análisis conjunto. *Estudios sobre Consumo*, Vol.28, pp. 27-44.

Saba, A. & Di Natale, R. (1998). Attitudes, intention and habit: their role in predicting actual consumption of fats and oils. *Journal of Human Nutrition and Dietetics*, Vol. 11, No. 1, pp. 21-32.

Saba, A., Vassallo, M. & Turín, A. (2000). The role of attitudes, intentions and habit in predicting actual consumption of fat containing foods in Italy. *European Journal of Clinical Nutrition*, Vol. 54, No.7, pp. 540-545.

Sandalidou, E. & Baourakis, G. (2002). Customers perspectives on the quality of organic olive oil in Greece, *British Food Journal*, Vol. 104 Nos. 3/5, pp. 391-406.

Sandalidou, E.; Baourakis, G. & Siskos, Y. (2002). Customers' perspectives on the quality of organic olive oil in Greece. A satisfaction evaluation approach, *British Food Journal*, Vol. 104, No. 3/4/5, pp. 391-406.

Sandalidou E.; Baourakis G.; Grigoroudis E. & Siskos Y. (2003). Organic and conventional olive oil consumers: a comparative analysis using a customer satisfaction evaluation approach, in A. Nikolaidis *et al.* (eds), *The Market for Organic Products in the Mediterranean Region*, Cahiers Options Mediterraneennes, Vol. 61, pp. 265-276.

Santosa, M. (2010). *Analysis of sensory and non-sensory factors mitigating consumer behaviour: a case study with extra virgin olive oil*, PhD Dissertation, Food Science, University of California, Davis.

Santosa, M. & Guinard, J-X. (2011). Means-end chains analysis of extra virgin olive oil purchase, *Food Quality and Preference*, Vol. 22, pp. 304-316.

Sanz, J. & Macías, A. (2005). Quality certification, institutions and innovation in local agro-food systems: protected designations of origin of olive oil in Spain, *Journal of Rural Studies*, Vol. 21, pp. 475-486.

Scarpa, R. & Del Giudice, T. (2004). Market segmentation via mixed logit:extra-virgin olive oil in urban Italy. *Journal of Agricultural &Food Industrial Organization*, Vol.2, No.1, pp.

Scarpa, R., Philippidis, G. & Spalatro, F. (2005). Product-country images and preference heterogeneity for Mediterranean food products: a discrete choice framework, *Agribusiness*, Vol. 21, No.3, pp. 329-349.

Schiffman, L. & Kanuk, L. (2001): *Comportamiento del consumidor*, Prentice Hall, 7ª ed.

Solomon, M.; Bamossy, G. & Askegaard, S. (1999). *Consumer Behaviour. A European Perspective*, Prentice Hall Europe.

Soons, L. (2004). The olive oil market of Mainland China, Master Thesis, Lund University, available at:
http://lup.lub.lu.se/luur/download?func=downloadFile&recordOId=1331369&fileOId=1331370 (accessed August 3, 2011)

Steenkamp, J.B. (1997). Dynamics in consumer behaviour with respect to agricultural and food products, in B. Wierenga, A. Van Tilburg, K. Grunert, JB. Steenkamp y M. Wedel (eds.), *Agricultural Marketing and Consumer Behaviour in a Changing World*, Kluwer Academic Publishers, Boston, pp. 143-188.

Tamsin, R.; Walshe, D.; Logstrup, S.; Giorgio-Gerlach, F. & Craplet, M. (2005). Destroying myths and misunderstandings of the claims regulation, European Community of Consumer Cooperatives, Brussels.

Thompson, K., Haziris, N. & Alekos, P. (1994). Attitudes and food choice behaviour. *British Food Journal*, Vol. 96, No.11, pp. 9-13.

Vanhonacker, F.; Lengard, V.; Hersleth, M. & Verbeke, W. (2010). Profiling European traditional food consumers, *British Food Journal*, Vol. 112, No. 8, pp. 871-886.

Van der Lans, I.A.; Van Ittersum, K.; de Cicco, A. & Loseby, M. (2001).The role of the region of origin and EU certificates of origin in consumer evaluation of food products. *European Review of Agricultural Economics*, Vol.28, No.4, pp. 451-477.

Van der Pol, M. & Ryan, M. (1996). Using conjoint analysis to establish consumer preferences for fruit and vegetables. *British Food Journal*, Vol.98, pp. 5-12.

Walley, K.; Parsons, S. & Bland, M. (1999). Quality assurance and the consumer: a conjoint study. *British Food Journal*, Vol.101, No.2, pp. 148-161.

Zampounis, V. (2006). Olive oil in the world market, in D. Boskou (Ed.) *Olive oil chemistry and technology*, AOC Press, pp. 21-39.

12

Quality Assessment of Olive Oil by ^{1}H-NMR Fingerprinting

Rosa M. Alonso-Salces[1,*], Margaret V. Holland[1],
Claude Guillou[1] and Károly Héberger[2]
[1]European Commission - Joint Research Centre,
Institute for Health and Consumer Protection,
Physical and Chemical Exposure Unit, Ispra,
[2]Chemical Research Center,
Hungarian Academy of Sciences, Budapest,
[1]Italy
[2]Hungary

1. Introduction

Olive oil is the oil extracted exclusively from the fruit of *Olea europea* L. only by means of mechanical methods or other physical procedures that do not cause any alteration of the glyceric structure of the oil thus preserving its characteristic properties. Olive oil is a highly appreciated edible oil, which is an important component of the Mediterranean diet, and is recognized for its potential health benefits.

The International Olive Council (IOC) establishes the definitions and classes of olive oils, based on methods of production and the free acidity of the oil, as well as the trade standard for their commercialization (International Olive Council, 2011). Much analytical work has been done on the authentication and quality assessment of this high added value agricultural product, as well as on the detection of its adulteration for both economic and health considerations (Frankel, 2010; Guillen & Ruiz, 2001). However, these issues continue to be major analytical challenges. In this context, the European Commission launched the TRACE project (http://www.trace.eu.org/) through the Sixth Framework Program under the Food Quality and Safety Priority with the aim of providing reliable analytical strategies to address this kind of problem.

This chapter reports research work on the use of ^{1}H-NMR fingerprinting, combined with pattern recognition techniques, for the quality assessment of olive oil. Two major issues have been studied, the geographical origin of virgin olive oil (VOO) and the stability of VOO at room temperature.

---

* Corresponding Author
The information contained in this chapter reflects the authors' views; the European Commission is not liable for any use of the information contained herein.

1.1 Geographical origin of virgin olive oil

At present, 75% of the global production of olive oil takes place in the Mediterranean basin, mainly in Spain, Italy and Greece (International Olive Council, 2011). VOO is permitted to be marketed under a Protected Designation of Origin (PDO), Protected Geographical Indication (PGI), or Traditional Specialty Guaranteed (TSG) label, on the basis of its area and method of production [Council Regulations (EC) No 510/2006 and No 509/2006]. In this context, the characterization of the geographical origin of VOO is becoming increasingly important. According to the EU definition, PDO products are most closely linked to the concept of *terroir* — a sense of place discernible in the flavor of the food. PDO products must be produced, processed and prepared in a specific region using traditional production methods. The raw materials must also be from the defined area whose name the product bears. The quality or characteristics of the product must be due essentially or exclusively to its place of origin, i.e., climate, the nature of the soil and local know-how. Food products with a PGI status must have a geographical link in at least one of the stages of production, processing or preparation. The European Commission has already registered in the "Register of protected designations of origin and protected geographical indications" 86 PDO and 15 PGI olive oils, produced in Italy, Greece, Spain, France, Portugal and Slovenia. As can be expected, given the financial benefits associated with these prestigious labels, it is very likely that economic fraud occurs (e.g. labeling a non-PDO product as a PDO one or adulteration with olive oils that do not fulfill the PDO requirements).

Another fraudulent practice is the mislabeling of the geographical origin of olive oils. The EU established new labeling rules that make origin labeling compulsory for virgin and extra virgin labeled olive oil [Commission Regulation (EC) No 182/2009]. So, oil produced from olives from just one EU Member State or third country has to be labeled with the name of the country of origin. VOO produced from olives from more than one EU Member State has to be labeled as a 'blend of Community olive oils', while oil produced using olives from outside the EU would be labeled as a 'blend of non-Community olive oils' or 'blend of Community and non-Community olive oils', or a reference to the EU and/or third countries of origin. Therefore, analytical methods are urgently needed to guarantee the authenticity and traceability of PDO and PGI olive oils, as well as the country of provenance, to help prevent illicit practices in this sector, and to support the antifraud authorities dealing with these issues.

VOO is made up of triglycerides (more than 98%) and minor components (about 1-2%) such as squalene, α-tocopherol, phytosterols, phenolic compounds, carotenoids, and aliphatic and terpenic alcohols, which constitute the unsaponifiable fraction of the oil (Bortolomeazzi et al., 2001; Harwood & Aparicio, 2000). The chemical composition of this fraction may vary both qualitatively and quantitatively depending on vegetal species, climatic conditions, extraction and refining procedures and storage conditions (Canabate-Diaz et al., 2007; Harwood & Aparicio, 2000), which also greatly influence the organoleptic quality and stability of the oil. The diversity and interdependence between all these factors makes it highly unlikely that these influences would be the same in different regions. Hence, the geographical characterization of VOO addresses all these agronomic, pedoclimatic and botanical aspects which are unique to the oil of each origin (Aparicio et al., 1994).

A considerable number of sensorial (Pardo et al., 2007), physical (F. Marini et al., 2004; Federico Marini et al., 2006) and chemical (Aguilera et al., 2005; Boggia et al., 2002; Lanteri et al., 2002; Federico Marini et al., 2006; Federico Marini et al., 2007) approaches combined with statistical analysis have been used to distinguish olive oils from different types, botanical, geographical origins and pedoclimatic conditions. For this purpose, fatty acids (Matos et al., 2007; Ollivier et al., 2006), triglycerides (Aranda et al., 2004), sterols (Alves et al., 2005; Matos et al., 2007; Temime et al., 2008), phenolic compounds (Lopez Ortiz et al., 2006; Vinha et al., 2005), and pigments (Cichelli & Pertesana, 2004) have been analyzed by conventional methods that usually require time-consuming pre-treatment methods (solvent extraction, isolation and/or derivatization) followed by chromatographic techniques (Aparicio & Aparicio-Ruiz, 2000) such as GC-MS and/or GC-FID (Bechir Baccouri et al., 2007; Haddada et al., 2007; Temime et al., 2006; Vichi et al., 2005) and HPLC-MS (Canabate-Diaz et al., 2007; Lopez Ortiz et al., 2006). PDO olive oils were distinguished using physicochemical parameters of the oils and chemometric class-modeling tools (Federico Marini et al., 2006), sensory parameters and fatty acid profiles of the oils (Ollivier et al., 2006), or the oil sterol composition (Alves et al., 2005). IRMS measurements of the alcohol and sterol fractions of olive oil also proved to be useful for its geographical characterization (Angerosa et al., 1999).

Fingerprinting techniques such as NMR (Mannina & Segre, 2002), NIR (Mignani et al., 2011), MIR (Reid et al., 2006), fluorescence (Kunz et al., 2011), FT-IR, FT-MIR and FT-Raman (Baeten et al., 2005; Lopez-Diez et al., 2003; Yang et al., 2005) spectroscopies, MS (Vaclavik et al., 2009), GC (Pizarro et al., 2011; Vaz-Freire et al., 2009) and DNA fingerprinting (Martins-Lopes et al., 2008; Ranalli et al., 2008) have been used for the determination of food authenticity (Reid et al., 2006). These types of techniques are particularly attractive since they are non selective, require little or no sample pre-treatment; use small amounts of organic solvents or reagents; and the analysis takes only a few minutes per sample. Chemometric analysis of NIR spectra of virgin olive oils allow us to determine its composition and geographical origin (Galtier et al., 2007). ^{1}H, ^{13}C and/or ^{31}P-NMR analysis of the bulk oil (Rosa M. Alonso-Salces et al., 2010b; Rosa M. Alonso-Salces et al., 2011b; Mannina et al., 2010; Petrakis et al., 2008; Rezzi et al., 2005) or the unsaponifiable fraction of olive oil (R. M. Alonso-Salces et al., 2010), in combination with multivariate techniques, have been used to distinguish VOOs from certain geographical origins. In section 3.2., the achievements of ^{1}H-NMR fingerprinting and chemometrics for the geographical characterization of VOO is reported. ^{1}H-NMR fingerprints of a statistically significant number of authentic VOOs from seven countries, namely Italy, Spain, Greece, France, Turkey, Cyprus and Syria and from three different harvests (2004/05, 2005/06 and 2006/07) were analyzed by pattern recognition and classification techniques, such as principal component analysis (PCA), linear discriminant analysis (LDA) and partial least square discriminant analysis (PLS-DA), to evaluate the best approach to identify the geographical origin at the national, regional and/or PDO level.

1.2 Stability of virgin olive oil

Another matter of major concern regarding the quality of edible oils is their oxidation, not only from the technological and economic point of view but also for safety reasons, due to the undesirable properties of some compounds produced during this process (Guillen & Ruiz, 2001).

The high resistance to oxidative deterioration of VOO is due to both its triglyceride composition, which is low in polyunsaturated fatty acids, and its antioxidant constituents, i.e. polyphenols and tocopherols. The oxidative stability of VOO has been evaluated by several methods: Rancimat test (Di Lecce et al., 2009; Esquivel et al., 2009; Kamvissis et al., 2008; Mateos et al., 2006; Platero-López & García-Mesa, 2007), oxygen stability index (OSI) (Carrasco-Pancorbo et al., 2007; Ceci & Carelli, 2010; Cercaci et al., 2007; Gómez-Caravaca et al., 2007; Márquez-Ruiz et al., 2008), peroxide value (Carrasco-Pancorbo et al., 2007; Di Lecce et al., 2009; Márquez-Ruiz et al., 2008), AOCS method (Diraman, 2008), conjugated dienes (Deiana et al., 2002) and conjugated trienes (Hrncirik & Fritsche, 2005) analyses, K_{232} and K_{270} UV indexes (Antolin & Meneses, 2000; Cañizares-Macías et al., 2004a; Márquez-Ruiz et al., 2008; Platero-López & García-Mesa, 2007), thermogravimetry analyses (Coni et al., 2004; García Mesa et al., 1993; Gennaro et al., 1998; Santos et al., 2002; Vecchio et al., 2009), differential scanning calorimetry analyses (Vecchio et al., 2009), ^{13}C NMR on chromatographically enriched oil fractions (Hidalgo et al., 2002), high-resolution chromatographic techniques coupled with UV spectrometry or mass spectrometry (B. Baccouri et al., 2008; Gallina-Toschi et al., 2005; Tena et al., 2009), capillary electrophoresis (Carrasco-Pancorbo et al., 2007; Gallina-Toschi et al., 2005), ultrasound-assisted method (Cañizares-Macías et al., 2004b; Platero-López & García-Mesa, 2007), microwave-assisted method (Cañizares-Macías et al., 2004a), chemiluminescence (Navas & Jiménez, 2007), electron paramagnetic resonance (Papadimitriou et al., 2006), and ORAC assay (Ninfali et al., 2002), among others. Fingerprinting techniques such as NMR (Alonso-Salces et al., 2011a; Guillen & Ruiz, 2001, 2006), FTIR (Guillen & Cabo, 2000), and fluorescence (Guimet et al., 2005; Tena et al., 2009) spectroscopies, DNA fingerprinting (Spaniolas et al., 2008), electronic nose (Lerma-García et al., 2009), and Oxitest method (Kamvissis et al., 2008; Mora et al., 2009) have been also used successfully to study edible oil stability.

Most of the studies on the oxidative stability of olive oil employed questionably high-temperatures which, unfortunately, cannot be considered reliable to predict the stability of olive oils at room temperature (Frankel, 2010), i.e. under normal storage conditions. This is due to the fact that the mechanism of lipid oxidation changes at the elevated temperatures at which these experiments were run. In this sense, the rate of lipid oxidation is independent of O_2 pressure at ambient temperatures; whereas it does become dependent on O_2 pressures at elevated temperatures due to the decrease in solubility of O_2. This causes the O_2 concentration to become a significant limiting factor that increases with the degree of oxidation. For this reason, in oxidative stability studies the use of several temperatures, in a range as low as practical, preferably at or below 60 °C, is an important consideration. Moreover, polymerization and cyclization of PUFA, which mainly occurs at elevated temperatures, are not significant at room temperature. Furthermore, volatile acids that are measured by the Rancimat and OSI methods are produced only at elevated temperatures (Frankel, 2010). For all these reasons, the results of the studies on the oxidative stability of olive oil at high temperatures are neither relevant nor can be extrapolated to normal storage conditions at room temperature. Olive oil stability at room temperature is of great interest, for instance, to know its storage shelf-life. Because VOO is relatively stable to oxidation due to its particular chemical composition, there has been apparently little or no control of its stability under normal storage conditions in the past. To provide some knowledge on this issue, the stability of olive oil at room temperature while protected from light by ^{1}H-NMR fingerprinting was studied (Alonso-Salces et al., 2011a). The ^{1}H-NMR spectra of the VOO aliquots kept under these conditions, over a certain period of time, were analyzed by principal component analysis (PCA).

2. Experimental

2.1 Chemicals and plant material

Deuterated chloroform for NMR analysis (99.8 atom %D) was provided by Sigma-Aldrich Chemie (Steinheim, Germany).

Virgin olive oils (963 samples) from seven countries of the Mediterranean basin, namely Italy (661 VOOs), Spain (144 VOOs), Greece (97 VOOs), France (39 VOOs), Turkey (14 VOOs), Cyprus (6 VOOs) and Syria (2 VOOs), were collected directly from the producers (olive oil mills) from most of the main producing regions of these countries during three harvests (2004/05, 2005/06 and 2006/2007). The sample collection was carried out with the collaboration of Dipartimento di Chimica e Technologie Farmaceutiche ed Alimentari - Università degli Studi di Genova (Italia), Laboratorio Arbitral Agroalimentario (Ministry of Agriculture and Fishery, Spain), General Chemical State Laboratory D'xy Athinon (Greece), General State Laboratory (Ministry of Health, Cyprus), Departamento de Química Orgánica - Universidad de Córdoba (Spain), Istituto di Metodologie Chimiche - Laboratorio di Risonanza Magnetica Annalaura Segre – CNR (Italy), Fondazione Edmund Mach - Istituto San Michele all'Adige (Italy), and Eurofins Scientific Analytics (France), in the framework of the EU TRACE project. The true type (virgin or extra virgin) and origin of the olive oils at the national, regional and PDO level were assured. The Italian samples were representative of the olive oil producing areas, which are markedly influenced by pedoclimatic factors from the North to the South of the country.

For the study of VOO stability, about a liter of VOO was divided into aliquots contained in dark glass 40mL-vials completely filled and kept at -20°C in a freezer. Each month, over a period of more than 3 and half years (samples for the months 20th, 32nd, 38th and 42nd are missing), one vial was taken from the freezer and stored at room temperature (r.t.) in a closed box. A preliminary supposition was made; this considered that the degradation of VOO at -20°C is not significant and thus the last aliquot taken out of the freezer was time 0. All aliquots were analyzed by ^{1}H-NMR once the last sample was taken from the freezer.

2.2 NMR analysis

Aliquots of 40 µL of each VOO were dissolved in 200 µL of deuterated chloroform, shaken in a vortex, and placed in a 2 mm NMR capillary. The ^{1}H-NMR experiments were performed at 300K on a Bruker (Rheinstetten, Germany) Avance 500 (nominal frequency 500.13 MHz) equipped with a 2.5 mm broadband inverse probe. The spectra of the samples used for the study of the geographical origin of VOOs were recorded using a 7.5 µs pulse (90°), an acquisition time of 3.0 s (32k data points) and a total recycling time of 4.0 s, a spectral width of 5500 Hz (11 ppm), 64 scans (+ 4 dummy scans), with no sample rotation. The spectra of the samples used for the VOO stability study were recorded using a 6.7 µs pulse (90°), an acquisition time of 3.5 s (50k data points) and a total recycling time of 7.0 s, a spectral width of 7100 Hz (14 ppm), 32 scans (+ 4 dummy scans), with no sample rotation. Prior to Fourier transformation, the free induction decays (FIDs) were zero-filled to 64k and a 0.3 Hz line-broadening factor was applied. The chemical shifts are expressed in δ scale (ppm), referenced to the residual signal of chloroform (7.26 ppm) (Hoffman, 2006). The spectra were phase- and baseline-corrected manually, binned with 0.02 ppm-wide buckets, and normalized to total intensity over the region 4.10-4.26 ppm (glycerol signal). TopSpin 1.3 (2005) and Amix-Viewer 3.7.7 (2006) from Bruker BioSpin GMBH (Rheinstetten, Germany) were used to perform the processing of the spectra. The region of the NMR spectra studied was 0-7 ppm for the geographical origin determination of VOOs, and 0-10 ppm in the VOO stability study. The data tables generated with the spectra of all samples, excluding the eight buckets in the reference region 4.10-4.26 ppm, were submitted to multivariate data analysis.

2.3 Data analysis

The data matrices, consisting of the ^{1}H-NMR buckets (variables) arranged in columns and VOO samples in rows, were firstly analyzed by univariate procedures (ANOVA, Fisher index and Box-Whisker plots), and afterwards, by the following multivariate techniques, already described in bibliography (Berrueta et al., 2007): unsupervised ones as principal component analysis (PCA); and supervised as linear discriminant analysis (LDA) and partial least squares discriminant analysis (PLS-DA). Statistic and chemometric data analysis were performed by means of the statistical software packages Statistica 6.1 (StatSoft Inc., Tulsa, OK, USA, 1984-2004), The Unscrambler 9.1 (Camo Process AS, Oslo, Norway, 1986-2004) and SIMCA-P 11.0 (Umetrics AB, Umea, Sweden, 1992-2005). Strategies used for variable selection in LDA and selection of the optimum number of PLS components in PLS-DA are described elsewhere (Rosa M. Alonso-Salces et al., 2010b).

For the geographical characterization of VOOs, the supervised techniques were applied to the autoscaled (or standardised) or Pareto-scaled data matrix of the VOO profiles following these steps: (*i*) the data set was divided into a training-test set and an external data set; (*ii*) the training-test set was subsequently divided into a training set and a test set several times in order to perform cross-validation; (*iii*) the training-test set was used for the optimization of parameters characteristic of each multivariate technique by cross-validation, for instance for variable selection in LDA or the number of PLS components in PLS-DA; (*iv*) a final mathematical model was built using all the samples of the training-test set and the optimized parameters; (v) this model was validated using an independent test set of samples (external data set), i.e. performing an external validation. During the parameter optimization step, the models were validated by 3-fold cross-validation (3-fold CV) or leave-one-out cross-validation (LOO). The reliability of the classification models achieved in the cross-validation was studied in terms of recognition ability (percentage of the samples in the training set correctly classified during the modeling step) and prediction ability (percentage of the samples in the test set correctly classified by using the models developed in the training step). The reliability of the final model was evaluated in terms of classification ability (percentage of the samples of the training-test set correctly classified by using the optimized model) and the prediction ability in the external validation (percentage of the samples of the external set correctly classified by using the optimized model) (Berrueta et al., 2007).

3. Results and discussion

3.1 ^{1}H-NMR spectra of virgin olive oil

^{1}H-NMR spectra of VOOs were recorded (Fig. 1). Olive oil is mainly made up of triglycerides, differing in their substitution patterns in terms of length, degree and kind of unsaturation of the acyl groups (Harwood & Aparicio, 2000). The chemical shifts of their ^{1}H signals are well known (Mannina & Segre, 2002; Sacco et al., 2000). However, the ^{1}H signals of the minor oil components, such as mono- and di-glycerides, sterols, tocopherols, aliphatic alcohols, hydrocarbons, fatty acids, pigments and phenolic compounds (Harwood & Aparicio, 2000), are only observed by ¹H-NMR when their signals do not overlap with those of the main components and their concentrations are high enough to be detected (R. M. Alonso-Salces et al., 2010a; Rosa M. Alonso-Salces et al., 2010b; D'Imperio et al., 2007; Guillen & Ruiz, 2001; Mannina et al., 2003; Sacchi et al., 1996). Table 1 gathers the common ¹H-NMR signals of the major and some minor compounds together with their chemical shifts and their assignments to protons of the different functional groups. Several signals of minor compounds were found in ¹H-NMR spectra recorded because they were not overlapped by those of the triglyceryl protons: cycloartenol at 0.318 ppm and 0.543 ppm, β-sitosterol at 0.669 ppm, stigmasterol at 0.687 ppm, squalene at 1.662 ppm, sn-1,2 diglyceryl group protons at 3.71 ppm and 5.10 ppm, and three unknown terpenes at 4.571 ppm, 4.648 ppm and 4.699 ppm.

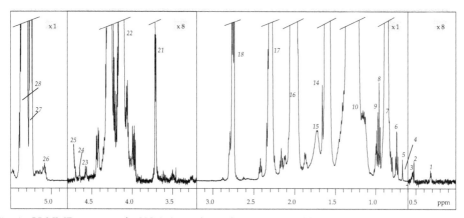

Fig. 1. ¹H-NMR spectra of a VOO (signal numbering, see Table 1).

3.2 Geographical origin of virgin olive oil

The large dataset of VOOs was studied regarding the situations that the antifraud authorities and regulatory bodies face. The PDO *"Riviera Ligure"*, some Italian regions and the main countries that produce VOOs were used as examples to prove the potential of the tools to detect the mislabeling of non-PDO oils as PDO VOOs, or the mislabeling of the provenance of VOOs at the regional or national level. With this purpose in mind, several multivariate data analysis techniques, datasets, types of data scaling and cross-validation were evaluated to attain the best classification models for each case study.

After removing 28 extreme samples, the dataset (935 x 342) was analyzed by PCA. The four first principal components, accounting for 63% of total system variability (TSV), showed that samples were distributed in a compact cluster. However, some overlapping sub-clusters due to the harvest year were observed in the score plot of the samples in the space defined by PC2 (13% TSV), PC3 (11% TSV) and PC4 (7% TSV). Taking into account that 70% of the samples were Italian and the rest from countries in the Mediterranean region, seasonal aspects seem to affect all samples in the same way, independently of their geographical origin. Therefore, in the modeling for the authentication of agricultural food products, it is important to include chemical data of several harvests in order to obtain general classification models that allow for the seasonal variability.

#	Chemical shift (ppm)	Multiplicity[a]	Functional group	Attribution
1	0.318	d	-CH_2- (cyclopropanic ring)	cycloartenol
2	0.527	s		
3	0.543	d	-CH_2- (cyclopropanic ring)	cycloartenol
4	0.669	s	-CH_3 (C18-steroid group)	β-sitosterol
5	0.687	s	-CH_3 (C18-steroid group)	stigmasterol
6	0.740	t	-CH_3 (^{13}C satellite of signal at 0.87 ppm)	
7	0.866	t	-CH_3 (acyl group)	saturated, oleic (or ω-9) and linoleic (or ω-6)
8	0.960	t	-CH_3 (acyl group)	linolenic (or ω-3)
9	0.987	t	-CH_3 (^{13}C satellite of signal at 0.87 ppm)	
10	1.19-1.37		-$(CH_2)_n$- (acyl group)	
11	1.243		-$(CH_2)_n$- (acyl group)	saturated (palmitic, stearic)
12	1.256		-$(CH_2)_n$- (acyl group)	oleic
13	1.288		-$(CH_2)_n$- (acyl group)	linoleic and linolenic
14	1.51-1.65		-OCO-CH_2-CH_2- (acyl group)	
15	1.662	s	-CH_3	squalene
16	1.96-2.07		-CH_2-CH=CH- (acyl group)	
17	2.26-2.32	m	-OCO-CH_2- (acyl group)	
18	2.72-2.82		=CH-CH_2-CH= (acyl group)	
19	2.754	t	=CH-CH_2-CH= (acyl group)	linoleic
20	2.789	t	=CH-CH_2-CH= (acyl group)	linolenic
21	3.69-3.73	d	-CH_2OH (glyceryl group)	sn 1,2-diglycerides
22	4.09-4.32		-CH_2OCOR (glyceryl group)	triglycerides
23	4.571	d		terpene
24	4.648	s		terpene
25	4.699	s		terpene
26	5.05-5.15	m	>CHOCOR (glyceryl group)	sn 1,2-diglycerides
27	5.22-5.28	m	>CHOCOR (glyceryl group)	triglycerides
28	5.28-5.38	m	-CH=CH- (acyl group)	

[a] Signal multiplicity: s, singlet; d, doublet; t, triplet; m, multiplet.

Table 1. Chemical shift assignments of ¹H-NMR signals of the main components in VOOs.

3.2.1 PDO virgin olive oils

Under the PDO of *Riviera Ligure*, only extra virgin olive oils produced in Liguria (Italy) that fulfill the PDO requirements related to olive varieties, farming practices, oil extraction procedures, bottling and labeling (Dossier Number: IT/PDO/0017/1540, Official Journal L22 24.01.1997) can be marketed. The ¹H-NMR dataset of VOOs from different geographical origins and PDOs was studied to create a classification model that differentiates between VOOs belonging to the PDO *Riviera Ligure* and those not belonging to this PDO.

Univariate data analysis (ANOVA, Fisher index and Box-Whisker plots) disclosed that any single variable could distinguish between Ligurian (belonging to the PDO *Riviera Ligure*) and non-Ligurian (not belonging to the PDO) samples. So, it was necessary to apply supervised pattern recognition methods to build classification models that can distinguish VOOs of this PDO from the rest. Several multivariate approaches (LDA and PLS-DA) were tested using balanced or unbalanced data sets, different cross-validation methods (LOO and 3-fold CV), different data scaling techniques (auto-scaling and Pareto-scaling) to find the best approach for the authenticity and traceability of PDO olive oils (Rosa M. Alonso-Salces et al., 2010b; Rosa M. Alonso-Salces et al., 2011b). Table 2 summarized the results of the best classification models achieved. Both supervised pattern recognition techniques performed better if using a balanced training-test set than an unbalanced data set (Rosa M. Alonso-Salces et al., 2010b; Rosa M. Alonso-Salces et al., 2011b). PLS-DA outperformed LDA. LDA achieved classifications of around 85% of hits for both categories. PLS-DA provided a model with 5 PLS components and the boundary at 0.540, that achieved slightly better results for the Liguria class (prediction ability in the cross-validation, 86-88%; classification ability of the final model, 92%; and prediction ability of the final model in the external validation, 88%) than for the non-Liguria VOOs (86-87%, 90% and 86% respectively). These results, together with the facts that in the cross-validation the recognition ability was higher but close to the prediction ability and the classification ability of the final model was also higher but close to prediction ability in the external validation, disclosed that the model achieved was feasible and not random, as well as being well-represented by the samples in the dataset.

Regarding the most important NMR variables on the classification models provided by these pattern recognition techniques, the variables selected in LDA were among the variables that presented the highest weighted regression coefficients (Esbensen et al., 2002) in the PLS-DA models (the larger the regression coefficient, the higher the influence of the variable on the PLS model). Thus, both pattern recognition techniques arrived at consistent results, and provided information about the most important features for the characterization of PDO *Riviera Ligure* VOOs. In this sense, the variables selected for the LDA model were five NMR buckets centered at the following chemical shifts: 6.61 ppm; 5.11 or 5.09 ppm; 4.57 ppm; 4.05 ppm; and 0.33 ppm. These buckets correspond to signals of the following VOO components: phenolic compounds and unsaturated alcohols, which present characteristic resonances in the spectral region 6-7.5 ppm (Owen et al., 2000) and 4.5-5 ppm respectively; sn-1,2-diglycerides (5.09-5.11 ppm) and sn-1,3-diglycerides (4.05 ppm), due to their CH glycerol protons; and cycloartenol (0.33 ppm), to the methylene proton of its cyclopropanoic ring (Sacchi et al., 1996).

In the PLS-DA models, the variables that presented the highest weighted regression coefficients were: 6.85-6.83 ppm, 6.75 ppm, 6.67 ppm, 6.59 ppm, and 6.23 ppm belonged to signals of phenolic compounds; 5.15-5.07 ppm were due to the CH glycerol protons of sn-1,2-diglycerides; 4.99 ppm to unsaturated alcohols; 4.71 ppm, 4.65 ppm and 4.57 ppm, to terpenes; 2.79 ppm, to diallylic proton of linolenic acyl group; 1.29 ppm, to methylene proton of linoleic and linolenic acyl group; and 0.33 ppm, to cycloartenol.

			Cross-validation				Model		External Validation	
			% Recognition		% Prediction		% Classification		% Prediction	
		N					132	135	67	601
	prior prob						0.49	0.51		
Technique	Miscellaneous	Validation	Lig	Non-Lig	Lig	Non-Lig	Lig	Non-Lig	Lig	Non-Lig
LDA[b]	5 NMR buckets selected: 6.61, 5.11, 4.57, 4.05 and 0.33 ppm; autoscaling	3-fold CV	84.1	85.9	84.1	83.7	82.6	85.2	86.6	79.7
PLS-DA[b]	5 PLS components selected; boundary: 0.540; autoscaling	3-fold CV	91.3	92.6	87.9	86.7	91.7	90.4	88.1	85.5
PLS-DA[c]	5 PLS components selected; boundary: 0.540; autoscaling	3-fold CV	-	-	86.4	85.9	91.7	90.4	88.1	85.5
PLS-DA[c]	5 PLS components selected; boundary: 0.540; autoscaling	LOO	-	-	87.1	85.9	91.7	90.4	88.1	85.5
PLS-DA[d]	5 PLS components selected; boundary: 0.540; autoscaling	3-fold CV/	-	-	-	-	91.7	90.4	88.1	85.5
PLS-DA[d]	4 PLS components selected; boundary: 0.520; Pareto scaling	3-fold CV/	-	-	-	-	87.1	83.0	80.6	81.0

[a] Abbreviations: N, number of samples; prior prob, prior probability; Lig, Liguria; Non-Lig, Non-Liguria; LDA, linear discriminant analysis; PLS-DA, partial least square discriminant analysis; Class codes: Liguria, 1; non-Liguria, 0.
[b] Statistica.; [c] The Unscrambler.; [d] SIMCA-P.

Table 2. Classification results obtained by supervised pattern recognition techniques for the authentication of VOO of the PDO *Riviera Ligure* using ^{1}H-NMR spectral data (balanced data set).[a]

3.2.2 Virgin olive oils from different regions

The large sample set of VOOs available was also studied from the point of view of the authentication of VOOs at the regional level, in particular, VOOs produced in certain Italian regions. The regions selected were those best represented in the dataset: Umbria (which is also a registered PDO: PDO *Umbria*), Sicily (6 PDOs: *Monte Etna, Val di Mazara, Valli Trapanesi, Valle del Belice, Valdemone* and *Monti Iblei*), Puglia (4 PDOs: *Terra d'Otranto, Collina di Brindisi, Dauno* and *Terra di Bari*), Lazio (3 PDOs: *Tuscia, Canino* and *Sabina*), Garda (3 PDOs: *Garda, Laghi Lombardi* and *Veneto Valpolicella, Veneto Euganei e Berici, Veneto del Grappa*), Campania (3 PDOs: *Peninsola Sorrentina, Colline Salernitane* and *Cilento*) and Calabria (3 PDOs: *Lametia, Alto Crotonese* and *Bruzio*). The binary classification models created for these regions were developed using an auto-scaled balanced training-test set by PLS-DA and LOO cross-validation (Rosa M. Alonso-Salces et al., 2010b). The final models were also evaluated by external validation. The results are summarized in Table 3.

Origin		Cross-validation		Model	External Validation		
Binary model[b]	N	prior prob	% Prediction	% Classification	N	prior prob	% Prediction
Umbria	35	0.45	71.4	82.9	12	0.014	50.0
Non-Umbria	43	0.55	74.4	79.1	845	0.986	74.8
Sicily	54	0.47	92.6	98.1	24	0.029	87.5
Non-Sicily	62	0.53	85.5	88.7	795	0.971	85.8
Puglia	47	0.42	68.1	72.3	22	0.027	81.8
Non-Puglia	64	0.58	62.5	71.9	802	0.973	65.1
Lazio	40	0.49	80.0	97.5	19	0.022	73.7
Non-Lazio	41	0.51	68.3	90.2	835	0.978	69.3
Garda	36	0.46	72.2	91.7	13	0.015	69.2
Non-Garda	43	0.54	74.4	90.7	843	0.985	80.1
Campania	21	0.43	71.4	81.0	7	0.008	57.1
Non-Campania	28	0.57	64.3	78.6	879	0.992	62.9
Calabria	17	0.38	70.6	94.1	5	0.006	60.0
Non-Calabria	28	0.62	85.7	96.4	885	0.994	79.9

[a] See abbreviations: Table 2; Models obtained by PLS-DA using autoscaling, LOO and The Unscrambler; Class codes: "Region", 1; "non-Region, 0.
[b] Binary models: Umbria vs. non-Umbria: 2 PLS components, boundary at 0.525; Sicily vs. non-Sicily: 3 PLS components, boundary at 0.460; Puglia vs. non-Puglia: 2 PLS components, boundary at 0.4435; Lazio vs. non-Lazio: 4 PLS components, boundary at 0.515; Garda vs. non-Garda: 3 PLS components, boundary at 0.555; Campania vs. non-Campania: 2 PLS components, boundary at 0.430; Calabria vs. non-Calabria: 3 PLS components, boundary at 0.445.

Table 3. Classification results obtained by supervised pattern recognition techniques for the authentication of VOO from certain Italian regions using ¹H-NMR spectral data.[a]

The model obtained to authenticate VOOs from Sicily recognized 98% of the Sicilian oils and 89% of the non-Sicilian ones and managed to correctly predict in the cross-validation step 93% and 86% of Sicilian and non-Sicilian oils respectively. Since this model achieved similar predictions in the external validation (higher than 85% of hits for both categories) to those in the modeling step, it can be considered stable and robust. In contrast, the models created for other regions such as Lazio, Garda and Calabria, were not so satisfactory: although the classification abilities were close to 90% of correct hits or even higher, the prediction abilities in the cross-validation were from 10 to 24% lower, which meant that the classification results were very dependent on the samples included in the training set in the modeling step. This also occurred for Umbria and Campania, but the models achieved about 80% of correct classification for the training set, and predictions on the test set were more than 10% lower, except for the oils belonging to the non-Umbria category (5% less). The external validation of some models (only 50% of Umbria, 57% of Campania, and 60% of Calabria VOOs were correctly predicted) confirmed that the classes were not well represented in the modeling step. Puglian VOOs, as well as non-Garda VOOs, were much better predicted in the external data set (82% of hits) than in the cross-validation (68% and 72% of hits respectively). This was probably due to the way samples were divided into the training-test set and the external set: the PCA scores of all the VOOs were regarded to select samples from the whole cloud of points including the borders. This procedure assured that the training-test set was representative of all the samples (at least of the 3 harvests studied), however the predictions on the external set could be overoptimistic.

Regarding the most influential variables, i.e. those with the highest weighted regression coefficients, on the binary PLS-DA models achieved for each region are the following. The signals due to cycloartenol (0.31-0.33 ppm) and *sn*-1,2-diglycerides (5.07-5.15-ppm) were important for all models except for Garda, as well as the resonances in the phenolic region at 6.73-6.79 ppm, which only did not influence the model for Sicily. The acyl group methylene protons of saturated fatty acids (1.23 ppm), ^{13}C satellite of signal at 4.09-4.32 ppm (α-methylene protons of the glyceryl group of triglycerides) at 3.97 ppm and the signal at 5.57 ppm were important specifically for the Umbria model; the signals at 0.53 ppm and 0.79 ppm, for the Sicily model; the methylic proton of the C18-steroid group of β-sitosterol (0.67 ppm) and the terpene signal at 4.57-4.59 ppm, for the Puglia model; the signal of the cycloartenol at 0.55 ppm, ^{13}C satellite of signal at 2.26-2.32 ppm (α-methylene protons of the acyl group) at 2.15 ppm, the glycerol proton of *sn*-1,2-diglycerides (3.71 ppm) and signals at 6.19 ppm and 6.15 ppm in the phenolic region, for the Lazio model; signals in the region 1.35-1.43 ppm, 2.35-2.39 ppm, and 4.33-4.35 ppm, the α-methylene protons of the acyl group (2.29 ppm and 2.33 ppm), the signal at 3.75 ppm, the α-methylene protons of the glyceryl group of triglycerides (4.27 ppm) and the signal at 6.15 ppm in the phenolic region for the Campania model; and the signal at 5.93 ppm for the Calabria model. The glyceryl protons of *sn*-1,3-diglycerides (4.05-4.07 ppm) and triglycerides (5.25 ppm, 5.29 ppm) were influent for the models of Umbria, Lazio, Umbria and Campania respectively; signals in the phenolic region at 6.25-6.29 ppm for the models of Puglia and Calabria; signals in the phenolic region at 6.63-6.65 ppm and 6.69-6.71 ppm for the models of Umbria and Campania ; signals in the phenolic region at 6.45-6.47 ppm for the models of Umbria and Garda.

These results disclosed that ^{1}H-NMR spectra of VOOs contained information related to the region of provenance of the oil, nonetheless further studies should be carried out with a considerably larger sample set for each region, and even for each of their PDOs, in order to guarantee the detection of counterfeit VOOs. In this regard, Sicily, which is an island at the southernmost part of Italy, produces an olive oil which is markedly influenced by pedoclimatic factors, in accordance with its geographical position. It is therefore coherent that the VOO produced on this island presents a characteristic chemical composition that allows one to distinguish it from all other VOO coming from different geographical regions.

3.2.3 Virgin olive oils from the main producing countries: Spain, Italy and Greece

The adulteration of VOOs from a certain country with VOOs produced in another country at a lower cost, or the false labeling of the VOOs as coming from a certain country when they were actually produced in another, are actual events that the antifraud authorities have to deal with regularly. The need for chemical approaches to detect these fraudulent activities is evermore apparent.

The ¹H-NMR data of the VOOs from the main olive oil producing countries, i.e. Spain, Italy and Greece, were analyzed by multivariate techniques with the purpose of creating classification models to distinguish the geographical origin of VOOs from these three countries (Rosa M. Alonso-Salces et al., 2010b). PLS-DA, using LOO cross-validation, was applied to the autoscaled data to provide binary classification models (country vs. non-country), which were also evaluated by external validation (Table 4). The model 'Greece vs. non-Greece' distinguishes Greek VOOs from all the rest of the VOOs; it classified properly more than 97% of the samples of both categories, Greece and non-Greece, and predicted correctly more than 90% of the samples in the test set of the cross-validation, as well as in the external validation. The binary models for Italy and Spain presented classification abilities of 89% for the Italian oils and the Spanish oils, 84% for the non-Italy category and 85% for the non-Spain category. The prediction abilities in the cross-validation for the model for Spain were ca. 80% of hits for both classes; whereas the predictions in the external validation were considerably different, for the Spanish VOOs it was overoptimistic (92%), and for the non-Spanish VOOs it was considerably low (67%). In the model for Spain, the variability of the non-Spain category was under-represented in the training-test sets. As a result, this model did not provide good predictions for this category in the external set. The model for Italy provided prediction abilities in the cross-validation of ca. 76% for both classes; and in external validation, close to this value. So, these predictions were substantially lower than the recognition ability of the model, indicating that the model was dependent on the samples included in the training set.

Origin	Cross-validation		Model	External Validation		
Binary model[b]	N	prior prob	% Prediction	% Classification	N	prior prob	% Prediction
Italy	72	0.35	75.0	88.9	568	0.78	75.7
Non-Italy	135	0.65	77.0	84.4	160	0.22	71.9
Spain	71	0.34	78.9	88.7	70	0.10	92.9
Non-Spain	136	0.66	80.9	85.3	658	0.90	67.2
Greece	64	0.31	92.2	98.4	31	0.04	96.8
Non-Greece	143	0.69	93.7	97.9	697	0.96	90.0

[a] See abbreviations: Table 2; Models obtained by PLS-DA using autoscaling, LOO and The Unscrambler; Class codes: "Country", 1; "non-Country, 0.
[b] Binary models: Italy vs non-Italy: 4 PLS components, boundary at 0.4020; Spain vs non-Spain: 3 PLS components, boundary at 0.3563; Greece vs non-Greece: 5 PLS components, boundary at 0.4725.

Table 4. Classification results obtained by supervised pattern recognition techniques for the authentication of VOO from the main producing countries, i.e. Italy, Spain and Greece, using ¹H-NMR spectral data.[a]

The most influential variables, i.e. those with the highest weighted regression coefficients, on the binary PLS-DA models obtained for each country are due to signals in the phenolic regions at 6.45-6.47 ppm and 6.83-6.85 ppm, which were important for the three models (Rosa M. Alonso-Salces et al., 2010b). In contrast, the model for Spain was particularly influenced by the methylic proton of the C18-steroid group of β-sitosterol (0.67 ppm), the β-methylene protons of the acyl group (1.59 ppm, 1.67 ppm), the allylic protons of the acyl group (1.99-2.07 ppm), the diallylic protons of the acyl group of linoleic (2.73-2.77 ppm) and linolenic (2.77-2.81 ppm), the glycerol proton of sn-1,2-diglycerides (3.71 ppm), sn-1,3-diglycerides (4.05-4.07 ppm) and triglycerides (5.25 ppm, 5.29 ppm), the olefinic protons of the acyl groups (5.37 ppm), the signals in the phenolic region at 6.37 ppm, 6.61 ppm and 6.71 ppm, and the signals at 0.53 ppm, 1.75-1.77 ppm, 2.35 ppm. Among the most important variables, those who only affected the model for Greece were the methylic proton of the linolenic acyl group (0.97 ppm), and the terpene signal at 4.55-4.57 ppm, and the signals at 0.77 ppm and 3.81 ppm. The resonances of cycloartenol (0.31-0.33 ppm and 0.55 ppm) and phenolic compounds at 6.23 ppm and 6.27 ppm were important for the models of Italy and Greece.

These results show that ^{1}H-NMR fingerprinting of VOOs can be a useful tool to assure authenticity and traceability of VOOs at the national level. From this study, a stable model was achieved to distinguish Greek VOOs from oils from other countries. However, for Italian and Spanish VOOs further studies should be performed with a larger balanced data set, in which all categories will be well represented, to obtain robust models. In the present data set, Spain was clearly under-represented, being the main producer (50% of EU production of olive oil); and Italy, even though it was quite well-represented, the number of samples were very unbalanced regarding the other countries, and so few Italian samples were used in the modeling step. The classification results might therefore be very dependent on the samples in the training-test set.

3.3 Stability of virgin olive oil

Regarding the importance of oil stability on its quality and nutritional properties, the stability of VOO was studied over a period of 43 months in the dark at room temperature by ^{1}H-NMR. The high stability of VOO is mainly due to its relatively low degree of fatty acid unsaturation and to the antioxidant activity of some of the unsaponifiable components. For instance, the oxidative susceptibility of olive oil is related to the antioxidant activity of α-tocopherol, which also showed a synergistic effect in association with some phenolic compounds with significant activity (Deiana et al., 2002).

In contrast with previous studies on the oxidative stability of edible oils, which were performed at high temperatures (Frankel, 2010); we studied VOO stability at r.t. (Alonso-Salces et al., 2011a). The ^{1}H-NMR spectra of 40 VOO samples in the spectral region 0-10 ppm (492 buckets) were analyzed by PCA. The samples represented in the PCA score plot of the first two principal components (Fig. 2) are distributed in the direction of the first principal component (PC1) in clusters, partially overlapped, according to the length of time they had been at r.t. So, PC1, accounting for 15% TSV, contains information related to the change and evolution of the chemical composition of VOO during storage in the dark at r.t., and hence, about the degradation of olive oil under these conditions.

The most influential features, i.e. buckets of ¹H-NMR spectra, on PC1 are those with the highest loadings in absolute value, and are shown in Table 5. Some of these chemical shifts correspond to ¹H-NMR signals of compounds involved in the hydrolytic and oxidative degradation of VOO. During the oxidation process, hydroperoxides (primary oxidation compounds) are produced (Guillen & Ruiz, 2001, 2006) , which may degrade into secondary oxidation products such as aldehydes, ketones, lactones, alcohols, acids, etc. The oxidation of edible oils is a matter of major concern also from a safety point of view because some oxidation products such as aldehydes are toxic (Guillen & Ruiz, 2001, 2006). Furthermore, several saturated and unsaturated aldehydes have been found to be responsible for rancid sensory defect in VOO (Morales et al., 2005), as well as for off-odours (Kalua et al., 2007), altering its organoleptic properties.

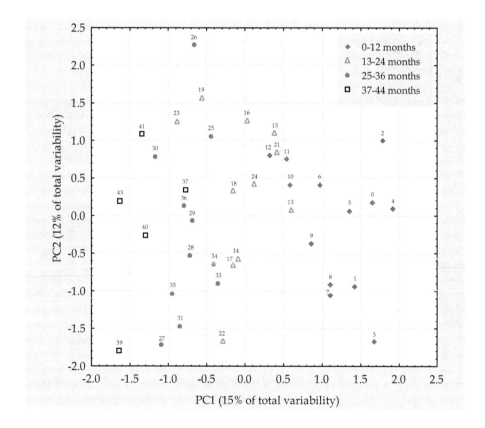

Fig. 2. PCA score plot of the samples used to study the stability of VOO on the space defined by the two first principal components. Samples are numbered according to the time (months) that they had been at r.t. in the dark before analysis.

Bucket (ppm)	Loadings	Multiplicity[a]	Functional group	Attribution
8.19	-0.807			
8.17	-0.860			
8.15	-0.823	broad signal	-OOH (hydroperoxide group)	hydroperoxides
8.13	-0.820			
8.11	-0.749			
8.09	-0.716			
6.97	0.766	s	-Ph-H (phenolic ring)	phenolic compounds
6.75	0.811	d	-Ph-H (phenolic ring)	phenolic compounds
6.57	-0.792			
6.55	-0.838	t	-CH=CH-CH=CH- (cis, trans diene	hydroperoxides
6.53	-0.834		system)conjugated	
6.51	-0.727			
6.01	-0.833			
5.99	-0.854	t	-CH=CH-CH=CH- (cis, trans	hydroperoxides
5.97	-0.860		conjugated diene system)	
5.95	-0.822			
5.57	-0.824	m	-CH=CH-CH=CH- (cis, trans	hydroperoxides
5.55	-0.776		conjugated diene system)	
5.25	0.880	m	>CHOCOR (glyceryl group)	triglycerides
4.45	0.726	m	-CH_2OCOR (^{13}C satellite of signal at 4.09-4.32 ppm, glyceryl group)	triglycerides
4.37	-0.782		>CH-OOH (methine proton of	hydroperoxides
4.35	-0.786		hydroperoxide group)	
4.27	0.770	m	-CH_2OCOR (glyceryl group)	triglycerides
4.09	-0.795			
4.07	-0.931	q	>CH-OH (glyceryl group)	sn 1,3-diglycerides
4.05	-0.875			
3.89	-0.745			
3.87	-0.802	broad signal		saturated alcohols
3.85	-0.715			
3.59	-0.708	broad signal		saturated alcohols
2.79	0.924	t	=CH-CH_2-CH= (acyl group)	linolenic
2.77	0.839		=CH-CH_2-CH= (acyl group)	linolenic and linoleic
2.75	0.870	t	=CH-CH_2-CH= (acyl group)	linoleic
2.21	-0.740			
2.19	-0.885	m	-OCO-CH_2- (^{13}C satellite of signal at	
2.17	-0.913		2.26-2.32 ppm, acyl group)	
2.15	-0.885			
2.03	0.792		-CH_2-CH=CH- (acyl group)	linoleic and linolenic
1.29	0.819		-(CH_2)_n- (acyl group)	linoleic and linolenic
1.27	0.852			
1.25	0.784		-(CH_2)_n- (acyl group)	oleic

[a] Signal multiplicity: s, singlet; d, doublet; t, triplet; q, quadruplet; m, multiplet.

Table 5. Stability of VOO: Loadings of the most influential variables on the first principal component, and chemical shift assignments of the ^{1}H-NMR signals.

The presence of hydroperoxides in the samples which had been stored at r.t. and protected from light for more than one year was confirmed by the ¹H-NMR signals at 8.09-8.19 ppm due to hydroperoxide protons; 6.51-6.57 ppm, 5.95-6.01 ppm and 5.55-5.57 ppm due to protons of the conjugated diene systems; and 4.35-4.37 ppm due to the methine proton of the hydroperoxide group, as observed by other authors (Guillen & Ruiz, 2001). All these signals presented very small intensities in comparison with characteristic VOO signals, indicating that the oxidative degradation was taking place at a very low rate and yield. This was also supported by the fact that characteristic resonances of aldehydes (9.3-9.9 ppm), the main secondary oxidation products, were not detected in the VOO over the 3 and half years of storage at r.t., so the secondary oxidation process had not yet occurred. These facts agree with the recognized high oxidative stability of VOO. Some minor signals at 6.97 ppm and 6.75 ppm were assigned to phenolic compounds (Owen et al., 2003). The decrease or disappearance, respectively, of these signals during storage at r.t. was in agreement with the role that these substances play as antioxidants during the oxidative degradation process of VOO.

During hydrolytic degradation of olive oil, triglycerides hydrolyze thereby increasing the content of free fatty acids and consequently, the acidity of the oil, which means deterioration in the oil quality. Several resonances indicated the occurrence of hydrolytic degradation. In this sense, slight changes in the intensity of the tryglyceride signals at 5.25 ppm, 4.45 ppm and 4.27 ppm and the α-methylene protons of the acyl group (¹³C satellite of the signal at 2.26-2.32 ppm) at 2.15-2.21 ppm were observed. Moreover, a slight decrease in the intensity of the signals at 2.75-2.79 ppm of the diallylic protons and at 2.03 ppm of the allylic protons of linoleic and linolenic acyl groups, and at 1.25-1.29 ppm of the methylene proton signal of oleic, linoleic and linolenic acyl groups, during storage at r.t., revealed that tryglycerides were degrading. The increase in the intensity of the signal at 4.05-4.09 ppm, assigned to the proton of the glyceryl group of sn-1,3-diglycerides, was indicative of the loss of quality and freshness of the VOO (Guillen & Ruiz, 2001). Young, good quality olive oils contain mainly native sn-1,2-diacylglyceride and only small amounts of sn-1,3-diacylglyceride. The increase in the latter was observed after one year of storage at r.t., which seems to be caused by intermolecular transposition and/or lipolytic phenomena (Sacchi et al., 1996). Moreover, in the samples stored for longer than 18 months, a broad signal also appeared in the region of saturated alcohols at 3.85-3.89 ppm, which can arise from lipolysis (Sacchi et al., 1996).

4. Conclusion

¹H-NMR fingerprinting of olive oil is a valuable analytical tool for the traceability of VOOs from different points of view, i.e. food authentication and food quality.

For authentication purposes, ¹H-NMR fingerprints of VOOs analyzed by supervised pattern recognition techniques allow the determination of their geographical origin at the national, regional and/or PDO level. PLS-DA afforded the best model to distinguish the PDO *Riviera Ligure* VOOs: 88% of the Ligurian and 86% of non-Ligurian oils were correctly predicted in the external validation. At the regional level, a stable and robust PLS-DA model was obtained to authenticate VOOs from Sicily, predicting well the origin of more than 85% of the samples in the external sample set. At the national level, Greek and non-Greek VOOs were properly classified by PLS DA: >90% of the oils were correctly predicted in the crossvalidation and external validation.

Regarding quality control, ^{1}H-NMR fingerprinting enables us to control the stability of VOO since this technique can detect its compositional changes due to oxidative and hydrolytic degradation. Under normal VOO storage conditions, i.e. at room temperature and protected from light, none of the signals present in the ^{1}H-NMR spectra of VOO at time zero disappeared or experienced significant decreases or increases over a period of more than 3 and half years. Only small changes in the signals and the appearance of some low intensity signals indicate that some oxidative and hydrolytic degradation of the VOO started after one year. These results confirm the high oxidative stability of VOO at r.t., and supports the best-before date for VOO that is normally between one and one and a half years, depending on the type of container and the olive variety used. Moreover, they show that VOO during this time period does not experience any significant changes which could render its consumption hazardous. In addition, aliquots (even small aliquots of 40 mL) can be preserved at r.t. in the dark (amber glass) until analysis for at least one year, which is of great interest to control laboratories of VOO with regard to storage space and expense. Furthermore, this research is a proof-of-concept that ^{1}H-NMR is a useful tool to study and evaluate the oxidative stability of edible oils in a quality control context at any temperature, since any toxic substances that may be generated during the degradation process can be detected and even quantified. Further studies would be needed to validate quantitative methods for this purpose.

5. Acknowledgement

The authors thank the research groups that participated in the collection of the olive oil samples: Dipartimento di Chimica e Technologie Farmaceutiche ed Alimentari - Università degli Studi di Genova (Italia), Laboratorio Arbitral Agroalimentario (Ministry of Agriculture and Fishery, Spain), General Chemical State Laboratory D'xy Athinon (Greece), General State Laboratory (Ministry of Health, Cyprus), Departamento de Química Orgánica - Universidad de Córdoba (Spain), Istituto di Metodologie Chimiche - Laboratorio di Risonanza Magnetica Annalaura Segre – CNR (Italy), Fondazione Edmund Mach - Istituto San Michele all'Adige (Italy), and Eurofins Scientific Analytics (France). The authors would like to acknowledge J.M. Moreno-Rojas for his help and useful remarks regarding the sampling, and N. Segebarth for sharing his wide knowledge on NMR with us.

6. Abbreviations used

VOO, virgin and extra virgin olive oils; PDO, Protected Designation of Origin; NMR, nuclear magnetic resonance; ANOVA, analysis of variance; PCA, principal component analysis; PC, principal component; LDA, linear discriminant analysis; PLS-DA, partial least squares discriminant analysis; TSV, total system variability; CV, cross-validation; LOO, leave-one-out cross-validation; r.t., room temperature.

7. References

Aguilera, M. P., Beltrán, G., Ortega, D., Fernández, A., Jiménez, A., & Uceda, M. (2005). Characterisation of virgin olive oil of Italian olive cultivars: 'Frantoio' and 'Leccino', grown in Andalusia. *Food Chemistry*, Vol.89, No.3, pp. 387-391, ISSN 0308-8146

Alonso-Salces, R. M., Heberger, K., Holland, M. V., Moreno-Rojas, J. M., Mariani, C., Bellan, G., Reniero, F., & Guillou, C. (2010a). Multivariate analysis of NMR fingerprint of the unsaponifiable fraction of virgin olive oils for authentication purposes. Food Chemistry, Vol.118, pp. 956-965, ISSN 0308-8146

Alonso-Salces, R. M., Holland, M. V., & Guillou, C. (2011a). 1H-NMR fingerprinting to evaluate the stability of olive oil. Food Control, Vol.22, No.12, pp. 2041-2046, ISSN 0956-7135

Alonso-Salces, R. M., Moreno-Rojas, J. M., Holland, M. V., & Guillou, C. (2011b). Authentication of Virgin Olive Oil using NMR and isotopic fingerprinting. In series: Food Science and Technology, Nova Science Publishers, ISBN 978-1-61122-309-5, New York

Alonso-Salces, R. M., Moreno-Rojas, J. M., Holland, M. V., Reniero, F., Guillou, C., & Heberger, K. (2010b). Virgin Olive Oil Authentication by Multivariate Analyses of 1H NMR Fingerprints and δ13C and δ2H Data. Journal of Agricultural and Food Chemistry, Vol.58, No.9, pp. 5586-5596, ISSN 0021-8561

Alves, M. R., Cunha, S. C., Amaral, J. S., Pereira, J. A., & Oliveira, M. B. (2005). Classification of PDO olive oils on the basis of their sterol composition by multivariate analysis. Analytica Chimica Acta, Vol.549, No.1-2, pp. 166-178, ISSN 0003-2670

Angerosa, F., Breas, O., Contento, S., Guillou, C., Reniero, F., & Sada, E. (1999). Application of stable isotope ratio analysis to the characterization of the geographical origin of olive oils. Journal of Agricultural and Food Chemistry, Vol.47, No.3, pp. 1013-1017, ISSN 0021-8561

Antolin, I. P., & Meneses, M. M. (2000). Application of UV-visible spectrophotometry to study of the thermal stability of edible vegetable oils. Grasas y Aceites, Vol.51, No.6, pp. 424-428, ISSN 0017-3495

Aparicio, R., & Aparicio-Ruiz, R. (2000). Authentication of vegetable oils by chromatographic techniques. Journal of Chromatography A, Vol.881, No.1-2, pp. 93-104, ISSN 0021-9673

Aparicio, R., Ferreiro, L., & Alonso, V. (1994). Effect of climate on the chemical composition of virgin olive oil. Analytica Chimica Acta, Vol.292, No.3, pp. 235-241, ISSN 0003-2670

Aranda, F., Gomez-Alonso, S., Rivera Del Alamo, R. M., Salvador, M. D., & Fregapane, G. (2004). Triglyceride, total and 2-position fatty acid composition of Cornicabra virgin olive oil: Comparison with other Spanish cultivars. Food Chemistry, Vol.86, No.4, pp. 485-492, ISSN 0308-8146

Baccouri, B., Temime, S. B., Campeol, E., Cioni, P. L., Daoud, D., & Zarrouk, M. (2007). Application of solid-phase microextraction to the analysis of volatile compounds in virgin olive oils from five new cultivars. Food Chemistry, Vol.102, No.3, pp. 850-856, ISSN 0308-8146

Baccouri, B., Zarrouk, W., Baccouri, O., Guerfel, M., Nouairi, I., Krichene, D., Daoud, D., & Zarrouk, M. (2008). Composition, quality and oxidative stability of virgin olive oils from some selected wild olives (Olea europaea L. subsp. Oleaster). Grasas y Aceites, Vol.59, No.4, pp. 346-351, ISSN 0017-3495

Baeten, V., Pierna, J. A. F., Dardenne, P., Meurens, M., Garcia-Gonzalez, D. L., & Aparicio-Ruiz, R. (2005). Detection of the presence of hazelnut oil in olive oil by FT-Raman and FT-MIR spectroscopy. *Journal of Agricultural and Food Chemistry,* Vol.53, No.16, pp. 6201-6206, ISSN 0021-8561

Berrueta, L. A., Alonso-Salces, R. M., & Héberger, K. (2007). Supervised pattern recognition in food analysis. *Journal of Chromatography A,* Vol.1158, No.1-2, pp. 196-214, ISSN 0021-9673

Boggia, R., Zunin, P., Lanteri, S., Rossi, N., & Evangelisti, F. (2002). Classification and class-modeling of "Riviera Ligure" extra-virgin olive oil using chemical-physical parameters. *Journal of Agricultural and Food Chemistry,* Vol.50, No.8, pp. 2444-2449, ISSN 0021-8561

Bortolomeazzi, R., Berno, P., Pizzale, L., & Conte, L. S. (2001). Sesquiterpene, alkene, and alkane hydrocarbons in virgin olive oils of different varieties and geographical origins. *Journal of Agricultural and Food Chemistry,* Vol.49, No.7, pp. 3278-3283, ISSN 0021-8561

Canabate-Diaz, B., Segura Carretero, A., Fernandez-Gutierrez, A., Belmonte Vega, A., Garrido Frenich, A., Martinez Vidal, J. L., & Duran Martos, J. (2007). Separation and determination of sterols in olive oil by HPLC-MS. *Food Chemistry,* Vol.102, No.3, pp. 593-598, ISSN 0308-8146

Cañizares-Macías, M. P., García-Mesa, J. A., & Luque De Castro, M. D. (2004a). Determination of the oxidative stability of olive oil, using focused-microwave energy to accelerate the oxidation process. *Analytical and Bioanalytical Chemistry,* Vol.378, No.2, pp. 479-483, ISSN 1618-2650

Cañizares-Macías, M. P., García-Mesa, J. A., & Luque De Castro, M. D. (2004b). Fast ultrasound-assisted method for the determination of the oxidative stability of virgin olive oil. *Analytica Chimica Acta,* Vol.502, No.2, pp. 161-166, ISSN 0003-2670

Carrasco-Pancorbo, A., Cerretani, L., Bendini, A., Segura-Carretero, A., Lercker, G., & Fernández-Gutiérrez, A. (2007). Evaluation of the influence of thermal oxidation on the phenolic composition and on the antioxidant activity of extra-virgin olive oils. *Journal of Agricultural and Food Chemistry,* Vol.55, No.12, pp. 4771-4780, ISSN 0021-8561

Ceci, L. N., & Carelli, A. A. (2010). Relation Between Oxidative Stability and Composition in Argentinian Olive Oils. *JAOCS,* pp. 1-9, ISSN 0003-021X

Cercaci, L., Passalacqua, G., Poerio, A., Rodriguez-Estrada, M. T., & Lercker, G. (2007). Composition of total sterols (4-desmethyl-sterols) in extravirgin olive oils obtained with different extraction technologies and their influence on the oil oxidative stability. *Food Chemistry,* Vol.102, No.1, pp. 66-76, ISSN 0308-8146

Cichelli, A., & Pertesana, G. P. (2004). High-performance liquid chromatographic analysis of chlorophylls, pheophytins and carotenoids in virgin olive oils: chemometric approach to variety classification. *Journal of Chromatography A,* Vol.1046, No.1-2, pp. 141-146, ISSN 0021-9673

Coni, E., Podestà, E., & Catone, T. (2004). Oxidizability of different vegetables oils evaluated by thermogravimetric analysis. *Thermochimica Acta,* Vol.418, No.1-2, pp. 11-15, ISSN 0040-6031

D'Imperio, M., Mannina, L., Capitani, D., Bidet, O., Rossi, E., Bucarelli, F. M., Quaglia, G. B., & Segre, A. (2007). NMR and statistical study of olive oils from Lazio: A geographical, ecological and agronomic characterization. *Food Chemistry,* Vol.105, No.3, pp. 1256-1267, ISSN 0308-8146

Deiana, M., Rosa, A., Cao, C. F., Pirisi, F. M., Bandino, G., & Dessi, M. A. (2002). Novel approach to study oxidative stability of extra virgin olive oils: Importance of α-tocopherol concentration. *Journal of Agricultural and Food Chemistry*, Vol.50, No.15, pp. 4342-4346, ISSN 0021-8561

Di Lecce, G., Loizzo, M. R., Boselli, E., Giomo, A., & Frega, N. G. (2009). Chemical and sensory characterization of virgin olive oils from Campania. *Progress in Nutrition*, Vol.11, No.3, pp. 170-177, ISSN 1129-8723

Diraman, H. (2008). A study on oxidative stability of Turkish commercial natural olive oils extracted from different systems. *Acta Horticulturae*, Vol.791, pp. 651-654, ISSN 0567-7572

Esbensen, K. H., Guyot, D., Westad, F., & Houmøller, L. P. (2002). *Multivariate data analysis in practice: an introduction to multivariate data analysis and experimental design* (5th ed.), Camo Process AS, ISBN 8299333032, Oslo, Norway

Esquivel, M. M., Ribeiro, M. A., & Bernardo-Gil, M. G. (2009). Relations between oxidative stability and antioxidant content in vegetable oils using an accelerated oxidation test - Rancimat. *Chemical Product and Process Modeling*, Vol.4, No.4, pp., ISSN 1934-2659

Frankel, E. N. (2010). Chemistry of extra virgin olive oil: Adulteration, oxidative stability, and antioxidants. *Journal of Agricultural and Food Chemistry*, Vol.58, No.10, pp. 5991-6006, ISSN 0021-8561

Galtier, O., Dupuy, N., Le Dréau, Y., Ollivier, D., Pinatel, C., Kister, J., & Artaud, J. (2007). Geographic origins and compositions of virgin olive oils determinated by chemometric analysis of NIR spectra. *Analytica Chimica Acta*, Vol.595, No.1-2, pp. 136-144, ISSN 0003-2670

Gallina-Toschi, T., Cerretani, L., Bendini, A., Bonoli-Carbognin, M., & Lercker, G. (2005). Oxidative stability and phenolic content of virgin olive oil: An analytical approach by traditional and high resolution techniques. *Journal of Separation Science*, Vol.28, No.9-10, pp. 859-870, ISSN 1615-9306

García Mesa, J. A., Luque de Castro, M. D., & Valcárcel, M. (1993). Determination of the oxidative stability of olive oil by use of a robotic station. *Talanta*, Vol.40, No.11, pp. 1595-1600, ISSN 0039-9140

Gennaro, L., Piccioli Bocca, A., Modesti, D., Masella, R., & Coni, E. (1998). Effect of Biophenols on Olive Oil Stability Evaluated by Thermogravimetric Analysis. *Journal of Agricultural and Food Chemistry*, Vol.46, No.11, pp., ISSN 0021-8561

Gómez-Caravaca, A. M., Cerretani, L., Bendini, A., Segura-Carretero, A., Fernández-Gutiérrez, A., & Lercker, G. (2007). Effect of filtration systems on the phenolic content in virgin olive oil by HPLC-DAD-MSD. *American Journal of Food Technology*, Vol.2, No.7, pp. 671-678, ISSN 1557-4571

Guillen, M. D., & Cabo, N. (2000). Some of the most significant changes in the Fourier transform infrared spectra of edible oils under oxidative conditions. *Journal of the Science of Food and Agriculture*, Vol.80, No.14, pp. 2028-2036, ISSN 0022-5142

Guillen, M. D., & Ruiz, A. (2001). High resolution H-1 nuclear magnetic resonance in the study of edible oils and fats. *Trends in Food Science & Technology*, Vol.12, No.9, pp. 328-338, ISSN 0924-2244

Guillen, M. D., & Ruiz, A. (2006). Study by means of H-1 nuclear magnetic resonance of the oxidation process undergone by edible oils of different natures submitted to microwave action. *Food Chemistry*, Vol.96, No.4, pp. 665-674, ISSN 0308-8146

Guimet, F., Boqué, R., & Ferré, J. (2005). Study of oils from the protected denomination of origin "Siurana" using excitation-emission fluorescence spectroscopy and three-way methods of analysis. *Grasas y Aceites*, Vol.56, No.4, pp. 292-297, ISSN 0017-3495

Haddada, F. M., Manai, H., Daoud, D., Fernandez, X., Lizzani-Cuvelier, L., & Zarrouk, M. (2007). Profiles of volatile compounds from some monovarietal Tunisian virgin olive oils. Comparison with French PDO. *Food Chemistry*, Vol.103, No.2, pp. 467-476, ISSN 0308-8146

Harwood, J. L., & Aparicio, R. (2000). *Handbook of olive oil: analysis and properties*, Aspen, ISBN 0834216337, Gaithersburg, MD

Hidalgo, F. J., Gómez, G., Navarro, J. L., & Zamora, R. (2002). Oil stability prediction by high-resolution ^{13}C nuclear magnetic resonance spectroscopy. *Journal of Agricultural and Food Chemistry*, Vol.50, No.21, pp. 5825-5831, ISSN 0021-8561

Hoffman, R. E. (2006). Standardization of chemical shifts of TMS and solvent signals in NMR solvents. *Magnetic Resonance in Chemistry*, Vol.44, No.6, pp. 606-616, ISSN 0749-1581

Hrncirik, K., & Fritsche, S. (2005). Relation between the endogenous antioxidant system and the quality of extra virgin olive oil under accelerated storage conditions. *Journal of Agricultural and Food Chemistry*, Vol.53, No.6, pp. 2103-2110, ISSN 0021-8561

International Olive Council. (2011). Designations and definitions of olive oils. Chemistry standards, In: *The Olive World. Areas of activity*, July 2011, Available from: http://www.internationaloliveoil.org/

Kalua, C. M., Allen, M. S., Bedgood, J. D. R., Bishop, A. G., Prenzler, P. D., & Robards, K. (2007). Olive oil volatile compounds, flavour development and quality: A critical review. *Food Chemistry*, Vol.100, No.1, pp. 273-286, ISSN 0308-8146

Kamvissis, V. N., Barbounis, E. G., Megoulas, N. C., & Koupparis, M. A. (2008). A novel photometric method for evaluation of the oxidative stability of virgin olive oils. *Journal of AOAC International*, Vol.91, No.4, pp. 794-800, ISSN 1060-3271

Kunz, M. R., Ottaway, J., Kalivas, J. H., Georgiou, C. A., & Mousdis, G. A. (2011). Updating a synchronous fluorescence spectroscopic virgin olive oil adulteration calibration to a new geographical region. *Journal of Agricultural and Food Chemistry*, Vol.59, No.4, pp. 1051-1057, ISSN 1520-5118

Lanteri, S., Armanino, C., Perri, E., & Palopoli, A. (2002). Study of oils from Calabrian olive cultivars by chemometric methods. *Food Chemistry*, Vol.76, No.4, pp. 501-507, ISSN 0308-8146

Lerma-García, M. J., Simó-Alfonso, E. F., Bendini, A., & Cerretani, L. (2009). Metal oxide semiconductor sensors for monitoring of oxidative status evolution and sensory analysis of virgin olive oils with different phenolic content. *Food Chemistry*, Vol.117, No.4, pp. 608-614, ISSN 0308-8146

Lopez-Diez, E. C., Bianchi, G., & Goodacre, R. (2003). Rapid quantitative assessment of the adulteration of virgin olive oils with hazelnut oils using Raman spectroscopy and chemometrics. *Journal of Agricultural and Food Chemistry*, Vol.51, No.21, pp. 6145-6150, ISSN 0021-8561

Lopez Ortiz, C. M., Prats Moya, M. S., & Berenguer Navarro, V. (2006). A rapid chromatographic method for simultaneous determination of β-sitosterol and tocopherol homologues in vegetable oils. *Journal of Food Composition and Analysis*, Vol.19, No.2-3, pp. 141-149, ISSN 0889-1575

Mannina, L., Marini, F., Gobbino, M., Sobolev, A. P., & Capitani, D. (2010). NMR and chemometrics in tracing European olive oils: the case study of Ligurian samples. *Talanta*, Vol.80, No.5, pp. 2141-2148, ISSN 1873-3573

Mannina, L., & Segre, A. (2002). High resolution nuclear magnetic resonance: From chemical structure to food authenticity. *Grasas y Aceites*, Vol.53, No.1, pp. 22-33, ISSN 0017-3495

Mannina, L., Sobolev, A. P., & Segre, A. (2003). Olive oil as seen by NMR and chemometrics. *Spectroscopy Europe*, Vol.15, No.3, pp., ISSN 0966-0941

Marini, F., Balestrieri, F., Bucci, R., Magrì, A. D., Magrì, A. L., & Marini, D. (2004). Supervised pattern recognition to authenticate Italian extra virgin olive oil varieties. *Chemometrics and Intelligent Laboratory Systems*, Vol.73, No.1, pp. 85-93, ISSN 0169-7439

Marini, F., Magrì, A. L., Bucci, R., Balestrieri, F., & Marini, D. (2006). Class-modeling techniques in the authentication of Italian oils from Sicily with a Protected Denomination of Origin (PDO). *Chemometrics and Intelligent Laboratory Systems*, Vol.80, No.1, pp. 140-149, ISSN 0169-7439

Marini, F., Magrì, A. L., Bucci, R., & Magrì, A. D. (2007). Use of different artificial neural networks to resolve binary blends of monocultivar Italian olive oils. *Analytica Chimica Acta*, Vol.599, No.2, pp. 232-240, ISSN 0003-2670

Márquez-Ruiz, G., Martín-Polvillo, M., Velasco, J., & Dobarganes, C. (2008). Formation of oxidation compounds in sunflower and olive oils under oxidative stability index conditions. *European Journal of Lipid Science and Technology*, Vol.110, No.5, pp. 465-471, ISSN 1438-7697

Martins-Lopes, P., Gomes, S., Santos, E., & Guedes-Pinto, H. (2008). DNA markers for Portuguese olive oil fingerprinting. *Journal of Agricultural and Food Chemistry*, Vol.56, No.24, pp. 11786-11791, ISSN 0021-8561

Mateos, R., Uceda, M., Aguilera, M. P., Escuderos, M. E., & Maza, G. B. (2006). Relationship of Rancimat method values at varying temperatures for virgin olive oils. *European Food Research and Technology*, Vol.223, No.2, pp. 246-252, ISSN 1438-2377

Matos, L. C., Cunha, S. C., Amaral, J. S., Pereira, J. A., Andrade, P. B., Seabra, R. M., & Oliveira, B. P. P. (2007). Chemometric characterization of three varietal olive oils (Cvs. Cobrancosa, Madural and Verdeal Transmontana) extracted from olives with different maturation indices. *Food Chemistry*, Vol.102, No.1, pp. 406-414, ISSN 0308-8146

Mignani, A. G., Ciaccheri, L., Ottevaere, H., Thienpont, H., Conte, L., Marega, M., Cichelli, A., Attilio, C., & Cimato, A. (2011). Visible and near-infrared absorption spectroscopy by an integrating sphere and optical fibers for quantifying and discriminating the adulteration of extra virgin olive oil from Tuscany. *Analytical and Bioanalytical Chemistry*, Vol.399, No.3, pp. 1315-1324, ISSN 1618-2650

Mora, L., Piergiovanni, L., Limbo, S., & Maiocchi, P. (2009). Evaluation of vegetable oils oxidative stability through the Oxitest reactor. *Industrie Alimentari*, Vol.48, No.495, pp. 51-56, ISSN 0019-901X

Morales, M. T., Luna, G., & Aparicio, R. (2005). Comparative study of virgin olive oil sensory defects. *Food Chemistry*, Vol.91, No.2, pp. 293-301, ISSN 0308-8146

Navas, M. J., & Jiménez, A. M. (2007). Chemiluminescent methods in olive oil analysis. *JAOCS*, Vol.84, No.5, pp. 405-411, ISSN 0003-021X

Ninfali, P., Bacchiocca, M., Biagiotti, E., Servili, M., & Montedoro, G. (2002). Validation of the oxygen radical absorbance capacity (ORAC) parameter as a new index of quality and stability of virgin olive oil. *JAOCS*, Vol.79, No.10, pp. 977-982, ISSN 0003-021X

Ollivier, D., Artaud, J., Pinatel, C., Durbec, J. P., & Guerere, M. (2006). Differentiation of French virgin olive oil RDOs by sensory characteristics, fatty acid and triacylglycerol compositions and chemometrics. *Food Chemistry*, Vol.97, No.3, pp. 382-393, ISSN 0308-8146

Owen, R. W., Giacosa, A., Hull, W. E., Haubner, R., Spiegelhalder, B., & Bartsch, H. (2000). The antioxidant/anticancer potential of phenolic compounds isolated from olive oil. *European Journal of Cancer*, Vol.36, No.10, pp. 1235-1247, ISSN 0959-8049

Owen, R. W., Haubner, R., Mier, W., Giacosa, A., Hull, W. E., Spiegelhalder, B., & Bartsch, H. (2003). Isolation, structure elucidation and antioxidant potential of the major phenolic and flavonoid compounds in brined olive drupes. *Food and Chemical Toxicology*, Vol.41, No.5, pp. 703-717, ISSN 0278-6915

Papadimitriou, V., Sotiroudis, T. G., Xenakis, A., Sofikiti, N., Stavyiannoudaki, V., & Chaniotakis, N. A. (2006). Oxidative stability and radical scavenging activity of extra virgin olive oils: An electron paramagnetic resonance spectroscopy study. *Analytica Chimica Acta*, Vol.573-574, pp. 453-458, ISSN 0003-2670

Pardo, J. E., Cuesta, M. A., & Alvarruiz, A. (2007). Evaluation of potential and real quality of virgin olive oil from the designation of origin "Aceite Campo de Montiel" (Ciudad Real, Spain). *Food Chemistry*, Vol.100, No.3, pp. 977-984, ISSN 0308-8146

Petrakis, P. V., Agiomyrgianaki, A., Christophoridou, S., Spyros, A., & Dais, P. (2008). Geographical characterization of Greek virgin olive oils (cv. Koroneiki) using ^{1}H and ^{31}P NMR fingerprinting with canonical discriminant analysis and classification binary trees. *Journal of Agricultural and Food Chemistry*, Vol.56, No.9, pp. 3200-3207, ISSN 0021-8561

Pizarro, C., Rodríguez-Tecedor, S., Pérez-del-Notario, N., & González-Sáiz, J. M. (2011). Recognition of volatile compounds as markers in geographical discrimination of Spanish extra virgin olive oils by chemometric analysis of non-specific chromatography volatile profiles. *Journal of chromatography. A*, Vol.1218, No.3, pp. 518-523, ISSN 1873-3778

Platero-López, J., & García-Mesa, J. A. (2007). Automated ultrasound-assisted method for the determination of the oxidative stability of virgin olive oil. *European Journal of Lipid Science and Technology*, Vol.109, No.2, pp. 174-179, ISSN 1438-7697

Ranalli, A., Contento, S., Marchegiani, D., Pardi, D., Pardi, D., & Girardi, F. (2008). Effects of "genetic store" on the composition and typicality of extra-virgin olive oil: Traceability of new products. *Advances in Horticultural Science*, Vol.22, No.2, pp. 110-115, ISSN 1592-1573

Reid, L. M., O'Donnell, C. P., & Downey, G. (2006). Recent technological advances for the determination of food authenticity. *Trends in Food Science & Technology*, Vol.17, No.7, pp. 344-353, ISSN 0924-2244

Rezzi, S., Axelson, D. E., Heberger, K., Reniero, F., Mariani, C., & Guillou, C. (2005). Classification of olive oils using high throughput flow ¹H-NMR fingerprinting with principal component analysis, linear discriminant analysis and probabilistic neural networks. *Analytica Chimica Acta*, Vol.552, No.1-2, pp. 13-24, ISSN 0003-2670

Sacco, A., Brescia, M. A., Liuzzi, V., Reniero, F., Guillou, C., Ghelli, S., & van der Meer, P. (2000). Characterization of Italian olive oils based on analytical and nuclear magnetic resonance determinations. *JAOCS*, Vol.77, No.6, pp. 619-625, ISSN 0003-021X

Sacchi, R., Patumi, M., Fontanazza, G., Barone, P., Fiordiponti, P., Mannina, L., Rossi, E., & Segre, A. L. (1996). A high-field H-1 nuclear magnetic resonance study of the minor components in virgin olive oils. *JAOCS*, Vol.73, No.6, pp. 747-758, ISSN 0003-021X

Santos, J. C. O., Dos Santos, I. M. G., De Souza, A. G., Prasad, S., & Dos Santos, A. V. (2002). Thermal stability and kinetic study on thermal decomposition of commercial edible oils by thermogravimetry. *Journal of Food Science*, Vol.67, No.4, pp. 1393-1398, ISSN 0022-1147

Spaniolas, S., Bazakos, C., Ntourou, T., Bihmidine, S., Georgousakis, A., & Kalaitzis, P. (2008). Use of lambda DNA as a marker to assess DNA stability in olive oil during storage. *European Food Research and Technology*, Vol.227, No.1, pp. 175-179, ISSN 1438-2377

Temime, S. B., Campeol, E., Cioni, P. L., Daoud, D., & Zarrouk, M. (2006). Volatile compounds from Chétoui olive oil and variations induced by growing area. *Food Chemistry*, Vol.99, No.2, pp. 315-325, ISSN 0308-8146

Temime, S. B., Manai, H., Methenni, K., Baccouri, B., Abaza, L., Daoud, D., Casas, J. S., Bueno, E. O., & Zarrouk, M. (2008). Sterolic composition of Chétoui virgin olive oil: Influence of geographical origin. *Food Chemistry*, Vol.110, No.2, pp. 368-374, ISSN 0308-8146

Tena, N., García-González, D. L., & Aparicio, R. (2009). Evaluation of virgin olive oil thermal deterioration by fluorescence spectroscopy. *Journal of Agricultural and Food Chemistry*, Vol.57, No.22, pp. 10505-10511, ISSN 0021-8561

Vaclavik, L., Cajka, T., Hrbek, V., & Hajslova, J. (2009). Ambient mass spectrometry employing direct analysis in real time (DART) ion source for olive oil quality and authenticity assessment. *Analytica Chimica Acta*, Vol.645, No.1-2, pp. 56-63, ISSN 0003-2670

Vaz-Freire, L. T., da Silva, M. D. R. G., & Freitas, A. M. C. (2009). Comprehensive two-dimensional gas chromatography for fingerprint pattern recognition in olive oils produced by two different techniques in Portuguese olive varieties Galega Vulgar, Cobrançosa e Carrasquenha. *Analytica Chimica Acta*, Vol.633, No.2, pp. 263-270, ISSN 0003-2670

Vecchio, S., Cerretani, L., Bendini, A., & Chiavaro, E. (2009). Thermal decomposition study of monovarietal extra virgin olive oil by simultaneous thermogravimetry/differential scanning calorimetry: Relation with chemical composition. *Journal of Agricultural and Food Chemistry*, Vol.57, No.11, pp. 4793-4800, ISSN 0021-8561

Vichi, S., Pizzale, L., Conte, L. S., Buxaderas, S., & López-Tamames, E. (2005). Simultaneous determination of volatile and semi-volatile aromatic hydrocarbons in virgin olive oil by headspace solid-phase microextraction coupled to gas chromatography/mass spectrometry. *Journal of Chromatography A,* Vol.1090, No.1-2, pp. 146-154, ISSN 0021-9673

Vinha, A. F., Ferreres, F., Silva, B. M., Valentao, P., Goncalves, A., Pereira, J. A., Oliveira, M. B., Seabra, R. M., & Andrade, P. B. (2005). Phenolic profiles of Portuguese olive fruits (Olea europaea L.): Influences of cultivar and geographical origin. *Food Chemistry,* Vol.89, No.4, pp. 561-568, ISSN 0308-8146

Yang, H., Irudayaraj, J., & Paradkar, M. M. (2005). Discriminant analysis of edible oils and fats by FTIR, FT-NIR and FT-Raman spectroscopy. *Food Chemistry,* Vol.93, No.1, pp. 25-32, ISSN 0308-8146

Permissions

The contributors of this book come from diverse backgrounds, making this book a truly international effort. This book will bring forth new frontiers with its revolutionizing research information and detailed analysis of the nascent developments around the world.

We would like to thank Boskou Dimitrios, for lending his expertise to make the book truly unique. He has played a crucial role in the development of this book. Without his invaluable contribution this book wouldn't have been possible. He has made vital efforts to compile up to date information on the varied aspects of this subject to make this book a valuable addition to the collection of many professionals and students.

This book was conceptualized with the vision of imparting up-to-date information and advanced data in this field. To ensure the same, a matchless editorial board was set up. Every individual on the board went through rigorous rounds of assessment to prove their worth. After which they invested a large part of their time researching and compiling the most relevant data for our readers. Conferences and sessions were held from time to time between the editorial board and the contributing authors to present the data in the most comprehensible form. The editorial team has worked tirelessly to provide valuable and valid information to help people across the globe.

Every chapter published in this book has been scrutinized by our experts. Their significance has been extensively debated. The topics covered herein carry significant findings which will fuel the growth of the discipline. They may even be implemented as practical applications or may be referred to as a beginning point for another development. Chapters in this book were first published by InTech; hereby published with permission under the Creative Commons Attribution License or equivalent.

The editorial board has been involved in producing this book since its inception. They have spent rigorous hours researching and exploring the diverse topics which have resulted in the successful publishing of this book. They have passed on their knowledge of decades through this book. To expedite this challenging task, the publisher supported the team at every step. A small team of assistant editors was also appointed to further simplify the editing procedure and attain best results for the readers.

Our editorial team has been hand-picked from every corner of the world. Their multi-ethnicity adds dynamic inputs to the discussions which result in innovative outcomes. These outcomes are then further discussed with the researchers and contributors who give their valuable feedback and opinion regarding the same. The feedback is then collaborated with the researches and they are edited in a comprehensive manner to aid the understanding of the subject.

Apart from the editorial board, the designing team has also invested a significant amount of their time in understanding the subject and creating the most relevant covers. They scrutinized every image to scout for the most suitable representation of the subject and create an appropriate cover for the book.

The publishing team has been involved in this book since its early stages. They were actively engaged in every process, be it collecting the data, connecting with the contributors or procuring relevant information. The team has been an ardent support to the editorial, designing and production team. Their endless efforts to recruit the best for this project, has resulted in the accomplishment of this book. They are a veteran in the field of academics and their pool of knowledge is as vast as their experience in printing. Their expertise and guidance has proved useful at every step. Their uncompromising quality standards have made this book an exceptional effort. Their encouragement from time to time has been an inspiration for everyone.

The publisher and the editorial board hope that this book will prove to be a valuable piece of knowledge for researchers, students, practitioners and scholars across the globe.

List of Contributors

Cinzia Benincasa, Kaouther Ben Hassine, Naziha Grati Kammoun and Enzo Perri
CRA-OLI Olive Growing and Olive Oil Industry Research Center Rende, Italy
Institut de l'Olivier, Sfax, Tunisia

Ana M. Costa Freitas, Maria J. B. Cabrita and Raquel Garcia
Escola de Ciências e Tecnologia, Departamento de Fitotecnia, Instituto de Ciências Agrárias e Ambientais Mediterrânicas ICAAM, Universidade de Évora, Évora, Portugal

Marco D.R. Gomes da Silva
REQUIMTE, Departamento de Química, Faculdade de Ciências e, Tecnologia/ Universidade Nova de Lisboa, Campus da Caparica, Portugal

Anna Grazia Mignani, Leonardo Ciaccheri and Andrea Azelio Mencaglia
CNR – Istituto di Fisica Applicata "Nello Carrara", Italy

Antonio Cimato
CNR – Istituto per la Valorizzazione del Legno e delle Specie Arboree, Italy

Sema Bağdat Yaşar, Eda Köse Baran and Mahir Alkan
Balıkesir University, Turkey

Ewa Sikorska
Faculty of Commodity Science, Poznań University of Economics, Poland

Igor Khmelinskii
Universidade do Algarve, FCT, DQF and CIQA, Faro, Portugal

Marek Sikorski
Faculty of Chemistry, A. Mickiewicz University, Poznań, Poland

Alessandra Bendini, Enrico Valli, Sara Barbieri and Tullia Gallina Toschi
Department of Food Science, University of Bologna, Italy

A. Lamagna
Comisión Nacional de Energía Atómica, Argentina
Escuela de Ciencia y Tecnología, Buenos Aires, Argentina

K. Pierpauli
Comisión Nacional de Energía Atómica, Argentina

C. Rinaldi
Comisión Nacional de Energía Atómica, Argentina
Consejo Nacional de Investigaciones Científicasy Técnicas, Argentina

M. L. Azcarate
Consejo Nacional de Investigaciones Científicasy Técnicas, Argentina
Centro de Investigaciones en Láseres y Aplicaciones CEILAP (CITEDEF-CONICET), Argentina

Zohreh Rabiei and Sattar Tahmasebi Enferadi
National Institute of Genetic Engineering and Biotechnology, Tehran, Iran

Ivonne Delgadillo, António Barros and Alexandra Nunes
Department of Chemistry, QOPNA Research Unit, University of Aveiro, Portugal

Rodney J. Mailer
Australian Oils Research, Australia

José Felipe Jiménez-Guerrero and Juan Carlos Gázquez-Abad
University of Almería, Spain

Juan Antonio Mondéjar-Jiménez
University of Castile La Mancha, Spain

Rubén Huertas-García
University of Barcelona, Spain

Rosa M. Alonso-Salces, Margaret V. Holland and Claude Guillou
European Commission - Joint Research Centre, Institute for Health and Consumer Protection, Physical and Chemical Exposure Unit, Ispra, Italy

Károly Héberger
Chemical Research Center, Hungarian Academy of Sciences, Budapest, Hungary